American Marine Engines
1885 – 1950

American Marine Engines
1885 – 1950

Stan Grayson

Devereux Books
Marblehead, Massachusetts

Published by Devereux Books
PO Box 503
Marblehead MA 01945

Internet address: www.devereuxbooks.com

Library of Congress Cataloging in Publication Data

Grayson, Stan, 1945-
American marine engines 1885-1950 / Stan Grayson
p. cm.
Includes bibliographical references and index.
ISBN 978-1-928862-13-0 (pbk. : alk.paper)
ISBN 978-1-928862-14-7 (hard cover : alk.paper)
1. Marine engines—United States—History. 2. Internal combustion engine industry—United States—History. 3. Inboard-outboard engine industry—United States--History. I. Title.
VM731.G6838 2008
623.870973'09041—dc22

200802485

Designed by Jon Albertson
www.albertson-design.com

Printed in the United States of America

On the cover: This beautiful and very rare engine was built by the Standard Motor Construction Co., once an important industry in Jersey City, New Jersey. It was probably constructed circa 1910–1916; the engine's basic design reflects patents granted to engineer C. C. Riotte in 1899. The stanchion base would have been familiar to a steam engineer and offered relatively light weight and easy accessibility. The engine has a 4" bore, 5" stroke, magneto and spark-plug ignition, and its original reverse gear. Starting was by compressed air. A 6-cylinder Standard won the first-ever Gold Cup motor boat race in 1904, and a smaller version powered the 26-foot tender to the presidential yacht *Sylph* last used by Woodrow Wilson. This antique Standard is owned by collector Harry Garman. Photography by Michael D. Gombert.

Contents

Acknowledgments

Special thanks are due those who critiqued portions of the manuscript: Dick Day, Dan Acierno, Jim Aamodt, and Ernie Darrow, and to Dave Birdsey, who loaned his magazine collection. Many others shared experience and information with me: Wayne Blake, George Boley, P. S. Brooke, Jr., Doug Brown, John Brown, Stephen Canright, J. B. Castagna, Walt Celley, Richard Durgee, Hugh Durgin, Joe Fossey, Harry Garman, Frank E. Goulde, John Griffin, Robert Henkel, Robert Hess, Norman Hoch, Joe Joyce, Donald Kitchin, Jr., Larry Healey, Mark Mason, Pierre McGraw, Andrew Menkart, Kate Mockler, Don Moyer, Duff G. Porter III, the late Charles F. Stokes, Doug Teel, Sharon Turk, and Myron Wininger. For anyone who assisted that I've overlooked, thanks!

Local history librarians throughout the country were diligent in their assistance. Among the public libraries, museums, and historical societies whose assistance was so important are the Algonac/Clay Township Historical Society, the Antique Boat Museum, the Ballston Spa Public Library, the Baltimore Museum of Science, the Bridgeport Public Library, the Brockville (Ontario) Public Library, the Carnegie Library of Pittsburgh, the Chesapeake Bay Maritime Museum, the Chicago Public Library, the Cincinnati Public Library, the Clyde (Ohio) Public Library, the Detroit Public Library, the Enoch Pratt Free Library, the Grand Rapids Public Library, the Gulf of Georgia National Historic Site of Canada, Laura Lamb at the Hamilton (Ontario) Public Library Special Collections, the Fall River Public Library, the Holyoke History Room at Holyoke Community College, the Joint Archives of Holland (Michigan), the Kenyon County (Kentucky) Public Library, the Kettering Archives, the Marshall (Michigan) District Library, the Michigan Historical Museum, the Milwaukee Public Library, the National Maritime Museum, the Newburgh (New York) Free Library, the Owls Head Transportation Museum, the San Francisco Public Library, the Onondaga County Public Library, the Oshkosh Public Library, the Public Libraries of Saginaw, the Pittsfield Public Library, the Portland (Maine) Public Library, the Providence Public Library, the St. John's Provincial Library, the St. Joseph (Michigan) Public Library, the St. Louis Public Library, the Simcoe County Archives, the Syracuse Public Library, the Timberland Regional Library (in Tumwater, Washington), the Troy (New York) Public Library, and the Westport Historical Society.

A Note on Prices as Reflected in This Book

Throughout this book, there are occasional references to the prices of marine engines or boats. Putting those prices into context for a contemporary reader can be a challenge because of inflation and because today's economy differs so dramatically from those of 1900–1920 and the years leading to World War II. In 1901 the average household income in America was $750. That was about the price of a 24-foot launch with an 8-hp, two-cycle engine. In 1914 a 2½ hp inboard "canoe motor" might be had for $35, a price likely to include a small fuel tank and propeller. A 4-cylinder, 10-hp Niagara — a very fine engine of the period — cost around $485. What did such prices mean to consumers of that era? In January 1914, after bitter labor disputes, Henry Ford startled industrial America by instituting his famous $5 per day wage. A union plumber or carpenter in New York City could expect to earn about $30 for a 44-hour week, about $1,500 per year. A Ford Model T Touring car cost $550. In Morristown, New Jersey, a nine-room house could be rented for $35 a month, and a fine seven-room house might cost some $3,200 to buy. An unpretentious wood-framed house in a working-class area of a city or in a rural area might be had for approximately $700 and up. Then as now, boats and engines would have been significant expenses. A general comparison of buying power for 1914 versus 2006 is that it would take about 700 2006 dollars to equal 35 1914 dollars.

Introduction

This book owes its existence to a suggestion made many years ago by Roger Taylor, then president of the International Marine Publishing Company in Camden, Maine. Roger had read an article about antique marine engines and was of the opinion that a book on the subject might find an enthusiastic audience. Given my experience researching and writing about the automobile industry, and my interest in boats, he asked if I'd be interested. That is how the first edition of *Old Marine Engines, the World of the One Lunger* came to be published in 1982.

Research included a trip to Nova Scotia to visit the Lunenburg Foundry, the last of all the companies still making a classic, two-cycle marine engine. I also toured the fascinating if dormant factory of Acadia Gas Engines in Bridgewater, and met several old-timers who well remembered the steady beat of a "one-lunger" engine. Throughout the research process, I met many able men dedicated to preserving the history of a dynamic industry that left a lasting technological and cultural legacy. After the writing and production was complete, graphic artist and replica ship designer Melbourne Smith created a dramatic dust jacket. The jacket proved eye catching, and the book received good reviews in both the general and enthusiast media.

Once *Old Marine Engines* was sold out, I began receiving sporadic calls from all over the U.S. and Canada. People reported that if the book could be had from sellers of rare books, it was priced at three times what it cost initially. Not a bad investment for those who owned a copy, but few were parting with theirs. Eventually, in 1994, under the Devereux Books imprint, I published a second edition to meet the ongoing demand. A third edition followed.

The decision to replace *Old Marine Engines* with a new book resulted in quite a challenge. What should the new book be? What subjects should take precedence? The primary goal was to include new information based on my ongoing research and to add more pictures. Of course, a balance had to be found. If it were to be reasonably priced, the book would have to be capped at a specific page count and budgets adhered to. As often happens with such projects, most of the original numbers were exceeded. Even then, during the writing, editing, and production process, I often worried as much about some material that could not be included as I did about the text and pictures that made the final cut.

Compared to the earlier book, this one has a larger page size to permit inclusion of more information. *Old Marine Engines* included a list of marine engine companies as an appendix; this book focuses on the company portraits. They are, however, prefaced by chapters that set the stage in terms of overall history and technology. While the focus of *Old Marine Engines* was the classic two-cycle "one-lunger," this book also encompasses the broader industry. To assist enthusiasts of various experience levels, a chapter has been devoted to engine collecting, including suggestions for identifying an engine and establishing its value. In this chapter, in particular, I have sought the input of several men with long experience in the collecting hobby and the marine engine restoration business.

One thing that has not changed since the writing of *Old Marine Engines* is the type of reference materials used. I have relied entirely on source material of the time. The bibliography gives a good idea of these materials. Reliance on such sources often turned up problems with previously published books, website articles, and papers in local historical societies written by well-intentioned individuals. I have done my best to resolve such matters. For what is presented on the following pages, the responsibility for a mistake of fact or an erroneous conclusion rests with me.

Finally, unlike *Old Marine Engines*, this book includes a chapter devoted to the research methods and materials that readers can use to learn more about a specific company or engine. It is my hope that those wishing to conduct such research, whether for their own information or to write about the industry, will find that chapter to be of particular benefit and serve as a guide. The subject deserves no less.

Stan Grayson
Marblehead, Massachusetts

May 2008

1928 Wolverine diesel. (Ed Malitsky photo)

Part I

Marine Engines Then and Now

CHAPTER ONE:

The Great Invention

Sometime in the year 1887 or 1888, a champion oarsman and commercial waterman in San Francisco named Henry Peterson was given an opportunity granted to few men. He glimpsed the future and made it his own. The future was a machine described by its inventor as a "vapor engine." The vapor came from a liquid, called gasoline, with a penetrating odor. Peterson learned that the vapor, when mixed with air, could be ignited by an electrically induced spark. The resulting explosion drove the engine's piston downward, turning the engine's flywheel. Both Peterson and the engine's creator, Daniel Regan, recognized that the flywheel's motion could, with some ingenuity, be transferred to a boat's propeller shaft.

At the time of Henry Peterson's revelation about internal combustion, he owned a fleet of five Whitehall boats — a type of fast rowboat that was then a staple of San Francisco's waterfront economy — and a shop on Folsom Street wharf. Peterson claimed that he installed a "primitive" engine in one of his Whitehalls in about 1887 and that, when he was able to keep the engine running, it effortlessly propelled the boat at a satisfactory speed. Still, it would be a decade before Peterson, presumably with significant assistance from Regan, finally had a boat that he believed would have a serious commercial application. This boat was named *Amy*.

A 40-foot launch with a 10-foot beam, *Amy* was little different in appearance from steam-

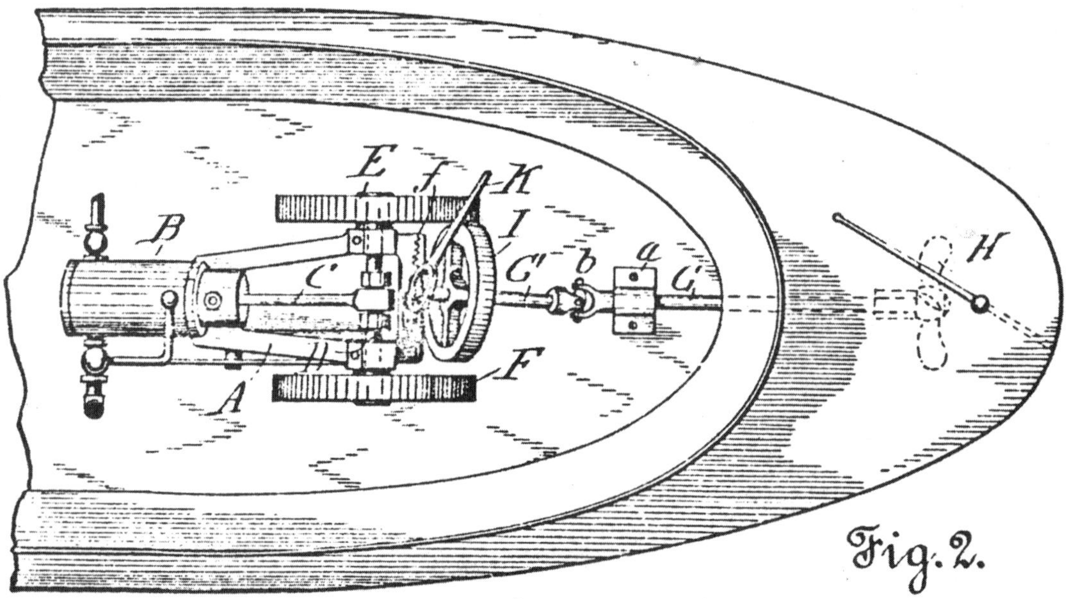

On May 17, 1892, inventor Daniel Regan was awarded a patent for his method of driving and reversing a boat's propeller shaft by a twin-flywheel, horizontal engine. The idea had merit because the engine's weight would have been located low in the hull, but vertical-style engines quickly proved most practical.

Among the most important American internal combustion engines was the Regan, which had an electric ignition system developed in the mid-1880s. The badges noting the engine's patent protection did not stave off copies. This Regan was restored by collector Bill Prine.

powered craft of her general size and shape. But instead of a boiler, steam engine, and a supply of wood or coal, there was a Regan engine mounted in *Amy*'s hull. In March 1897, before an audience of skeptical fellow boatmen who literally bet against him, Peterson eased Amy out of her berth at Pier 10 and then proceeded to tow 255 log pilings from Folsom Street to the Arctic Oil Works wharf.

"As no one supposed so small a boat could undertake such a job," the *San Francisco Chronicle* reported, "all sorts of bets were made that Peterson would fail in his attempt." But the bettors lacked Henry Peterson's vision. They lost. Peterson immediately ordered a bigger boat and built on his initial success. In the aftermath of the 1906 earthquake, Peterson recognized that an enormous building boom awaited San Francisco, and he added to his fleet. By 1916, the Henry Peterson Company owned 14 launches and tow-boats with engines of from 20 to 100 hp, and 11 barges of from 40 to 450 tons. By then, the Bay area was home to at least a half-dozen engine companies, some with impressive factories that were producing a wide variety of models for customers both in the immediate area and for export.

In the first decade of the 20th century, which we may aptly refer to as the "internal combustion century," inventive men like Daniel Regan and forward-thinking commercial watermen like Henry Peterson began popping up throughout America. As a matter of fact, Peterson was not alone in his experiments. According to engineer Charles Winslow, whose paper presented to a meeting of the Society of Automotive Engineers in 1955 ranks as one of the key documents on engine development on the West Coast, Capt. N. Anderson claimed he had installed a gasoline engine in a freighting sloop in Alameda in 1888. In 1890, another was installed in a stern-wheeler named *Ruth* operating on the Eel River in Scotia. Presumably, these were Regan engines although Winslow did not specifically say so.

American Marine Engines

Daniel S. Regan was likely the first American to develop a commercially viable gasoline engine — and almost certainly the first anywhere to market an engine with electric ignition — but his initial efforts were soon joined by others in Michigan, New England, and New York State. In each of these areas, there was a broad potential market for marine engines, and in each geographic area the work of one man sowed the seeds for growth that quickly spread. Besides Daniel Regan in San Francisco, there was Clark Sintz, an Ohioan who soon established himself in Grand Rapids, Michigan, and the brothers Frank and Ray Palmer in the Mianus/Cos Cob area of Connecticut. The seminal developments of these industry pioneers would profoundly influence many others, and the locales in which they worked quickly spawned other engine companies and became centers of the industry.

In San Francisco, when Regan's Vapor Engine Company and its successor failed in bitter lawsuits, the patents were acquired by the Union Gas Engine Company which, starting in 1892, grew to become a major force both in the U.S. and in export markets. Other San Francisco companies like Hercules, Standard, Atlas, and Imperial turned out heavy-duty, slow-turning four-cycle engines with marvelously conceived and executed make-and-break ignition systems. Such engines would forever be associated with California. They soon found their way north and became staples of the graceful Columbia River gill net boats, Bristol Bay fishing craft, and cannery tenders.

In New England, Palmer, Lathrop, Mianus, Bridgeport, and other Connecticut builders found a ready market as commercial fishermen quickly adopted the new technology. They installed a variety of two-cycle motors in Cape Cod catboats, Maine Hampton boats, Connecticut lobster boats and sharpies, dories, and other types. Many of the engines were single-cylinder models, and their distinctive exhaust note soon earned them the nickname "one lunger," a term that has survived ever since.

Thanks to its early date of commercial sale beginning in 1892, the Sintz was among the most influential marine engines ever built and did much to spur development of the industry in Grand Rapids, Detroit, and elsewhere in Michigan, Wisconsin, and Ohio. This 1899 Sintz is equipped with a reversible propeller. Pre-1900 marine engines can often be recognized because the mounting plate was beneath the crankcase rather than on the centerline of the crankshaft.

This circa 1910 'Frisco Standard single has had its side plates removed to show the construction method and connecting rod. The make-and-break mechanism is visible atop the cylinder.

In the Midwest, a childhood mechanical prodigy and inventive genius named Clark Sintz developed a two-cycle engine that caught the eye of a wealthy enthusiast who convinced Sintz to move from his birthplace in Springfield, Ohio, to Grand Rapids, Michigan. There the Sintz engine entered production and it immediately gained the attention of Detroit entrepreneurs, sparking the creation of new companies in what would soon become the most dynamic center on earth for marine engine development. (Although Detroit would always remain the hub of the industry in the Midwest, marine engine companies emerged elsewhere in Michigan, in Wisconsin — where Fairbanks-Morse would become among the country's largest manufacturers — Ohio, Iowa, and Kentucky.) Besides recreational motor boats, the marine engine immediately revolutionized the Great Lakes fishing industry where the Kahlenberg became a favorite for a wide variety of boat types. Kahlenberg, Fairbanks-Morse, and others also found their way into the "gas boats" that began quickly replacing steam vessels on America's great inland waterways.

That Michigan played such a key role in the marine engine industry was prompted, in large part, by its central importance in the automobile industry. During the first decade of the 1900s, it was not necessarily clear to either investors or

This four-cycle 3-hp "light duty" Regal was built in Coldwater, Michigan, in 1913. The timer and throttle control are located on the same shaft. The muffler is water-cooled. Founded in 1901, Regal offered light-, medium-, and heavy-duty models for both commercial and recreational applications.

Palmer's first engine, this little 1-hp, two-port, two-stroke was used for display at boat shows after it was found and refurbished at the company in 1940. As was often the case on pre-1900 engines, the crankcase had a "wine-bottle" shape evocative of a stationary steam engine. The engine has been fitted with a Palmer generating valve and has Palmer's original-style igniter. This historic little Palmer is now owned by collector Dick Day.

This round-bottomed skiff powered by a 1-cylinder engine was among the illustrations in an E. W. Roberts article about marine engines published in The Gas Engine *in January 1899. Unfortunately, the engine brand was not specified.*

inventors whether the automobile or the motor boat offered the most growth potential. For a period, engine companies often hoped to sell their wares for both applications, not to mention for use in motor trucks and farm tractors. Engine designers in Detroit like Alanson Brush (who would become famous for his work on the first Cadillac), Charles Brady King (a wealthy friend of Henry Ford who eventually started both a marine engine company and then an automobile company), and Joe Van Blerck (a Dutch immigrant who pioneered high-performance marine engines), to name but three, at one time or another hoped to find a market for their engines in land-bound applications as well as boats.

Later, as technology produced ever more reliable and robust automobile engines appropriate for marine conversion, the long-time linkage between the two industries continued. What's more, the ceaseless development of four-cycle engines by automakers had enormous impact on the marine engine. Looking back on the industry in 1920, Walter L. Fay of Fay & Bowen in Geneva, New York, reported that "the motor car was responsible for the rapid development of a four-stroke engine, lighter, faster, and of greater power in proportion to its weight, and the attention of marine engineers became directed toward producing engines which were to the boat what the auto engine was to the car." By the early '20s, Connecticut lobstermen who had previously used two-cycle engines had replaced them with heavy-

The Gasolene Engine Equipment Company was an important New York City agent for a variety of engine companies as this 1913 ad reveals.

duty four-cycle models from Lathrop, Palmer, Mianus, Bridgeport, and Frisbie.

While West Coast marine engine development remained centered in San Francisco and Oakland, in the East, companies began to flourish from New Jersey to Maine, with Connecticut and New York becoming the areas of greatest commercial activity. Connecticut had already been established as a center for mass production and the manufacture of arms and high-grade machine tools by the time the internal combustion movement gained momentum. With its long coastline and a skilled workforce of machinists and foundry men, the state became a natural center for engine manufacture. Its proximity to the investment capital and markets in the New York City area also helped. Central and upstate New York, meanwhile, became another hotbed of activity as boaters around Lake Erie, the Finger Lakes, and the St. Lawrence River provided a steady demand. The city of Buffalo, for example, was home to at least four important companies.

All this activity was, of course, noted in Canada. In Nova Scotia, the Hawboldt engine

The machine shop of the Hicks Gas Engine Works, circa 1908. James Hicks, wearing a vest, is seen at the right with a book in his hands. While the overhead belts were long-typical of such shops, the Hicks premises were well lit and comparatively neat.

The product of Canadian engine pioneer Forman Hawboldt, this 2-cylinder model was based on Hawboldt's first engine that he successfully demonstrated in Chester, Nova Scotia, in 1902.

emerged from its inventor's barn in 1902. In Lunenburg, the first Atlantic marine engine was created in 1907 by a company that would ultimately outlast every one of its contemporaries in the U.S. and Canada. Newfoundland became a large and enduring market for both Canadian and American engine makers and the single-cylinder two-cycle "one-lunger" with its water-resistant make-and-break ignition system became a part of Newfoundland commerce, lore, and culture for over a half-century. Far removed from Atlantic Canada, in British Columbia, four-cycle engines such as the Easthope would likewise play an important role as fishing boats, cannery tenders, and workboats of all types were adapted to gasoline power.

The Rise of the Marine Engine Industry Marine engine development progressed through a series of four reasonably well-defined eras. The Pioneer Era — 1884 to 1898 — was the period during which the internal combustion engine, while still primitive in its technology, was nonetheless developed into a commercially viable and reasonably practical machine, although a certain mechanical knack was required of owners. Once the basic technology was established, the pioneer companies began increasing production, and they were soon joined by many others, all competing for sales to markets that, at the time, appeared to be limitless — an appearance that would prove deceiving.

The First Production Era — 1899 to 1912 — saw engines being manufactured in ever-increasing numbers. One estimate placed the total production in 1911 alone at between 75,000 and 100,000, nearly twice what it would be some 20 years later. During these years, companies also began targeting specific markets from the cost-conscious to the high-end yacht and commercial user.

The decade beginning in 1895 saw engine builders struggling to overcome not only significant technological obstacles, but also the often negative perceptions of potential customers. Internal combustion needed to demonstrate that it was not only a viable alternative to steam but

It didn't take long for clever men to adapt the gasoline engine to drive a paddlewheel, thus eliminating the complexity, danger, and licenses involved in steam. This photo, possibly of a Gray installation, dates from 1912.

Visitors to the New York Boat Show in 1937 could view a comprehensive and well-lit display of fine Scripps engines. Closest to the camera is a pair of 678-cubic-inch, 200 Series engines. One has a head and cylinder removed so salesmen could demonstrate the construction. On the table is one of the cylinder heads and a manifold. A V-12 Scripps marine engine is in the middle of the first row while a Scripps-converted Lincoln Zephyr V-12 and Ford V-8 are at the end.

superior in all regards. That meant stressing the absence of fire and sooty coal and convincing customers that a gasoline engine was reliable enough to get one home safely, and that the volatile gasoline wouldn't explode.

Ease of operation was recognized early as a key factor to acceptance. In May 1898, *The Gas Engine* noted that "The perfect gasoline engine for marine purposes will be one which you can actually stop the engine itself exactly as it is done with steam and start it again by as simple a process and that too in either direction ahead or astern." These matters — easy starting, easy stopping, and reversing — would continue to pose challenges for the following 10 to 20 years. Reported one observer in August 1910: "With the advent of the gasoline motor for marine use, there appeared a problem which was of little consequence in connection with its more flexible predecessor, the steam engine. The problem was that of reversing the direction of the boat's motion..." Companies producing marine transmissions like Paragon, Snow & Petrelli (Joe's Gears), and many others emerged to meet the need, as did many engine producers themselves.

As for easy starting, while fishermen eventually became adept at starting their two-cycle marine engines — and at reversing them "on the switch" — such skills remained elusive for the weekend boater. It was not until the development of compact four-cycle engines with electric starting that the motorboat was finally recognized as practical enough to be operated safely and reliably by pleasure boaters both male and female.

Even with such obstacles, the motorboat industry experienced enormous growth. "It has taken some years to educate the public to the fact that the motor boat is safe, efficient and reliable," reported *American Industries* magazine in 1910. "It is remarkable how few accidents occur in consideration of the large number of these vessels in use, the number being estimated at 250,000." That Americans were a potentially large and lucrative market for motors and boats had been proved as early as 1904 when some 90,000 people attended what was then known as the Annual Sportsman's Show at Madison Square Garden in the spring of 1904. Of the show, *Motor Boating* reported that "It was plainly evident at that time that motor boating was destined to become one of the chief sports and pleasures of the water-loving Americans, as well as an important factor in the business of many who had heretofore depended on wind or steam."

The Sportsman's Show soon evolved and was joined by annual shows at New York, Boston, and elsewhere devoted exclusively to motor boats and marine engines. Whatever were the challenges posed by operating a marine engine in the first decade of the 20th century, demand seemed to grow inexorably. Said *American Industries* in its 1910 article on the subject: "Every little town bordering on a waterway has its motor factory, some, to be sure, turning out only twenty or thirty a year, but all tending to swell the total." That year, it was estimated there were between 700 and 800 builders of marine gasoline engines in America.

Boating magazines followed engine development trends carefully, recording the evolution of the marine engine industry and its products. For example, during the period from 1909 to 1914, two-cycle motors dropped from 65 percent of production to 51 percent, as four-cycle models were consistently improved. During that same brief period, two-port, two-cycle motors fell from a predominant 54 percent share of the market, and three-port motors had assumed that majority position by 1914. High-tension ignition systems like the Atwater Kent had become the norm by 1914 except on the heavy-duty two-cycle motors favored by fishermen and the commercially oriented four-cycle engines built in California, which retained make-and-break igniters. In 1916, *Power Boating* identified 59 manufacturers of two-cycle engines who offered a total of 518 models compared to 65 four-cycle manufacturers offering 500 models. Then the market changed abruptly. By the time America entered World War I in 1917, two-cycle engines accounted for only about two percent of the total market, while 4-, 6-, and 8-cylinder, four-cycle models dominated sales.

The Maturation Era — 1913 to 1929 — saw engine companies learning from their own mistakes and those of competitors while also gaining a greater understanding of market demands and how to meet them. The marine engine had lost its novelty. Engines had long since become a necessity, and the drive was on to produce more refined, reliable engines at ever lower prices. America's entry into World War I prompted a focus in the boating press on state-of-the-art marine engines. The straight-8 Duesenberg and Van Blerck V-12 — engines suitable for installation in patrol boats and submarine chasers — were described in great detail, and suppliers of high-grade components for the engines advertised their wares in the press.

The end of World War I marked the beginning of what would become a profound change in the marine engine industry. Government contracts for war-related materiel combined with financial unease in the markets led to a dramatic drop in demand. Materials shortages and labor difficulties also didn't help. When production finally began increasing in the '20s, it was for ever more compact and less expensive engines, a need increasingly met by companies, like Kermath and Universal, that focused on standardized models that could be mass-produced and sold at attractive prices.

Increasingly, too, marine engine companies took advantage of technical advances made by the automobile where volumes were high and the resulting income could finance intense development programs. Among the leaders in the trend was the Gray Motor Company in Detroit, which began adapting Studebaker and Pontiac engines to marine use. At the same time, the upper echelon market segment remained strong, and some of the finest of all high-end marine engines from Sterling and Scripps began appearing during these years. *Speejacks*, a 98-foot, Winton-powered motor yacht, became the first such boat to complete a circumnavigation in 1923.

The difficulties of the post-World War I period also marked a technological dividing line that at last ended the two-cycle versus four-cycle arguments that had always been a part of the marine engine industry. Two-cycle popularity had peaked by about 1912, and it was all downhill thereafter. Said Fay & Bowen's Walter L. Fay: "The old war cry of the two-cycle man against the cams, valves and gears of the four-cycle type soon lost its force as the automobile gained in favor. Admitting that the two-cycle type is the simpler, it still would be an insult to the present intelligence of the public to claim that there is anything in the modern four-cycle pleasure boat engine to prevent its operation with entirely satisfactory results by the ordinary man or woman."

Fay went on to state what by the start of the Roaring Twenties was fast becoming obvious. The electric starter had finally overcome one of the chief remaining obstacles to successful operation of a motor boat by the typical owner. Operating a boat was now little more challenging than operating a car. Production-built runabouts equipped with windshields and electric lights completed the favorable comparison. From now on, engine makers hoping to improve sales volumes, let alone survive, would focus on improving power-to-weight ratios, increasing reliability and longevity, and developing engines capable of volume production so that prices could meet market expectations. Barely a decade later, as the Roaring Twenties were about to give way to the Great Depression, it was estimated that the marine engine industry had sold 46,410 motors in 1929, employed 9,170 people, and had 1,330 distributors and 9,100 dealers.

Agent's ad placed by Horace J. Conley, 1905

Sales and Marketing Although word-of-mouth experience was a key manner in which some engine companies increased sales to specific markets, specifically fishermen, an engine company's primary sales and marketing tools were magazine ads and brochures. Of course, the principal boating magazines including first *The Rudder* and then *MotorBoat*, *Motor Boating*, and *Power Boating* featured dozens of ads. But the larger engine makers, or those with an expanded understanding of the market's potential, advertised in general interest publications, too. *The Saturday Evening Post*, *Everybody's Magazine*, and *Harper's* were a few of these magazines. As for the brochures, a number of companies produced lavish catalogs featuring artwork showing the engines and their components. Explanatory copy was sometimes extensive as makers tried to impress potential customers with what set their products apart from others. Usually, the brochures were printed on good quality, coated paper, and the cover stock was frequently embossed. Such publications have themselves now become collectibles.

Except for strictly local engine builders who sold directly to customers based on word-of-mouth recommendations, marine engine manufacturers sold their products through sales agents. Agents had headquarters in strategically located cities such as New York, New Orleans, Detroit, Minneapolis, Chicago, Boston, Seattle, Portland, Maine, and Portland, Oregon, Baltimore, and other potentially lucrative markets. The engine companies, of course, had one or more sales executives whose responsibility it was to develop the sales network and to provide marketing support. The latter proved necessary, especially in the many cases where local outlets proved to be managed by men better suited to the mechanical aspects of the trade rather than business and customer care.

Typically, the sales agents handled a variety of different makes so they could meet the needs of boaters with varying application needs and budgets. The various transactions of these old marine engine dealers make for fascinating reading because of the variety of brands they recommended. Following are sample sales transactions reported in January 1920 by Walter H. Moreton, a well-known marine engine dealer on State Street in Boston who represented several manufacturers: a Sterling 125-hp Model F for a 44-foot high-speed cruiser designed by John G. Alden and built at Coombs's yard in Belfast, Maine; a 12-hp Universal for a yacht tender; a 20-hp Red Wing for a 24-foot runabout, and a 16-hp Frisbie for a Boston Harbor police launch.

The Industry's Decline Despite periods of steady and occasionally dramatic growth, the marine engine industry was never an easy business. There were hundreds of companies and, except for those that did business with some success on a purely local level, the important makers battled to dominate market niches. Ultimately, good marketing supported by service and a good reputation were vital to success. Relatively few companies could put all those elements together.

Economic factors always played a role. When gasoline rose from some five cents a gallon around 1910 to 10, then 15 cents and more during the mid-teens, and 23 cents in the early '20s, builders of large commercial gasoline engines either developed a diesel or failed. Most companies offered engines with optional equipment that included a special carburetor adapted to kerosene, which was significantly cheaper than gasoline. Other companies offered engines designed to run on kerosene or fuel oil with a "hot bulb" ignition system that didn't need a spark plug or mechanical igniter. Such an engine, often called "semi diesel," became an important factor on the commercial market.

The immediate post–World War I era brought with it shortages of materials and also labor strife. When things got back to "normal," normal had changed and the market began shifting. Certainly, there were customers to support the fine Sterling, Scripps, and a few other hallmark brands, but the growing clamor was for affordable engines that were easily operated by men, women, and even children. Kermath, Universal, and Gray were comparatively well positioned, but everyone faced demands for lower prices. Companies emerged that weren't really marine engine builders at all but firms who marinized Ford's ubiquitous Model T (and later the Model A and V-8) and sold their versions at prices lower than that of a true marine engine of the same horsepower rating. A number of companies designed engines to use Model T parts.

Then came the Great Depression. The marine engine industry, like virtually all other American business, suffered during this economic

This is an example of the late '30s Lincoln Zephyr V-12 as converted for marine applications by the Scripps Motor Co. The Lincoln V-12 was a 267.3-cubic-inch, 110-hp engine that was essentially the Ford V-8 with four additional cylinders set at 75 rather than 90 degrees. It was priced well below the Scripps V-12 marine engine.

disaster. The number of people able to afford the luxury of a new motor boat or new engine could support only a portion of the available suppliers. At Chris-Craft, to cite but one example, production of the top-line 27-foot runabout dwindled: Only 62 were sold from 1932 to 1941. The entire production boat-building industry was hit hard, and many companies failed, reducing demand for engines.

Even as marine engine sales volumes plummeted, demand increased for affordable yet durable engines. Many, if not most, new marine engines were now based on automobile engines converted with a water-cooled exhaust manifold, geared water pump, and different front and rear housings. Kermath's V-12 was a version of the Cadillac V-12. Both Kermath and Scripps marinized the Ford V-8 and Lincoln V-12. Industrial blocks, especially those from Hercules and Waukesha, found their way into many boats under the label of one marine engine maker or another. While marine engine companies touted the features of these engines, they never advertised their origins. The industry looked forward to a "new day" when people again had jobs and money. Many companies that had persevered through the post–World War I downturn now closed.

As for the general business trend, it was not until around 1936 that executives began speaking confidently about the worst being over. General Motors, which had already invested years of effort in the development of a high-speed diesel that would replace the cast-iron stalwarts then available, still pressed ahead under the direction of the brilliant Charles Kettering. By the late '30s, big GM diesels had reached the point where the U.S. Navy began installing them in submarines. The smaller GM models, developed in-house and then handed over to Gray for marine conversion, emerged in production form by the end of the decade.

World War II proved to be both a bonanza and then a final bust for the marine engine industry. Government orders for all sorts of marine engines — from little 4-cylinder Palmers destined for lifeboats to big Sterlings, Packards, and Scripps installed in rescue and torpedo boats, and Gray diesels used in landing craft — kept the plants busy. Factories were sometimes enlarged. Work shifts were added. Gray, Kahlenberg, General Motors, and others proudly flew the "E Pennant" banners awarded by the war department for their contributions in terms of production.

Like other marine engine companies, Sterling built engines for the armed forces during World War II. This ad from Yachting *in 1944 tells of the company's work and included the famous E Pennant awarded to firms whose performance was recognized as outstanding by the government.*

Even as the plants were working overtime, however, engine company executives knew that the war, which represented a bonanza in terms of production and revenues, wouldn't last forever. As early as 1942, in an ad depicting its Admiral V-12, a marinized version of Vimalert's aero engine, the Sterling Engine Co. of Buffalo, New York, looked ahead: "When the war is over, we will be able to give to marine and industrial users of engines the same assurances of our skill as engine builders." For Sterling, as for others, however, that opportunity never came. Instead, the world changed.

The end of World War II spelled the end for the purpose-built marine engine as it had been known for roughly five decades. In the postwar period, the great volume of business trended toward outboard motors and mass-produced fiberglass runabouts. By about the mid fifties, comprehensive model lineups that had long sustained Kermath, Chrysler, Universal, and Gray no longer seemed needed. Even Chris-Craft's initially promising postwar production couldn't support the remaining engine builders, and the company, which had tended to downplay the potential impact of fiberglass, finally had to reinvent itself.

The postwar period could be summed up succinctly for marine engine manufacturers —

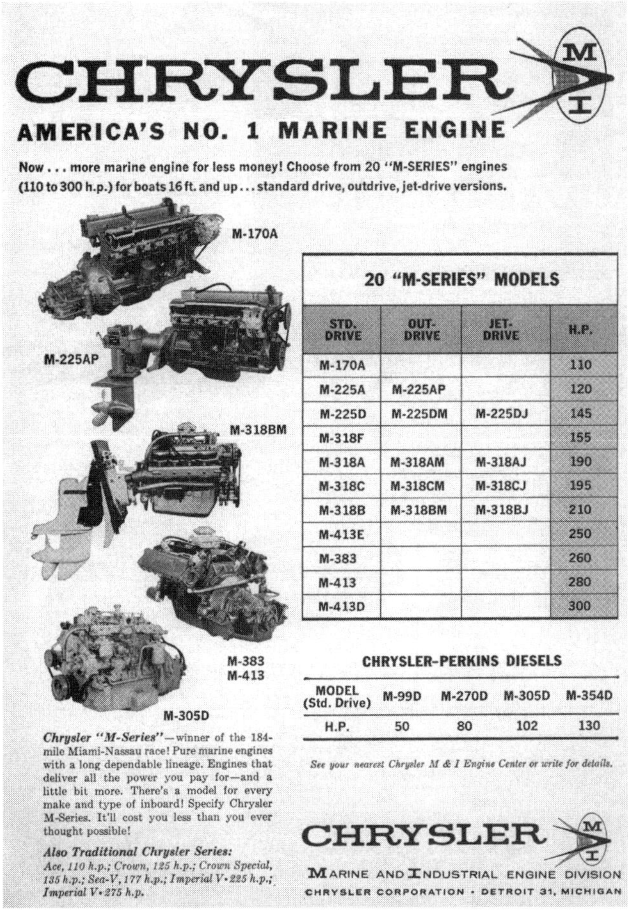

Chrysler ad, 1961.

merge, sell, or die. Gray executives made the stunning decision to reject becoming part of General Motors and instead opted for takeover by Continental. When Continental failed and corporate conglomerates became involved in what remained, Gray became nothing more than a once-grand name now viewed largely for its value in terms of spare parts. Palmer Brothers, sold to a company with no interest in innovating, failed and finally closed when an attempted revival by former employees didn't work out. Packard's marine engines died when the company, once America's premier automaker, ceased production. Universal did better than most and managed to soldier on well into the postwar era with a variety of engines. Eventually, even its tried and true Atomic Four finally fell victim to comparably sized, modern diesels.

The makers of slow-turning diesels, like Atlas, whose development dated back to the mid teens, also failed. Once the answer to high-priced gasoline for commercial applications, the classic marine diesel with its common rail injection — the great innovation that replaced cumbersome, leak-prone air injection systems — enjoyed a reign that ended abruptly after the war. Atlas diesels, the old Wintons, and Superiors, were now entirely replaced by compact, mass-produced, high-speed GM diesels, while engines from Cummins, Fairbanks-Morse, Buda, and a few others accounted for the remainder. The grand old diesel engine makers' only value now was to a company that wished to corner the market on spare parts.

Packard ad, 1947.

In only one area did the classic marine engine manage to hang on, if in a small way. That was Canada. In Newfoundland, for decades an important market for a variety of engine makers, the provincial government supported the development of a make-and-break ignition, two-cycle engine introduced in 1945 by the E. F. Barnes Company. Sales continued into the 1960s. In British Columbia, the Easthope engine continued to find a small market into the 1970s. In Nova Scotia, at the Lunenburg Foundry, the famous Atlantic two-cycle engine remained in production to at least some degree into the '80s.

"The trend," said the current president and CEO of the company, Peter Kinley, "was for more outboards to be used in the small boat fishery, and successful fishermen wanted more power than the Atlantic two-cycle could produce. Government policy also favored diesel fuel engines to eliminate the explosion hazard of gasoline engines. Fishermen could not get loans from the loan board for installation of an inboard gasoline engine." Finally, even the stalwart Atlantic passed into history, and Peter Kinley remembers well the day that event was recognized.

"In 1986 I was invited to send a sample engine to Vancouver for display in the Canadian National Pavilion at the Worlds Fair, Expo '86," he said. "It was to be included in an exhibit on marine transportation." When Kinley and his wife visited the Fair, however, they discovered the engine was displayed not in an area of the country's up-to-date technology but in an exhibit of the anthropology of coastal peoples. "The Atlantic engine was considered as a piece of history," he remembered, "as important to the development of the country as the birch bark canoe, but all the same, a technology of yesterday."

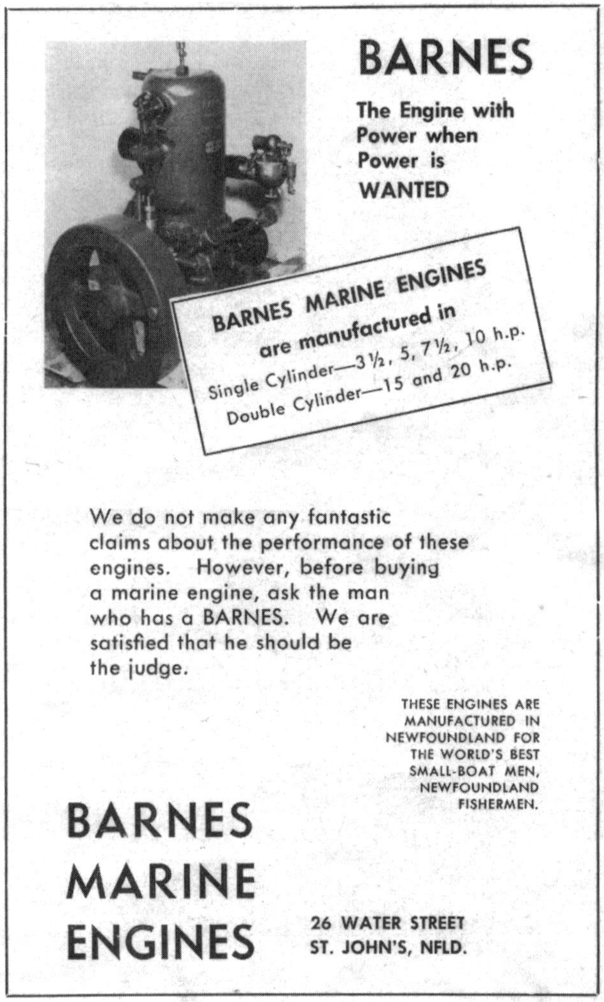

Barnes ad, circa 1946.

CHAPTER TWO:

Technologies, an Overview

Marine engines of the period from roughly 1885 to 1905 relied on technology that was usually developed by inventive men with practical experience but little or no formal engineering training. Born in the period just before or after the end of the Civil War, those who first created and refined the internal combustion engine possessed natural mechanical gifts. Many were American-born; others were immigrants from Germany, England, Scotland, Canada, and elsewhere. Some had backgrounds in steam engines. Many were skilled machinists. Several had gained wealth and experience in one of the great boom industries of that time, bicycles.

How did they do it? Many were mechanical prodigies. Most seem to have taken careful note of the work of others. Some of the pioneers noted that their lives were profoundly changed by the Otto engine they saw at the Philadelphia Exposition in 1876 or the engines at Chicago's Columbian Exposition in 1893. Having seen one or more engines, the would-be inventors applied their own ideas for improvement, and then proceeded on the principle of, to use a term familiar in their own time, "by guess and by luck." Soon enough, however, college-educated mechanical engineers entered the picture, and graduates of Cornell, the Massachusetts Institute of Technology, the University of Michigan, the University of Ohio, and many other schools found employment throughout the industry.

Whether or not the pioneers possessed engineering degrees, their work method was to draw, build, test, evaluate, and modify. Despite formidable obstacles, progress was inexorable and even, in hindsight, fast. Writing on the occasion of *Motor Boating* magazine's 10th anniversary

Engine builders always tried to present their factories as modern and sprawling, whatever the reality. This illustration was included in the 1911 catalog of the C. T. Wright Engine Co. of Greenville, Michigan.

issue in December 1917, C. B. McCuaig of the Buffalo Gasolene Motor Co. noted, "Ten years ago we were in an age of brick and mortar. Today we are in an age of steel and concrete. In 1907 the Marine Engine Industry had not even begun to realize the possibilities of its field."

By the time McCuaig wrote that, the biggest hurdles faced by the industry's pioneers had been surmounted, and key questions had been resolved. In the Midwest and New England, the sometimes heated debate about whether two-cycle or four-cycle engines would prevail had been answered in favor of the latter. Ignition — what the famous German automobile pioneer Karl Benz once called "the problem of problems" — had, with the exception of the California marine engine, resolved itself into an almost universal adoption of spark plugs fired by battery or battery and magneto systems. After needed refinement, the carburetor had become the accepted method of feeding the air and fuel mixture into the engine.

As early as about 1906, many of the pipes and fittings that might remind one of plumbing equipment began to disappear from engines, replaced by flanged connections and cast manifolds. New engines with individual cylinders and exposed mechanical components had, with the exception of heavy diesel and semi-diesel engines, largely disappeared by 1917. By then, the conventional idea that separate cylinders were an advantage if catastrophic damage occurred to one cylinder, had considerably faded. Engines with pairs of cylinders cast as one — "en bloc" — began to appear, and they were followed by sleek multi-cylinder engines with precisely cast and machined one-piece blocks. With such construction came internal oil passages instead of exposed brass lines, and mechanical elements, often including even the flywheel, shrouded from view.

Today, it is well over a century since pioneers like Daniel Regan, Clark Sintz, the Palmer brothers, Edmund Wilson Roberts and their contemporaries blazed the often treacherous trail toward successful internal combustion. Their inventions now seem both wonderful and instructive to those who grew up long after the basic challenges had all been met and the goal became refinement and ever-improved performance. It is the now-primitive technology inherent in the early marine engine that makes it fascinating to collectors and, it must be said, sometimes frustrating to all but the most expert mechanics when an engine cannot be made to start or run reliably. This chapter provides an overview of some key early marine engine technologies and components.

Two-Cycle Engine Types Although apparently simple in terms of mechanics, the two-stroke possessed its own inherent peculiarities, which should not be underestimated as stumbling blocks to successful operation. Such engines enjoyed their greatest success in the hands of commercial fishermen. Of course, it took some time before even those stalwarts learned the occasionally mysterious ways of their "one lungers."

Pleasure boaters took longer to master the engine, and there were doubtless many who never became comfortable with machines that had to be started by rocking the flywheel and kept running by adjusting the mixing valve, grease cups, and ignition timer. Today, a collector seeking to operate such an engine must hope that he will have to learn only its special traits and that mechanical wear or a botched detail in some prior rebuild hasn't added to the challenge.

The basic types of two-cycle motor were the two-port and the three-port. The former took in fuel and air through a check valve in the base. Once the mixture had been drawn in by the partial vacuum created during the piston's upward stroke, the check valve closed. The mixture was then compressed on the piston's down stroke and forced through the transfer port into the combustion chamber. The three-port engine eliminated the check valve. It took in mixture from a carburetor through a port (the third port, cast into the cylinder wall) that was uncovered on the piston's upstroke.

The two-port's check valve was required to maintain crankcase compression as the piston descended, preventing the mixture from being forced back out of the base. Pulled open by vacuum against the pressure of a spring and then closed by the spring, the check valve tended to limit rpm potential. (Restorers should be aware that, if replacing a check valve's spring, its tension can't be so strong that an engine with somewhat diminished compression won't properly open the valve.) Discussing the subject of check valves in 1912, engineer A. E. Potter noted that "there is a limit to the number of times that the tappet of a check valve can be lifted from its seat and become reseated within the space of one minute. There is also a considerable frictional resistance of the gas to be overcome in high-speed motors, which a check valve at such high speed materially increases."

This inherent aspect of a two-port engine — how fast the check valve could be reliably opened and closed together with efficient flow of mixture — wasn't a concern to the commercial fisherman, workboat operator, or even some recreational boaters. Two-port engines like the Palmer, Lathrop, and Mianus ran reliably at lower rpm, helped by a flywheel that tended to be heavier than a three-port engine's, and also at the higher end of the engine's speed range. Owners did not expect precise throttle control and instead adapted their operating methods and propeller size to deliver the needed rpm. Still, the check valve could pose operational problems, especially if it failed to seat fully and permitted some of the mixture to be forced back past it, diminishing

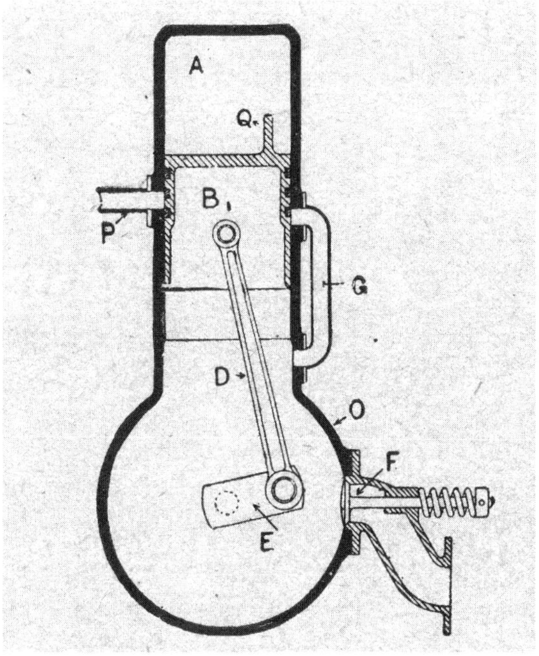

The drawing above shows a typical two-port, two-cycle engine. The check valve (F) feeding mixture into the base is clearly indicated.

The drawing below is of a three-port, two-cycle Detroit. The inlet port is located in the cylinder wall and a carburetor is fitted.

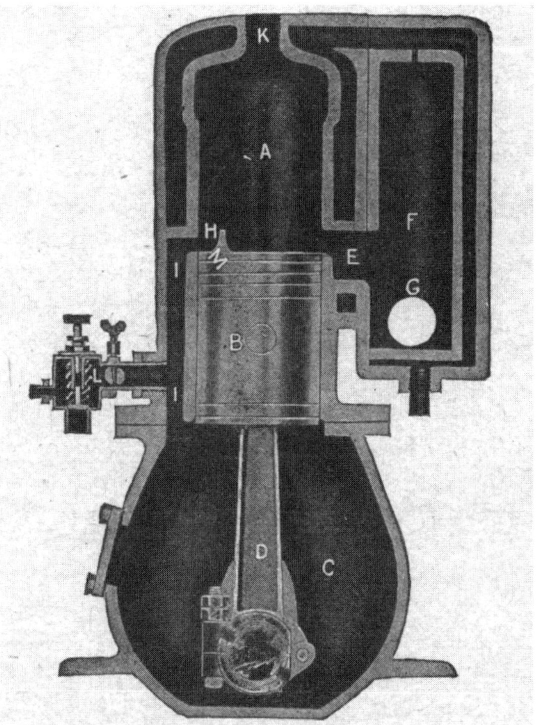

crankcase pressure. Today, those hoping to get a two-port motor to run reliably must be sure that the piston and rings provide adequate compression, and that the crankshaft bearings are tight enough for sufficient vacuum to be maintained in the base. The check valve should be inspected for good sealing, and if the engine won't run, to ensure it has not been installed backwards.

In 1910, at the Boston Boat Show, 119 models of two-port engines and 86 three-port models were on display. That ratio, however, was then well on the way to being reversed. The two-port's initial dominance had been caused in part by the need to circumvent the patent on the three-port, two-cycle engine that was held by Englishman Joseph Day, who had acquired it from inventor Frederic William Caswell Cock of Bath. The patent did not expire until October 1906. After that, more three-port engines were introduced.

With two- and three-port motors being sold at the same time, debates between makers of both types were heated. Each manufacturer cited special attributes for its own method. A principal argument was whether the partial vacuum created by the piston's upstroke on a three-port motor, and the resultant strong puff of indrawn mixture, was preferable to the less energetic but more prolonged intake process of the two-port.

In the end, the issue resolved itself by the two-port motor's being favored for work and heavier-duty applications, and the lighter, more compact, three-port models, for recreational and performance applications. In the period from about 1910 to 1916, at the same time as hydroplane-style hulls were being developed, lightweight, high-performance, multicylinder two-cycle engines were offered by such makers as Pierce-Budd, Elbridge, Emerson, Roberts, and Termaat & Monahan.

As if to add further complication to the argument about two- versus three-port motors, some makers ran ads classifying their products as a two-/three-port or a three-/four-port! The additional port — often called an "accelerator port" or "auxiliary air valve" — could be adjusted to admit fresh air to the combustion chamber. The purpose was to increase power by helping drive out the burnt exhaust gases. On a two-port motor, an accelerator port might be added to either the base or the transfer port. On a three-port, the accelerator was usually mounted near the top of the transfer port. Operators of such engines were advised that the port "should be closed when starting or running slowly, but when running at high speed, it is gradually opened until the best possible results are obtained."

Finally, on this matter of two or three ports, some engines had a base and cylinder casting with removable plugs so the engine could be operated in either fashion — port-controlled or with the mixture drawn into the base. Because of the higher vacuum pressure required by a three-port, the engine would become impossible to start or throttle if problems developed with crankcase leaks through the bearings or piston rings. Changing it to a two-port by adding a check valve could often put things right until the problems became so serious that new bearings or rings were needed.

Delivering Fuel and Air "There is no use in denying the fact that the carburetor, vaporizer, mixing-valve, or whatever the apparatus may be called that gasifies the liquid fuel, is one of the prime factors or secrets of the efficiency and reliability of the internal combustion engine," noted an article in *The Gas Engine* in November 1905. And that was putting it mildly. These various devices were key to an engine operator's success, but the terminology applied to them can now seem confusing. Even at the time, terminology regarding carburetors, vaporizers, generator valves, and mixing valves routinely overlapped.

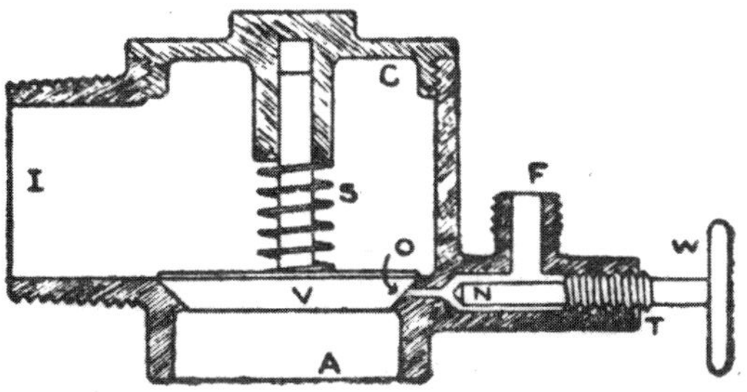

This primitive generator valve shows the needle valve N. Fuel entered the generator valve body at F. Air entered through the opening A. O is the orifice through which fuel was fed into the air stream. The valve threaded into the crankcase and mixture entered at I. No adjustment for valve lift is included in this design.

Among the very first devices intended to deliver vapor to an internal combustion engine were small fuel tanks, often no larger than a gallon in size, in which cotton waste was suspended into the gasoline. The fuel was absorbed by the fabric "wicks." Air entering the tank passed over the wicks and the resultant vapor left the tank through a pipe that led to the engine. There, the vapor was drawn through a check valve into the crankcase, if a two-cycle, or into the combustion chamber of a four-cycle. In his catalog of 1891, Daniel Regan labeled his tank a "carburetor"; the term was also used by Regan's competitor, the Pacific Engine Company. Such a device would generally be called a "vaporizer," but by the early 20th century, that term was not always consistently defined. Back east, in Connecticut, Ray Palmer named his version of an apparatus similar to Regan's a "gasometer." Such a device was better suited to stationary applications than to a boat. According to Frank Palmer, the boat in which the first Palmer engine was installed had to be rocked from side to side to slosh the gasoline around and keep the fuel/air mixture sufficiently rich to maintain operation. It was obvious that something more sophisticated was needed.

Mixing Valves Beginning around 1895, mixing valves — also routinely called vaporizers and generating valves (devices to "generate" vapor) — began appearing on engines from coast to coast. Engine makers including Palmer, Union, Kahlenberg, and Globe manufactured their own mixing valves, but many used a bought-out valve such as those manufactured by Lunkenheimer or Monarch. Whatever the brand, all vaporizers worked pretty much in the same way. They admitted gasoline along a needle valve adjustable by the operator. The check valve halted the flow of fuel until the valve was opened on the upstroke of the piston. Air drawn past the needle-valve orifice then picked up the fuel, and the resulting mixture entered the base as a mist. When the piston reached the top of its stroke, suction in the base ceased and the valve was returned to its seat by spring pressure.

The simplest vaporizer's only adjustment was the needle valve itself. "There is no regulation of air," read Palmer's operating booklet circa 1895. "Regulation is by changing quantity of Gasoline." This was done using a knurled wheel kept from rotating on its own by a ratchet. Later, most vaporizers also permitted valve lift to be regulated as a means of throttle control. Some models included a throttle plate as well. Usually included was a rod that permitted the valve to be manually operated and that would, it was hoped, dislodge anything that might be keeping the valve from seating properly. Before 1909 when it became common to mix oil with the gasoline, oil residue in the crankcase could gum up the valve. The valve seat itself was subject to wear and required regrinding at regular intervals. Dirt particles could clog the needle valve and promptly shut down the engine. Finally, as the tank emptied, fuel pressure varied. This, coupled with the motion of the boat, forced the engine operator to try to compensate by making sensitive adjustments.

Most vaporizers, if not all, came in standard pipe sizes, the sizes determined by the engine maker according to the requirements of an engine's cylinder. A few were as small as a half-inch. The more common sizes were ¾-inch, 1-inch, 1¼ -inch, and 1½-inch, and a few were 2-inch. Engineer E. W. Roberts reported that, "Generally speaking it will be found that an engine will run better, giving more power, at a higher speed if the size of the mixing valve is slightly larger than the inlet pipe leading to the engine." The valves were manufactured with National Pipe Thread (NPT) threads. The few that were not NPT typically had special SAE-style two-bolt flanges that mated with the maker's engine but not other engines except by happy coincidence.

Carburetors The idea was grand: Create a device that would automatically provide the correct fuel/air mixture under varying loads and temperatures with little or no adjustment by the operator. The invention that promised to do this was the float-feed carburetor that would keep consistent the amount of vapor available to the engine. Not surprisingly, early efforts to produce such a device encountered problems. Among other things, the first carburetors proved wasteful of fuel because they managed to vaporize only what E. W. Roberts called "its lighter constituents." The result was gummy residue that could clog up passages.

All involved recognized that much work needed to be done, and that efforts to adapt the carburetor from an automobile engine to a marine engine seldom resulted in success. In 1905, the magazine *Power Boat News* put it simply: "... attention could be diverted toward developing a carburetor designed especially for marine engines." The magazine also noted that cost was a factor in the choice between vaporizers and carburetors. "To make the 2-ported, 2-cycle gasoline engine sell within a price favorable to all, the vaporizer is yet a noticeable feature, and the cause of most of the trouble consistent with this type."

The principal company that gradually solved the problems involved in creating a float-feed carburetor that worked reliably afloat was formed by Frank Wheeler and George Schebler. Wheeler was the financial expert who would later become cofounder of the Indianapolis Motor Speedway. Schebler was the engineer. George Schebler's carburetor developments began as early as 1901, but the business really started in 1904 with the encouragement of Harry Stutz. Stutz had already recognized the potential of the automobile industry and understood the enormous appeal to manufacturers of a carburetor that would work reliably on a variety of engines. For a period of about one year, Stutz was Wheeler-Schebler's chief salesman. By 1911, Stutz's hunch

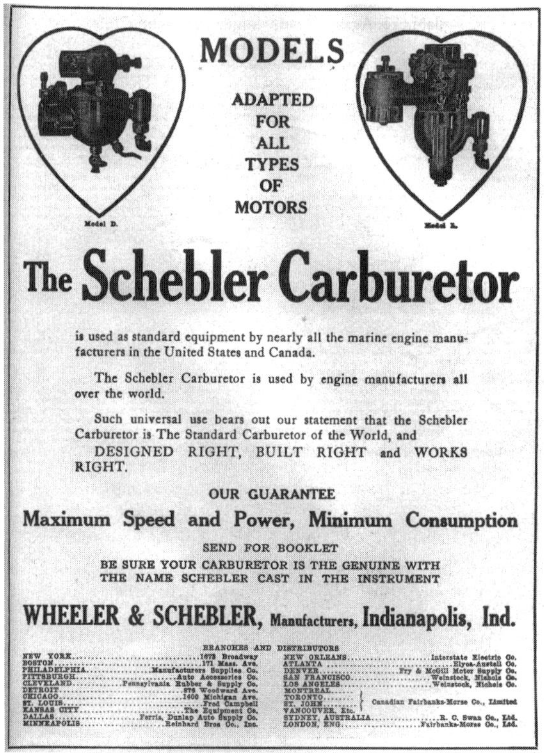

This 1915 Schebler ad continued the long-standing theme of enclosing the carburetor in the outline of a heart, suggesting buyers would love their new acquisition.

about the potential of carburetors had been proved correct, and Wheeler-Schebler built a landmark factory in Indianapolis where a variety of carburetors were mass produced.

In 1924, the company's leading sales executive was George T. Briggs, who had joined Wheeler-Schebler in 1905. Briggs recalled the company's first customer was the Truscott Boat and Engine Company of St. Joseph, Michigan, an important pioneer engine maker. Soon enough, other marine engine companies adopted what Schebler called its Model "D" carburetor. They included Lathrop, Palmer, Sterling, Scripps, and Buffalo. The "D" was available in a half-dozen sizes, and a check-valve could be fitted if the carburetor was to be used on a two-port, two-cycle engine.

The Schebler Model D was designed for either side-draft or updraft installation without modification, a necessary feature that was soon copied by all competitors' models. If the Model "D" was a big step toward reliable operation, it still required adjustment, and an instruction sheet accompanied each carburetor that described preliminary, low-, and high-speed adjustments. One owner, writing about the topic of adjustment in *The Rudder*, advised: "The needle valve should be adjusted for slow speeds and the auxiliary air valve for high speeds. Once adjusted, let both alone: if there is trouble, hunt for it somewhere else [scale from the tank, dirt or sand in the fuel, etc.]. Carburetors do not get out of adjustment unless aided by someone's fingers."

Watermen, however, became adept at adjustments. They would set the needle valve slightly differently for wet and dry days and also became expert in paring down the needle valve to save fuel. Before any adjustment at all was undertaken, the owner was advised by Schebler's instruction sheet as follows: "Connect the carburetor to the intake as close to the motor as possible and so that the gasoline connection is about

This Breeze ad appeared in 1906.

eight inches below the bottom of the gasoline tank. Clean the tank and the gasoline line thoroughly. Use copper or brass pipe, with a free passage of three-sixteenths inch. Do not use rubber tubing for gasoline connection. See that the ignition is properly timed and that the spark plugs are clean and the points have the proper gap. This precaution is most important because proper carburetion cannot be obtained if the ignition is faulty."

Carburetors from Schebler, Krice, Monarch, Kingston, Holley, Breeze, and other suppliers were used by most engine makers, but sometimes an engine builder would develop its own unique carburetor. The Camden Anchor-Rockland Machine Company, maker of Knox engines, was

This Krice ad appeared in 1913.

one such, Ferro another. These were all updraft carburetors mounted low on the engine where they were sure to be fed by gravity from the boat's fuel tank.

Typically, a small pan approximately 3 inches deep by 4 inches wide by 6 inches long, with a brass fine screen to ward off a backfire would be hung beneath the carburetor(s) to catch overflow caused by choking, and this fuel would then be deposited overboard. The safety issue of both carburetors and vaporizers was not to be taken lightly. Predictable leakage of gasoline from a worn mixing valve into the bilge was certain if a boat owner forgot to shut off the fuel line.

Although the updraft and side-draft carburetor remained a typical installation right up until World War II, the performance benefit of a downdraft carburetor was well understood years earlier. In 1929, Scripps introduced a new series of engines equipped with a Holley downdraft carburetor noting "the advantage of permitting a cooler mixture to enter the cylinder in this way handling a larger charge with a consequent increase in power." Later, in 1934, Chrysler offered two versions of its popular 224-cubic-inch, 6-cylinder Crown. The version equipped with a downdraft Holley carburetor instead of a conventional Carter model developed an additional 4 horsepower, 86 in all.

Whatever performance advantages were possessed by downdraft carburetors, development of "safety," non-drip carburetors kept the updraft models in use on most engines. These carburetors typically had the intake throat tilted upward about 30+ degrees and included a pick-up tube that, when the throttle was opened, picked up any pool of fuel in the bottom of the throat. In the 1920s makers began to apply backfire traps to marine carburetors, and marine engine makers began supplying their engines with these devices. They were not, however, required until the passing of the Safe Boating Act in 1940.

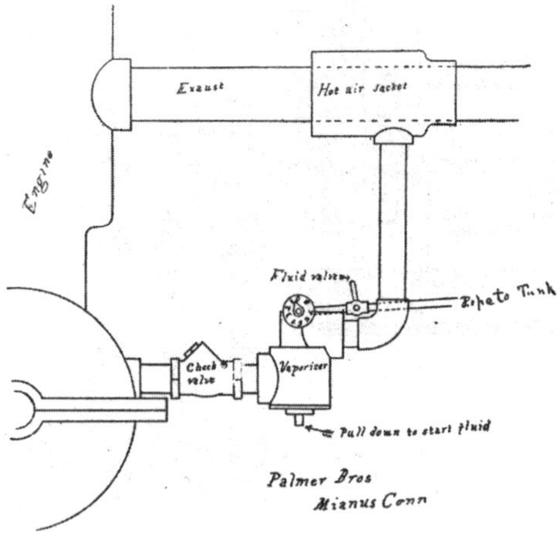

This Palmer layout drawing, circa 1900, shows how air heated by the exhaust pipe was to be directed into the vaporizer to warm the mixture. (Courtesy of Dick Day)

Warming the Mixture Because the process of changing liquid gasoline to a vapor results in a loss of heat, frost can form near and in a mixing valve or carburetor. Any water sucked into a jet might form frost even on a hot day, causing rough or slow running or stopping the engine entirely. In cold weather, ice might form. Everyone recognized that warming the intake air was desirable. In 1905, engineer E. W. Roberts wrote of the need to draw air "from around some hot portion of the engine as from a perforated sleeve surrounding the exhaust pipe leading through the carburetor or jacketing the carburetor with [warmth] from the water jacket of the engine."

In San Francisco, the Union Gas Engine Company noted in its 1905 catalog that "air, before being drawn into the vaporizer and combined with the gasoline, is slightly heated, the result being a dry, warm, explosive gas, requiring a very small spark to ignite, and which is practically at the same degree of temperature in all weather." The key word here is *before*. Many other makers' engines had combined intake/exhaust manifolds, but these generally heated the air downstream of the mixing device. Still, the 1911 Ferro catalog touted the virtue of its combined intake/exhaust manifold noting that the "ingoing gas is carried at an even temperature of about 78 degrees. This prevents the ingoing gas from condensing."

Doubtless the most efficient method of warming the intake air followed Roberts's theory and Union's practice. When Lozier introduced its fine two-cycle engine in 1901, the muffler supplied with the engine had an auxiliary outlet pipe that was to be connected directly to the vaporizer. A section of the outlet pipe was perforated and, in wet or foggy weather, a sliding sleeve could be positioned to cover the holes and increase the volume and temperature of the air. By the time Palmer introduced its ZR engine in the early '20s, it had engineered a two-part casting that fit over the exhaust pipe and fed hot air from its outlet through a flexible metal hose to the carburetor intake. Such a system was widely used by most engine companies. Chesapeake watermen, at least, referred to such warming components as the "oven."

Ignition: Flame and Hot-Tube Engineer C. E. Dawson described the principle of a flame-ignition engine in *The Gas Engine* in October 1906. "The flame is kept alight in a pocket situated in a slide or the plug of a cock; this system was the only one considered reliable in the early days of the compression four-stroke cycle and it did great service for many years."

Keeping alight a flame presented obvious problems in a moving vehicle and left the pioneer designers to come up with another method. Of course, ignition timing in such engines was also a problem. It remained so after the first hot-tube ignition systems entered production. Dawson described this method as "a simple tube connecting with the combustion space of the correct length to produce explosion at the highest point of compression, or it is often combined with a timing valve, which is open at the time when ignition is required." A lot of engines were built with hot-tube ignition, but they still required the use

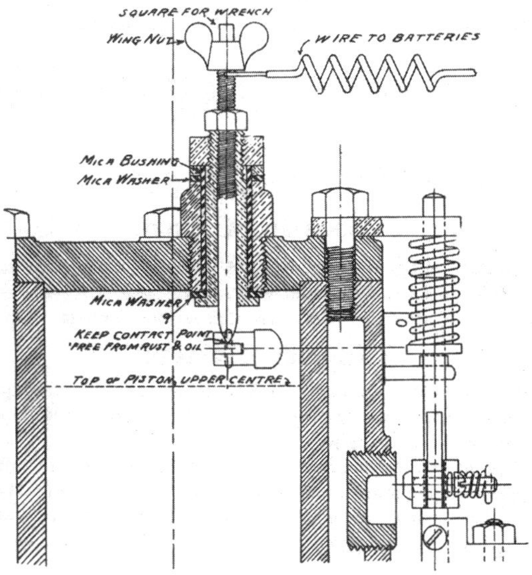

Inside a make-and-break engine: This cutaway of a circa 1898 Palmer shows the arrangement of the stationary electrode with its insulating washers, the movable contact, and related components. (Courtesy of Dick Day)

of fire to ignite the burner, one per cylinder, that heated the tube itself. In the first years of the 20th century, makers of engines with electric ignition were quick to point out that a "spark cannot be extinguished by the wind."

Make-and-Break Ignition Make-and-break ignition systems have no spark plugs, devices that underwent their own development period relating to insulating materials and metallurgy. Instead, a pair of electrodes within the cylinder — one stationary and one movable — was mechanically opened and closed by a rod driven off the crankshaft. The igniter mechanism included such levers, cams, triggers, springs, and set screws as were needed to permit electrodes in the cylinder to remain open until it was time to snap them closed and then back open. When that happened — the action timed to the

Make-and-break igniter: This one was developed and used by Ferro. "A" is the movable electrode, "B" the stationary.

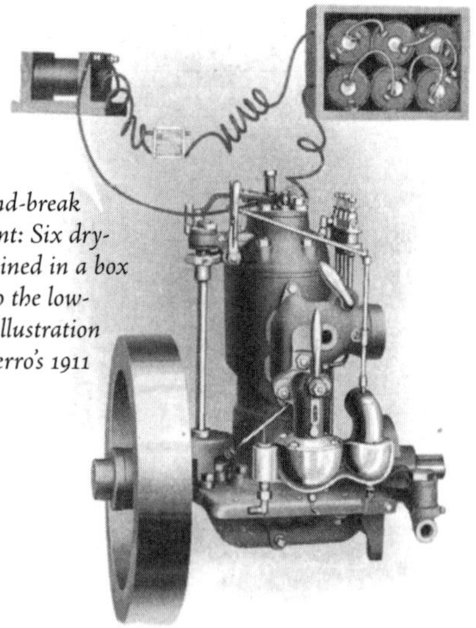

A typical make-and-break wiring arrangement: Six dry-cell batteries contained in a box supplied current to the low-tension coil. This illustration was included in Ferro's 1911 catalog.

desired instant of ignition — an arc generated by a low-tension coil was drawn across the electrodes' gap. (Think of the arc as the same thing an electric welder produces. A spark contains very low current but very high voltage, some 10,000 to 15,000 volts.)

Engineer Dawson described the make-and-break as a system "in which a current of low voltage is employed in combination with a plain spark coil [an inductance coil] whose function is to increase the spark, which is produced when any electric circuit is broken; this break is made by the action of the so-called hammer break [contact points] operated by the engine."

Being a low-tension (meaning low voltage but high current) system, the make-and-break was largely impervious to moisture. To demonstrate, salesmen would splash water over a running engine. Such an act would stop a jump-spark (spark-plug) ignition engine immediately unless the spark plug was protected by a completely waterproof cover. Protection was also needed for the coil and battery.

Despite the make-and-break system's advantages, producing an igniter that operated reliably, didn't lose its spring tension, and permitted some timing adjustment was a stumbling block for all the early would-be engine makers. Not only were there the mechanical aspects of the igniter to be considered, but the whole "sparking outfit" including the coil and batteries had to be carefully integrated. How long the contact points remained closed — technically known as "dwell" — could dramatically affect battery life. Linked to this matter was whether an igniter could also provide enough sparks as rpm increased.

"The lack of uniformity in the time length of contact on gas and gasoline engines of the make-and-break type seems to furnish a subject for discussion that should result in material benefit to all parties interested … including the manufacturers and users of engines as well as the manufacturers of and dealers in ignition supplies," wrote E. E. Hudson in a paper read before the National Gas and Gasoline Engine Trade Association at Chicago in December of 1909.

Some had done better at solving the problem than others. As early as 1895, Ray Palmer recognized that then-available make-and-break igniters suffered from a long dwell time, which chewed up battery life, and a slow break which did not take advantage of the collapsing field of a high-current inductance coil. Palmer invented an igniter based on two springs and a latch mechanism that proved to be particularly successful.

The reality of make-and-break igniters, at least as applied to the classic two-cycle one-lunger, was that they all had their idiosyncrasies. "Father had good luck with his Lathrops," Martha's Vineyard fisherman Oscar Pease told me once, "but when he got a boat with a Mianus, he was always having trouble making it run right. It had a different make-and-break system that was very tricky to get adjusted. And just about the time you did, it would wear a little and go out of time."

The mechanical complexity of the igniter mechanisms, as adapted to multicylinder, four-cycle engines built in California, now impresses one as being remarkable examples of engineering design and casting and machining skills. In fact, these make-and-break ignition systems contribute mightily to the fascination and desirability of those engines. Such marvels were not limited to the West Coast. In 1908, Buffalo showed a four-cycle, 4-cylinder engine with make-and-break igniters operated by an overhead shaft.

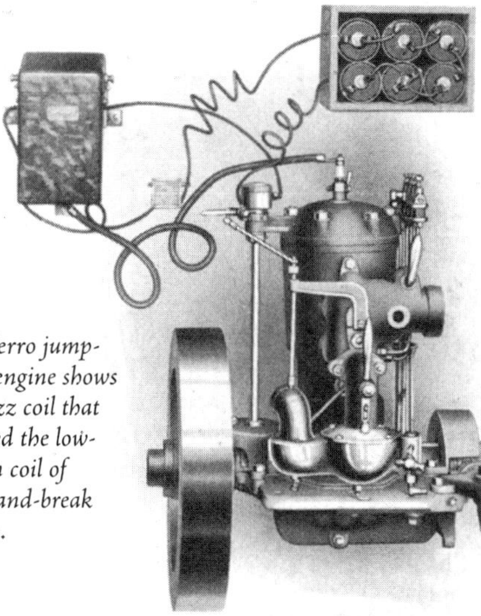

This Ferro jump-spark engine shows the buzz coil that replaced the low-tension coil of make-and-break models.

What kept the make-and-break system viable for so long on marine engines, especially the medium- and heavy-duty models used commercially, was the fact that these remained relatively slow-speed machines that operated at speeds of 350 to 500 rpm. Although the systems were subject to mechanical wear and could require rather frequent adjustment, a man who used his motor on a daily basis became quite expert at recognizing when adjustment was needed and how to do it. Some manufacturers continued to offer make-and-break ignition as an option even after the high-tension, jump-spark system had been made largely reliable.

Jump-Spark Ignition In the jump-spark system, as succinctly defined by C. E. Dawson, "A spark of sufficiently high tension to jump across a gap in the combustion space is generated by means of an induction coil whose primary winding is operated by a convenient source of electric current and the circuit is completed at the moment when combustion is needed." A vibrating "buzz" coil was often used to fire the plugs despite the trouble they were known to represent in terms of adjustment and trouble-shooting. Jump-spark systems with non-vibrating coils would ultimately replace the varnished box containing the buzz coil.

The great advantage offered by the jump-spark ignition system was its better adaptability to higher engine speed and elimination of the igniter's expense and complexity. Most makers advised that jump-spark engines produced more power than make-and-break models.

Ignition was adjusted by the engine's timer. Timers were driven by gears off the crankshaft. The timer usually consisted of a brass case in which the "timer spool" was mounted. The spool was composed of fiber insulation material except for a brass segment that came in contact with a terminal mounted on the case. As the spool revolved, contact would be made and broken, resulting in a spark at the plug. Multisegment timers were used on multicylinder engines.

Ignition was advanced or retarded by moving the timer handle from left to right or up and down. It was typical on most timers to have a hard rubber cut-out button on the handle. The

Inside a timer: This Ferro timer was typical. The cover has been removed to show the spool. The handle was moved to adjust timing and is equipped with a cut-out button.

button was connected to the ground wire, and pushing it could quickly stop the engine without resorting to opening the ignition switch. The engine could also be reversed by pushing the button and, just before the engine stopped, moving the timer handle from the advance to the retard position.

Current Sources The wet battery, essentially a jar containing zinc and copper plates immersed in sulfuric acid and water, was used during the early internal combustion period. Although such batteries could be renewed with a fresh shot of chemicals, they were cumbersome and had obvious limitations aboard smaller boats. Instead, dry cells, storage batteries, and magnetos soon gained prominence. Eventually, the Edison storage battery and others like it would become typical although one observer called storage batteries "the best misunderstood piece of electrical apparatus on the boat."

To capitalize on real and perceived problems involving batteries, magneto manufacturers quickly advanced their own agenda. The Electric Goods Manufacturing Company, which had locations in Boston and Canton, Massachusetts, claimed that sales of its Perfex Waterproof Power-adding Ignition were five times larger in 1910 than 1909 and looked ahead to even bigger gains. It claimed that half the engine makers exhibiting at that year's boat show in New York had used its magneto because "power boaters in general are fast coming to the point where they will not buy or

A wet-cell battery of 1911 looked like nothing so much as a crockery jar with removable lid that permitted the electrolyte to be replenished.

A typical magneto, this one supplied with Ferro engines.

use an engine that is not equipped with ignition that is absolutely waterproof... and [the Perfex magneto] is being welcomed by all battery users as the solution of their troubles." It was not unusual for magnetos to also be referred to as "sparkers."

Whatever the advantages of a magneto, the coil that fired the one-lunger's make-and-break igniter or spark plug was usually connected to a group of dry-cell batteries yielding a minimum of 6 volts. These batteries, often Columbia N° 6 cells made in Cleveland by the National Carbon Company, were about 6 inches high and weighed about 3 pounds. The 1½-volt dry cells were connected in series; it was typical to mount them together in a box for convenience.

Dry-cell life was a concern to all. The issue was topical enough for engineer and engine designer Ernest Graef to market the "Graef Dry Battery." This was a foot-long box containing 8 cells. The idea was that when the first four cells weakened, one could conveniently switch to the remaining four. To prolong dry-cell life, a magneto was sometimes used. At Lozier, on 5-hp models

Columbia dry-cell batteries, and others like them, were familiar to thousands of boat owners in the pre–World War I period and were still being advertised regularly into the late '20s.

and larger, a low-tension magneto was mounted on the hand-hole plate. The magneto was driven by a friction wheel off the flywheel. "The instant the engine is started, the magneto is switched into circuit by means of a simple switch, and the batteries, therefore, will always be at rest, except when the engine is being started, and will last an indefinite length of time."

On multicylinder, four-cycle engines, spark sources and the plugs themselves were always redundant. Not only was this felt necessary for reliability, but the firing of two or more plugs was needed to get the most power from the low-octane gasoline of that era in combustion chambers such as the T-head, which were inherently inefficient. In its circa 1912 catalog, the Loew Manufacturing Company of Cleveland repeatedly stressed that its Loew-Victor motors were furnished with "two separate systems of ignition operating two separate sets of spark plugs and deriving current from two separate sources, battery and magneto." Dual ignition, at least, would be a staple of most multicylinder engines for years, from the fisherman's Palmer to the pleasure boater's Peerless or Sterling.

Dual ignition was not always deemed sufficient. In 1922, Peerless explained its Philbrin Double Triplex ignition system that "consists of three separate systems, all of which are connected to double spark plugs located on opposite sides of the cylinders. The Philbrin system has a very decided advantage over the other two spark battery systems, owing to the fact that it is synchronized electrically instead of mechanically." Such engines, however, required a level of involvement and understanding that boat owners a decade later had no interest in acquiring.

This Packard Electric Co. illustration of 1915 shows a capable technician attending to the electrical system, always a potential trouble source.

Circa 1905 ad of The Induction Coil Co.

Cylinder Manufacture With the exception of larger companies like Camden Anchor, Rockland Machine, Lozier, the Lunenburg Foundry in Nova Scotia, the Standard Gas Engine Company of San Francisco, and others that possessed both their own machine shops and foundries, the castings for cylinders (and usually the crankshaft forgings and other components) were supplied by vendors. The C.T. Wright Engine Company catalog of 1910 was not inaccurate when it stated, "The average engine builder does not really build engines. He assembles them. He has the castings made and then buys Smith's commutator, Jones' carburetor, Brown's propeller wheel, etc."

The casting and boring of a one-piece cylinder for a two-cycle engine or a four-stroke's T-head required great skill. Not only did the cylinder water jacket need to be formed, but so did the water passages inside the integral head. Such castings required the pattern maker to know his craft thoroughly, and the patterns and cores were built to minute tolerances. The sand used for the intricate molds required special handling

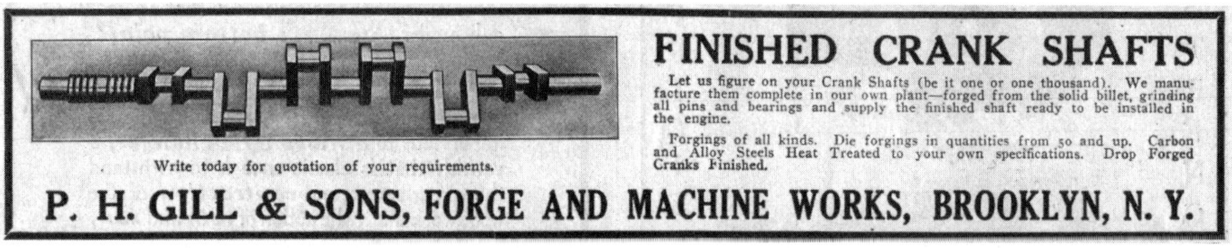

This 1913 ad touted Gill as a supplier of forged crankshafts. There were many such suppliers.

and was not a material simply picked up off the beach. Any mud or salt in the sand rendered it unusable for casting. Sand that became known as "Albany molding sand" was used by many Eastern foundries.

The final product's quality relied on the skill of the molder who would mix the sand with water until he judged, from experience, when the perfect consistency had been reached. Then the sand, mixed with core oil, was packed into the core molds and pounded firm. The molds were then baked at 250 to 300 degrees Fahrenheit until they had the consistency of a cookie. Finally, the cores could be assembled, and the pouring could commence. The alloy used was generally referred to as "gray iron," this being an alloy that combined long wearing properties with a density that, in theory, would result in a nonporous casting that would allow no oil to migrate from the engine's inside right through to the exterior surfaces. Fresh castings were allowed to "cure" for up to three months, heated naturally during the day and cooled at night, to relieve stresses in the metal.

Until en bloc cylinder blocks became typical around 1918, multicylinder engines were based on individual cylinder castings secured to the crankcase. The weight of the castings was recognized as a disadvantage by a few makers of high-performance engines. In Detroit, the T-head cylinders of Van Blerck's racing and performance models were cast with a large portion of the water jacket removed. This method permitted a casting of more uniform thickness to be achieved because the core positions could be better controlled. The open spaces were later covered with a brass sheet secured with machine screws. The whole was then pressure-tested against leakage.

Many engines of the period from 1900 to 1910 did not have detachable cylinder heads because gasket material (also referred to as "packing") was prone to leaks. Makers of "headless" or "dome-head" engines, and they were many, were quick to point out the head gasket joint as a potential headache. The issue as presented by the Acadia Gas Engine Company was typical of all who chose a headless design. "The cylinders are individually cast of a mixture of high grade nickel grey iron, with head integral, doing away with joints and packing."

Those planning to purchase an engine with a removable head were advised that ⅝-inch head studs — the threaded rods extending up from the block — were the minimum acceptable and that ¾-inch studs were preferable. For years, head gaskets were typically made of 1/32-inch asbestos. Usually, the area in which the head mated with the cylinder was machined so that the asbestos gasket was protected from direct exposure to the combustion chamber.

Experienced mechanics learned to soak the gasket in boiled linseed oil to improve its durability without stiffening it as happened when coated with shellac. To make gasket removal possible without destroying the gasket, it was suggested that graphite be thoroughly rubbed into one surface. Those who took the time to read instructions or follow advice were helped in avoiding gasket problems by retightening the head bolts after a new engine had been run or a new gasket installed. Although one occasionally reads of "asbestos and copper wire" being used to form gaskets, the most reliable gasket type didn't appear until Ford created a one-piece compressible copper and asbestos gasket for the Model T.

Make-and-break engines also needed something to seal the joint where the igniter entered the cylinders. Often, this was a metal-to-metal "ground joint" — surfaces machined perfectly flat — that avoided a gasket altogether. Others, especially after the copper/asbestos gasket had been perfected, used one for the igniter as well as the head/cylinder joint and exhaust manifold gaskets. "The [igniter] joint is made tight with a copper asbestos gasket..." said Fay & Bowen of its two-cycle, make-and-break engine in 1912.

Finally, regarding cylinders, although the great majority of engines were designed with cylinder centered on the crankcase, a few had the cylinder offset. Usually seen in two-cycle engines, the offset cylinder was said to equalize side thrust on the cylinder wall on the power stroke, wasting no energy and reducing shock loads on the bearings. Real benefits were, however, probably minimal.

Four-Cycle Cylinder Head Designs

If port layout was the trickiest part of two-cycle design, valve layout and operation occupied a similar if not quite so mysterious level of importance for the four-cycle designer. The performance limits imposed by early intake valves that opened by suction as the piston descended were well recognized. Mechanically operated valve systems of different types soon predominated. Among the most popular, particularly for higher-end engines, was the T-head.

The T-head got its name because the cylinder head overhangs the cylinder on both sides, forming a shape that reminded some of the letter "T." The layout placed the intake valve on one side of the head and the exhaust valve on the other. This resulted in a combustion chamber that was wide and flat, something not in its favor, which was recognized even at that time. "It is considered good practice to have the incoming charge sweep over the exhaust valve and so help cool it," wrote George Crouch in an analysis of valve layout in 1910. "The T-head motor does not comply with this requirement. Neither does it comply with the thermodynamic principle that economy of fuel is gained by reducing the surface of the combustion chamber to a minimum in order to keep the heat loss to water jacket as small as possible." Crouch might have added that the T-head required two camshafts, which added to its overall complexity and cost.

Whatever its shortcomings, the T-head needs to be viewed in the context of its era for, in those terms, it offered several advantages. Among other things, there was plenty of space around

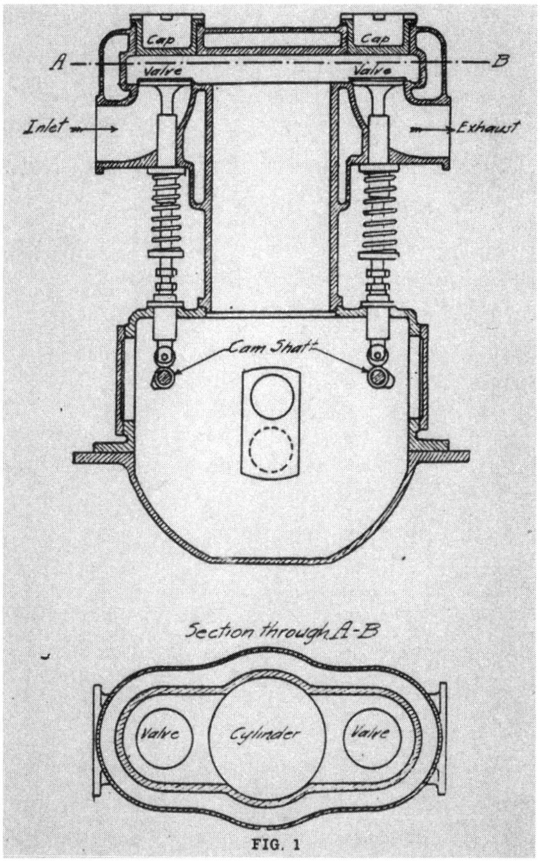

FIG. 1

T-head cylinder

each valve, so that a valve could be easily accessed, an important consideration in those days of rapid carbon build-up and, sometimes, burned valves. Further, there was little wear on the valve train, which was driven by cams fully enclosed in the crankcase where they were well lubricated. The valve springs were easily tended to and the valves easily adjusted. Also, with such a wide combustion space, two or more spark plugs were easily accommodated, and at least one or two of the plugs could be located by the inlet port to promote quick ignition of the mixture.

Finally, it was recognized that there was some cross-flow effect created by the incoming mixture as its inertia tended to carry it across the head and purge the exhaust gas. Widely used during the era before World War I, the T-head survived afterward only in some heavy-duty duty models such as those from Lathrop and from Palmer, whose T-head ZR-1 was known around the Chesapeake Bay as the Cadillac of the workboat engines. The ZR-1 was produced from the early 1920s through 1962, and the last one operating on the Chesapeake didn't turn its final revolution until 2001.

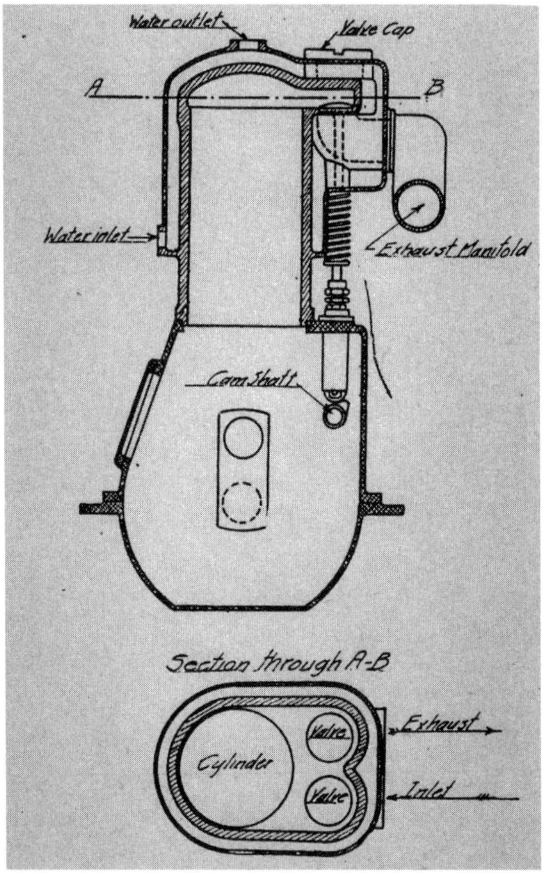

L-head cylinder

The L-head or "side valve" engine was the type that came to dominate the market. In the early production period leading up to World War I, when the vaporization of fuel was an ongoing concern, George Crouch pointed out that an L-head's side-by-side valves could contribute to the "warming of the intake pipe [manifold] so that vaporization is more nearly complete." In addition, the close proximity of the intake valve to the exhaust meant that the cool air-fuel mixture might reduce exhaust valve temperatures and contribute to exhaust valve life.

In the L-head, a single camshaft drove intake and exhaust valves mounted on the same side of the block. These engines were less costly to manufacture than T-head types. Savings were due to the elimination of a camshaft and its drive gear and bearings. This was certainly one reason why so much attention was given to overcoming the L-head's inherent weakness in terms of combustion chamber efficiency. Among the leaders in this research in England was the Cambridge-educated Dr. Harry R. Ricardo. His experiments led to the Ricardo-type cylinder head in which gas flow and engine efficiency were much improved.

By 1924, Chrysler and Packard had introduced models with Ricardo-inspired cylinder heads, and they were followed in 1926 by Pontiac. But everyone was aware of such developments. While marine engine manufacturers before World War I, such as Speedway, Scripps, Red Wing, Clifton, and many others, had L-head engines, they were joined after the war by companies as diverse as Gray and Sterling (the Petrel), Lycoming, Fay & Bowen, and many more. In 1927, when he surveyed the state of marine engine development for *The Rudder*, engineer Harry Lowe Brownback noted, "The 'L' head engine leads in all but the highest powers, but the study of high turbulence heads has boosted the performance of this type of cylinder, so that it is beginning to approach the efficiency of the more complicated and costly overhead rig."

"Among the advantages of the valve-in-head motor," wrote George Crouch in 1910, "may be noted the compact combustion chamber with small wall surface, and a direct gas passage to and from the cylinder. These are important features, as they tend to give increased power on a given cylinder size, as well as a decreased gasoline consumption ... [T]he combustion chamber approaches more nearly the hemispherical form which would give the least possible surface ..." In 1921 Palmer introduced its long-lived, overhead valve engine, the YT.

The Standard Gas Engine Company of San Francisco pointed out in its 1915 catalog "that the four-cycle, overhead-valve type of internal combustion engine produces the greatest amount of power for the amount of gas consumed." Standard even spoke to what would be referred to by engineers as "flame propagation," the manner in which the air/fuel charge was burned in the combustion chamber. "The ability of the Standard overhead valve engine to most rapidly burn the charge of gas is therefore an important advantage."

Writing about overhead valves in 1910, George Crouch assumed that such valve mechanisms might always be afflicted by noisy operation as, in the 'Frisco Standard's layout, where the valves were not enclosed. Such designs, acceptable in fishing or work boats, were quickly replaced for general consumption. By the early

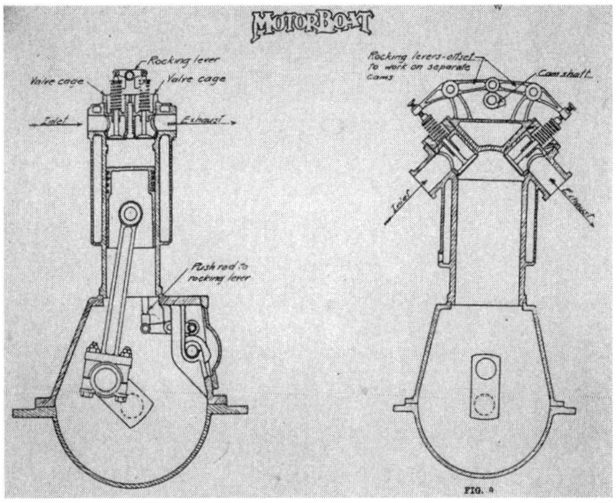

Overhead valve cylinder head (l.) and overhead cam cylinder head (r.)

'20s, fully enclosed valve trains were expected, whether overhead valve or L-head.

The Camshaft Beginning with the pioneers, it was recognized that the camshaft was, in a large sense, responsible for an engine's performance characteristics and smooth operation. But development of the camshaft presented a multitude of problems that included metallurgy, manufacture, and, of course, design. A typical camshaft of the pre–World War I era was made up of individual cam lobes that were milled to shape and then slid onto a keyed steel shaft. They were then "pinned" in place. The Sterling Engine Company, which never built anything but high-grade engines, described this method in its 1911 catalog. "Cams are of hardened steel ground to shape, and are designed to obtain the maximum power and speed from the engine. [Lobes are] Pressed on the cam shaft, keyed, and pinned in place."

In Petaluma, California, where the Corliss Gas Engine Company made a line of heavy-duty marine engines for several years in the period leading up to 1916, a company catalog described its camshafts as being made of "machine steel carefully milled and hardened." The lobes were "fastened to the cam shaft with gib head keys." As was typical of marine engines, these cams were enclosed and ran in a continuous oil bath. The valve lifters were also made of hardened steel and even had roller followers to reduce wear and noise.

As early engine designers gained experience, they concluded that "pinning" cam lobes to the cam shaft left much to be desired. It added to manufacturing time and costs, and all those lobes added to the parts total. Beginning in the period around 1912 to 1915, one-piece camshaft forgings began to be manufactured.

It was recognized, of course, that the shape of the camshaft's lobes was a key factor in determining an engine's performance characteristics. E. W. Roberts was sharing advice about valve size, timing, and lift by 1904 and advising that for a high-speed engine — already the subject of increasing rpm was quite topical —"it is better to use a larger opening and a smaller lift, because a small lift is easier to handle, especially on a large, high-speed engine."

Lubrication Before it became common, in 1909, to lubricate a two-cycle engine with an oil/gasoline mixture — Pierce-Budd may have been the first to prove the method —manufacturers used a variety of lubricating devices. Most common were drip-feed oilers and mechanically driven pumps. Force-feed systems were comparatively expensive, and the various brass oil lines were subject to damage. As for the drip-feed oilers, they required adjustment so that a feed rate of about 10 to 20 drops per minute was achieved. Blue smoke showed too much oil was being fed, which could also foul a spark plug or the igniter contacts. Eventually, except for the rare holdout such as Lathrop or makers of engines to be run on kerosene to which oil could not be added, two-cycle manufacturers standardized an oil/fuel mixture. The most typical was 1 pint to 5 gallons. A high-grade 30-weight motor oil was (and is) best used.

The two-cycle engine's crankshaft was sealed by grease forced in from brass lubricator cups. At the time, grease known as "Albany" or simply as "bearing" grease was recommended. Today, restorers use wheel-bearing grease to seal the shaft or, if the shaft exhibits excessive wear, water-pump grease.

Four-cycle lubrication systems evolved from splash-fed types that basically relied on crankshaft "slingers" to lubricate the bottom end and gravity or force-feed oilers to feed oil to the pistons. Gradually, high-grade engines were equipped with full-pressure lubrication. This eventually became standard practice.

Impurities in the oil were recognized as a threat to engine life, and such makers as Van Blerck had begun installing a "strainer" in the oil system by 1912. Peerless and Sterling were other pioneers in the effort to filter and even cool the circulating oil. Even the finest mesh strainer would not filter all the impurities. The first full-flow oil filter was almost certainly designed and sold commercially by California engineer Charles Winslow, who, together with Elbert Hall (of Hall-Scott), had patented such a filter in 1924.

Reversing Gear Backing up had never represented a problem for an easily reversed steam engine, but making a motor boat back up was a different story. Propellers with reversible-pitch blades and reverse gears were two expedient means of addressing the problem. Such companies as C.T. Wright, Barber Brothers, Wolverine, and the Camden-Anchor-Rockland Machine Company made their own reversing outfits. In addition to their ability to reverse the boat's direction, reversing propellers also could be varied in pitch. Despite these advantages, as automobile transmissions began to be developed, the marine gear began to be made ever more reliable.

At the Paragon Gear Works in Taunton, Massachusetts, the first marine gear was built

Who was Joe? The "Joe" of Snow & Petrelli's ad was a marketing creation intended to humanize an important but otherwise seemingly humdrum product. This ad is from 1911.

by Henry Tuttle in 1899. The original sales were made to the engine buyer who was purchasing accessories. (It had not yet occurred to the engine companies to offer complete packages.) When industry observers looked back on the early days of marine engines and marine gears, they viewed the Paragon as a device that "may well be classed with the pioneer devices of the motorboat industry." The Paragon was a planetary-type transmission with a multiple-disc friction clutch. It operated positively and required little maintenance, which was fortunate because, as some said, "while the motor would have the best of care the reverse gear many times is beneath the flooring entirely out of sight and very many times is preserved in bilge water."

Paragon soon had plenty of company. Snow and Petrelli, makers of "Joe's Famous Gears" in New Haven, Connecticut, had entered the field in 1906. The first Joe's gear was based on a friction clutch and three gears, the motion imparted to the gears through an eccentric and the whole transmission enclosed in an oil-tight case. In 1912 Joe's Duplex Drive Gear was placed on the market with double clutches, and Joe's Husky Gear for diesel, semi-diesel, and oil engines was introduced in 1919.

Other reverse gear makers included the Smith and Baldridge Machine Company, Harthan, and Holliday, all makers of planetary-type reverse gears; Gies made a transmission that included both chain- and gear-drive; and Carlisle Johnson made a fully enclosed, constant-mesh transmission that was very popular.

What about New England's two-cycle pioneers, Palmer, Lathrop, and Mianus, whose make-and-break engines could be reversed "on the switch" by opening the ignition switch and closing it at just the right instant in the crank's rotation? Some fishermen became adept at this skill, but its drawbacks outweighed the advantages for most. Aside from the fact that misjudgment in just when to open and close the switch could cause a serious collision, the absence of a neutral position that permitted warming up and adjusting the engine was an inconvenience. Palmer included a reverse gear as early as 1902 when it introduced its Model K. In the Midwest, two-cycle manufacturers, including Gray, Ferro, Detroit, and Michigan, were among those who equipped their engines with reverse gears. Others, including Buffalo and Sterling, soon adopted a similar strategy.

Reduction Gears In 1916, when Joe Van Blerck introduced his magnificent V-12 "Twin Six" engine, its list price of $7,250 (about $1.3 million in today's dollars) included an optional reduction gear unit. Van Blerck was a pioneer in the technology. Brilliantly, he recognized that a lightweight engine running at comparatively high rpm while turning a big propeller at a comparatively low rpm was the key to overall performance and packaging. The concept worked on boats as diverse as racing hydroplanes, fast runabouts, and a 109-foot Herreshoff schooner named *Mariette*. Despite its comparatively compact dimensions and light weight (versus, for example, a contemporary Winton), the Van Blerck eight-cylinder in *Mariette* could produce a sustained 10-mph cruise speed while taking up relatively little space.

As engine development progressed, rpm and power outputs rose ever higher, making it possible to, among other things, use smaller displacement and lighter engines in ever-larger boats. Now, increased attention was paid to harnessing this horsepower, and reduction gears and V-drives became ever more prevalent. Chrysler's marine engine manager, H. E. Fromm, summed things up when he noted, "The development of the small bore engine with reduction gear has made it possible for the industry to offer popular-sized cruisers with unusual accommodations at unheard-of prices." It was recommended that a reduction ratio of 1½:1 was appropriate for cruising motor boats while ratios as low as 7:1 would permit a high-speed engine to turn a workboat's big propeller.

A *Motor Boating* report on new Lycoming engines in 1934 noted: "The new built-in reduction gear models give the boat owner and builder greater opportunity to utilize the advantages of high speed engines in large boats requiring large diameter propellers. Greater economy in fuel consumption, smoother operation and lower initial cost are among the outstanding advantages to be realized in the use of high-speed engines working through reduction gears."

The Marine Engine Comes of Age Bit by bit, component by component, the marine engine matured, becoming ever more reliable and ever easier to start and run. It must have been a great relief to boat owners, both novice and experienced, when electric starters and generators began to be introduced as options on higher-grade motors by 1914. Such equipment became common options or else standard equipment on Scripps in 1915 and on many others during the early '20s.

By the late '20s, the "modern" marine engine had emerged. With the exception of some old-style, heavy-duty, low-rpm models, the trend was always to higher rpm, lighter weight, and, of course, electric starting. The "auto-boat" had finally become a reality under the impetus of Chris-Craft, Dodge, and others. In 1930, the Lycoming 4-cylinder engine that powered a Dodge 16-foot runabout produced 40 horsepower at 3,200 rpm and weighed only 390 pounds with its electric starter and mechanical fuel pump. It was just three feet long.

Motor Boating concluded its 1934 issue introducing that year's crop of marine engines with these words: "We live in a wonderful age… Never content with things as they are, we seek constantly for improvement, more power, more speed, more of anything and everything… New and better engines is the watchword. Our fine engines of only a few years ago are no longer supreme. Science has done things to the new products to make them better and more powerful." The magazine pointed to new metal alloys that permitted weight reductions, new carburetors and improved fuels that helped improve operation and power, improved lubricants that reduced friction, and design refinements that continued to increase power output and fuel efficiency.

Those who had begun their working careers in the marine engine industry at the turn of the century had lived through a period of ceaseless experimentation and dramatic advances. They had witnessed the progression of the marine engine from a slow-turning, single-cylinder, cast-iron machine of great weight and sometimes problematical operation to ever more reliable, compact, multicylinder engines with horsepower-to-weight ratios that would have been unthinkable to any but the most farsighted of the pioneer developers.

CHAPTER THREE:

Collecting Antique Marine Engines

"An engine for yachts should be as beautiful as the handsomest steam engine built for that purpose," opined *The Gas Engine* in the spring of 1905. "The owner should be able to take his friends to the engine room with as much pride as he has in any other part of his little floating palace." An antique marine engine of that period, or roughly the time from 1890 to 1916, remains a fascinating, even artful mechanical presence. Like any antique, it inspires the imagination. What would it say if it could talk? Who conceived it? What does it tell us about a previous generation's understanding of technology? Who were the men who built, sold, and used it before it was finally set aside to gather dust and rust?

Technology's march was good for boat owners but unkind to today's marine engine connoisseurs. The primitive engines of the early years gave way to ever more sophisticated models. What were in their time the most exotic of engines became obsolete when something better came on the market. The question then was simply how to dispose of a heavy, dirty old engine, not whether to store or even carefully preserve it. Instead, when time passed an engine by, it was usually scrapped (The oldest motors now in exis-

Longtime collector and Palmer specialist Dick Day (l.) confers about an engine during an antique engine meet at Mystic Seaport. (Stan Grayson photo)

tence had to survive two world war scrap drives.), abandoned, or converted to some other application. The latter, at least, ensured that a marine engine might continue to provide useful work and receive a measure of care long after its siblings had ceased to exist.

Once finally abandoned, often literally buried in the backwoods or abandoned in lakes and rivers from Maine to Alaska, old marine engines still remain to be discovered and retrieved. (Unlike saltwater engines, those found in lakes or rivers are often in very restorable condition because cast iron doesn't deteriorate rapidly in fresh water.) One thing is sure: While some might make an heirloom out of a piece of furniture or, occasionally, a boat, marine engines were seldom viewed in a similar manner. What's more, although stationary engines were produced in the thousands for a nationwide market, marine engine production, even at its sometimes impressive peaks, was small by comparison. Any marine engine, particularly one in intact original condition, is now an unusual to genuinely rare item.

Who Collects Antique Marine Engines Old marine engines appeal to engine hobbyists who collect, restore, and display their motors; maritime and regional museums seeking to preserve these evocative examples of American ingenuity and industry; and owners of antique boats wanting to create a truly authentic restoration of a race boat, runabout, cruiser, or, on occasion, workboat.

Individual collectors have always played a pivotal role in the preservation of old marine engines for they, more than any institution, have developed an appreciation for these objects. Working alone or as members of clubs, collectors have managed to find, restore, preserve, and research a variety of marine engines that might otherwise have been lost forever. Many collectors are men of immense, even artistic mechanical ability, able to discern and solve perplexing problems and create elegant solutions to challenges as diverse as how to sleeve the ported cylinder of a worn-out two-cycle motor or adapt modern electrical components to a make-and-break igniter.

Museums, of course, also play a role in the old engine world. That said, the acquisition, preservation, and interpretation of marine engines by museum professionals has for years lagged behind more conventional artifacts. Unfortunately, many donated engines have been ruined because their water jackets were not filled with antifreeze to halt destructive erosion caused by salt deposits. Progress has been made in some cases. Today, maritime museums throughout the U.S. and Canada occasionally add marine engines to their collections. So do local history museums that celebrate the culture and industry of a specific city or town.

The best museum displays are those that make an effort to interpret marine engines in a way that's meaningful to a general audience. As an artifact, the marine engine must be presented in terms of both its human background and technical significance, subjects not uniformly well understood by all curators. Who were the men who created it? What was the company's place in the local economy and overall industry? What engineering features set the engine apart from its competitors? How was it used by the commercial watermen or recreational boaters of its time? Only in the answers to such questions will the old marine engine speak to new generations of museum patrons.

Antique Engines for Antique Boats As this is written, those long involved in the old engine hobby report that more and more collectors are choosing to use at least one of their old marine engines for its intended purpose. Restored marine engines of the pre-'teens, 'teens, '20s, '30s, and '40s give owners of antique boats the chance to keep their boats original in aura, sound, and performance. Among the most unusual of such boats would be a launch from the pioneer or early production eras of marine engine development. A Lozier, Fay & Bowen, or Truscott launch, or a Toppan dory powered with its original engine, or a restored version, is as exciting as it is rare. Occasionally, too, a cruiser emerges with its original T-head Sterling still firmly bolted in place atop the engine beds. Although boats like that are certainly unusual, they do exist and show up with some regularity at antique boat shows when their owners judge enough progress has been made to share their treasure with the public.

Reproduction boats are part of the scene, too. An engine collector who well understands his 1- or 2-cylinder two-cycle may be prompted to build an appropriate hull in which to show it off. Such projects are not only great fun, but they also demonstrate to owners of modern boats, and boat show attendees, just how practical an old marine engine can be. That said, when it comes to an antique two-cycle engine, it pays to be on intimate terms with the needs of its ignition system and vaporizer or carburetor. Given that, one can be confident in the ability to easily start the motor, tend to any needs that may arise during the boat's use, and, in the absence of a reversing mechanism, reverse it safely "on the switch." Such skills are not required, of course, with the turn-key-style operation of the Chrysler, Lycoming, Gray, or Kermath typically found in vintage runabouts.

A boat owner's performance expectations also impact whether or not to use an antique engine in an antique boat. A 100-hp Chrysler Imperial may barely have enough oomph to get a 24-foot runabout on plane with four people aboard. A 26- or 28-footer is likely to need 200 to 300 hp to perform in a way that will satisfy some owners. Boaters expecting 21st century performance from a 75-year-old marine engine might better look to repowering. Today, the trend may be toward modern replacement engines for those who regularly use their boats as opposed to those who launch them for the occasional boat show or weekend and then return them to storage.

When an old marine engine is put back into use as a "runner," its long-term survival may well be jeopardized. Too often, a boat owner may acquire several identical old engines and then cannibalize the best parts from each to make a single "good one." While the practical advantage of such an approach is understandable, some of these old engines are rare enough that each should be cared for, not diminished or ruined. Even worse, careless maintenance practices on an in-use old engine, such as forgetting to winterize it so the water jackets freeze and crack, can easily ruin a rare antique.

A Challenging Vocation As hobbies go, collecting old marine engines poses some special challenges. The always exciting chase for a new acquisition is likely to involve the time and expense of considerable travel. What's more, unless the motor in question is a single-cylinder model, the logistics of moving, storing, and displaying it must also be resolved. While a collection of small engines can be housed in the garage, bigger, heavier engines usually require a small barn with wide access doors or a purpose-built structure with a concrete floor. Most collectors with such buildings also have found they need a forklift truck to move things around and, of course, a rugged flat-bed trailer for transporting.

Finding an intact marine engine with its original mixing valve or carburetor, lubrication system, and ignition equipment intact is a rare and always exciting event. These days, such finds are few and far between. Instead, the engine is likely to be in some stage of deterioration and is likely to be missing key parts. Finding or making key components and fittings and getting an old engine running reliably can be a daunting and expensive project although the advent of Internet sites devoted to the field has made things easier. The skills involved in restoration, however, are those of the machinist and welder and may well also demand ability as pattern maker and rigger. While a few hobbyists possess the know-how and shop resources to handle all aspects of a restoration, others must farm out some of the needed work.

Identifying an Engine Of all the thousands of marine engines once manufactured, an unfortunately high percentage of survivors lack the all-important original builder's plate that positively identifies them. Not only can identification be a challenge in its own right, but the similarity between engines of different brands can further complicate the process. Sometimes, such similarities can be traced to the fact that engines sold by different companies were built by the same maker who then affixed a nameplate appropriate to a specific customer. The badge-engineering so predominant in the automobile industry was alive in the marine engine industry decades before there was a Ford and Mercury or a Chevrolet and GMC.

Identifying a re-badged engine (assuming an unscrupulous collector hasn't created a fake by knowingly attaching a badge to an unrelated engine) can be easy compared to identifying two engines that possess suspicious similarities but are otherwise more different than alike. Such cases, however, should not take one entirely by surprise. The prolific designer, writer, and mechanical engineer E. W. Roberts estimated that between 1903 and 1908 alone, he personally had designed about 90 two-cycle engines, most for companies other than his own. Many of these would possess what we might now call the "Roberts touch." Engineer Stanley Du Brie was consulted rather widely, and doubtless there were other mechanical engineers who designed for a variety of clients.

Given such situations, the first question of the collector who has decided to acquire an unknown engine is almost certain to be, "What is it?" Today, by posting a photograph of a mystery engine on an engine-related Internet chat forum — www.oldmarineengine.com is the premier such site at this time — one can take advantage of the entire world of collectors. Quite possibly, someone may spot the picture and know immediately what he's looking at. If not, the collector's only source of identification is the time-consuming but enjoyable perusing of magazine ads or company brochures from the period when the engine is assumed to have been built. Owning an engine built by a small company that didn't advertise can make identification even more problematic and is one more reason why engines possessing the original nametag are so desirable. The accompanying table offers suggestions that may help in linking an existing engine to one in an old picture or maker.

What Year Was It Built? Once an engine is identified, the next question often involves the year it was produced or what the serial number means. In the case of such large companies as Lathrop, Gray, Universal, Kermath, Palmer, and others, the meaning of a serial number can usually be deciphered with the help of experts in the brand. The more information one can gather from owners of similar engines, the more likely one is to make progress on relating serial numbers to dates.

When nothing more specific exists, certain details can suggest at least a general time frame. An engine with a "wine-bottle" crankcase is likely to date to the late 19th century. So, too, is an engine whose mount is at the bottom of the crankcase. Engine mounts at the center line of the crankshaft are almost certainly of a later era. Another clue can be gained from checking whether the engine has flanges or pipe fittings (see chart on p. 41).

Identifying an Old Marine Engine

Component	Identification Hints	Comment
Flywheel	Solid or spoked? How many and what size/shape holes? How many spokes? Is inside of rim ribbed to provide a handhold? Split hub? Gib key or retaining nut?	
Cylinder	Dome head or removable head? Painted cylinder casting or brass jacket? In multi-cylinder – are cylinders individual, cast in pairs, or is the block a solid casting?	
Igniter	Location of the stationary electrode? General configuration of mechanism?	
Timers	Swing arm or rotary? Gear or eccentric driven? Handle shape? Cut-off switch? Location?	
Base	Shape? Unusual features? Clean out plate shape and lettering?	Pre-1900, the mounting plate was usually cast at the bottom of the crankcase. After 1900, mounting flanges were typically cast at the center line of crankshaft except in a few cases of very tiny engines.
Intake and exhaust manifolds	Unique shape, combined intake/exhaust, or other features? Water heated intake manifold?	After about 1905, SAE-style flanges tended to be used to isolate the pipe fitting from the cylinder casting to avoid destruction of these valuable castings when trying to replace a water or exhaust pipe. The pipe fitting for oil/grease feeds didn't receive this treatment as they rarely caused serious rusting problems. In fact most were brass or bronze. Post 1909 many of the threaded oil feed holes on two-cycle engines were shipped with steel NPT pipe plugs as makers did not want to guarantee their product unless oil was mixed in the fuel. Grease cups for crankshafts remained on two strokes. Closed crankcase four strokes eliminated the need for crankshaft grease cups by about 1908.
Internal parts	Connecting rod marking?	Forgings such as connecting rod webs sometimes give name of maker and model.
Oilers	Number and location? Force-feed or drip?	After about 1909 makers continued for a few years to leave the threaded holes for pipe thread connections but stopped providing oilers except at customer request.
Water pump	Location and style? NPT attached to cylinder or SAE style flange? Two-bolt or three-bolt? Gear or plunger? Location behind flywheel or aft of cylinder?	
Carburetor or vaporizer	Is the brand and style or model consistent with literature or photos of the possible match?	Care is needed with this component. Many owners changed carburetors or valves, particularly if NPT was used to connect to engine. Most collectors accept these items if they fit the period of the engine. Unique flanged attachment units are considered probably original.
Serial number	A serial number can be a bonus once an engine has been identified. For major brands such as Palmer, Gray, and some others, the number can provide a clue to the year of manufacture.	Serial numbers don't mean much unless tied to model by years of production.

Desirability For many collectors, a marine engine of the period 1890 to 1916 would represent the find of a lifetime. Engines of that epoch are usually based on marvelously cast individual cylinders, and boast a profusion of brass lubricating devices, water pipes, pumps, and manifolds. Add to those features mechanical igniters that may range from crude to incredible engineering and machining feats, spark coils in their original varnished boxes, and other wondrous features not seen in more than a century, and the antique marine engine becomes an evocative artifact.

Such engines — especially those few that have survived with the patina of original paint, highlighted by original brass — are wonderful to behold. These engines have personality! In their place came ever more refined engines that were easier to operate and maintain, more powerful, cleaner, and more efficient but lacking in individuality by comparison. As the industry matured after World War I, marine engines tended to look more and more alike. For that matter, many differed only in key details — intake and exhaust manifolds, water pump, front and rear covers, and the reverse gear — from the engine under the hood of an automobile. Whatever mechanical advances or unusual features they contained were hidden within the cylinder block and head.

This is not to say that a specific engine of a later period may not be collectible or intensely sought after. A Scripps F-6, a Sterling Sea Gull, a Kermath Sea Wolf, a Speedway, an N-Model Van Blerck, Chris-Craft V-8, Chrysler Crown, or Packard Eight: Any of them may be highly valued, but usually by someone with an antique boat of the appropriate vintage and size for such an engine. As in any collecting endeavor, personal taste and need play an important role.

Determining Value Among the very first questions that arise regarding an old marine engine is, "What's it worth?" Many antique objects, whether duck decoys, Windsor chairs, toy trains, or antique automobiles, have established track records thanks to widely publicized and well-attended auctions. There is, however, no similar published scale of values for antique marine engines. Only the advent of eBay has resulted in a limited and fleeting dissemination of prices. Overall, the comparative rarity of old marine engines and a market that, while sizable, is but a fraction of that for more generally understood antique objects, has rendered assessing values quite challenging.

What's more, with the exception of some highly knowledgeable men, marine engine collecting currently remains a comparatively unsophisticated avocation in terms of what engines and what periods are more desirable than others. That, however, is bound to change. Most involved believe that it is only a matter of time before the marine engine version of "Foos fever" — the sudden rise in demand and escalation of prices that saw Foos stationary engine values shoot from perhaps $2,000 to $25,000 — strikes.

"The market," noted one long-time collector, "can be driven nuts. When things get crazy for a specific brand or engine, then more everyday versions get to be better appreciated and then their value starts to rise."

As this is written, establishing the value of a marine engine remains a guessing game although there are some general guidelines. For one thing, with the exception of a few once-popular brands built in comparatively large quantities, any marine engine is a rare item. Some, of course, are more rare than others; rarity and condition are the key factors determining worth. Complete engines in original condition including the transmission are the rarest of all. Even an intact, original engine, however, will have a value that is dependent almost entirely on the perceptions of a relatively small number of interested parties. How much does rarity impact value? The mere fact that an engine is incredibly rare usually doesn't mean it's worth a fortune because the market itself is not large. In fact, an apparently one-off survivor may be seen as an "oddball" and have less value than a popular, widely recognized model by an important company since the latter is sought after by a greater number of collectors. Among the few generally applicable rules that many collectors agree on is that price is inversely proportional to weight — with the exception of high-performance engines in excellent to outstanding condition with a solid provenance.

The value of an old marine engine, then, is similar to that of any other antique object: it is largely in the eye of the beholder. In the still rather small world of antique marine engine collectors and "dealers," pricing will range from honest estimates based on years of experience to rather transparent efforts to escalate values. Generally speaking, a price that causes one to raise eyebrows in surprise is, in the absence of established standards, likely to be inflated. Engines that are offered on occasion by antiques dealers are almost certain to be aggressively priced, which is not to say, necessarily, that they are over-priced. Once an engine enters the "antiques world," however, potential buyers may find themselves in the position of paying a price that "makes the market," which means the price may be higher than anyone has previously paid for that particular engine. Engines offered by ethical men who have made a career in the engine restoration business are almost certain to be priced appropriately to a market they have been involved with, and have helped make, for years.

At this point, there aren't so many potential buyers that values of most one-lungers or pre–World War I multicylinder engines have risen to the levels seen in the world of antique stationary engines. Big money, however, is involved in some racing-oriented marine engines. A Packard Gold Cup model, the rare Sterling, or converted aviation engines like the Liberty or Hispano-Suiza V-8 can be costly because increased interest in vintage boat racing has resulted in increased demand.

Such machines, of course, are rare. There is a romance attached to them that is absent from more pedestrian models. The enthusiasts who happen to have taken up vintage boat racing and collecting also generally happen to be those who can well afford a lavishly expensive antique or reproduction race boat. They can also afford to pay what it takes to get and restore one of the few, generally dilapidated engines appropriate for powering it.

Those involved in this end of the market caution that prices have risen ever higher not only because of demand for the engines themselves, but also because of escalating restoration costs. Said one: "As the old-time tool and die makers and machinists who would work for an hourly rate reasonable in their time pass away, costs have sky-rocketed with some shops approaching $100 an hour. The ultimate result may be the replacement of antique engines with modern ones as a matter of necessity."

The desire to keep a runabout or cruiser authentic may also lead to higher prices for a specific engine. The owner of a restored runabout that may once have been powered by a Sterling, Scripps, a Kermath Sea Wolf, a Chrysler Imperial, or a Hall-Scott LM-6 would certainly pay a premium if a similar engine (especially with some spare parts) could be found and put into reliable operating condition.

As this is written in the spring of 2008, the market value of antique marine engines ranges from $100 to $1,500 or so for a single-cylinder two-cycle in poor to fair condition. The majority of one-lunger marine engines of the period 1902 to 1918 will probably trade in a range of $2,000 to $7,500 with the highest prices commanded by intact, solid original engines complete in every respect including paint. These wonderful engines are not only important historically, but they are fun to display, and are not too big and heavy to be impractical.

Pre–World War I multicylinder motors, such as a Niagara, Jencick, Sterling, or Van Blerck — to cite a few rare but highly desirable brands — would doubtless press the envelope in terms of pricing and, one suspects, would be worth every dollar. Depending on condition, a multicylinder T-head engine would likely command a price ranging from $1,500 to $15,000, perhaps even more in isolated cases. These marine engines, with their individual cylinders and external valves, water pumps and piping, and lubrication systems, are simply amazing objects. Finding one in good original condition would be a dream for any collector, and sometimes the dream comes true.

The biggest money is currently being spent on such models as a recently restored Packard Model 2025 V-12, which changed hands for a six-figure price. Such engines, however, fall outside the general scope of marine engine collecting. That Packard was one of only nine similar engines originally intended for air racing before it was installed in a once-famous racing boat. An engine like this, with its twelve individual cylinders, each equipped with four exposed valves, is such a magnificent accomplishment that its value might be expected to rise further in coming years.

Impact of Locale on Values Inevitably, perhaps, collectors of any antique object often are drawn to those made in their own general area. In the marine engine field, that means collectors may be willing to pay a premium for an engine built in their own state, city, or, failing that, region. The fact is that a California-built marine engine should, and doubtless will, one day become an object of desire for a collector in Connecticut as it now is for a California enthusiast. And vice-versa. It is certainly easy to envision a collector in New England being drawn to the fascination of a heavy-duty California engine with its marvelous make-and-break igniter, sometimes novel crankcase shapes, and other intriguing features. Likewise, collectors throughout the country may find themselves drawn to one of the many two-stroke engines manufactured in Michigan, a heavy-duty engine from Connecticut, or one produced by the many companies in the Finger Lakes region of New York.

The accompanying table, created with the input of long-experienced collectors and restorers, is an effort to take a step toward defining value criteria. While it cannot pretend to attach a dollar figure to a specific motor, it does provide a guide to what factors make a particular engine more or less desirable, and that is a big step toward deciding what it's worth.

Color Painting an engine in its original color is always desirable, but determining color can be a challenge. Part Two of this book includes color information for specific brands whenever it was available. Sometimes, careful removal of paint layers will reveal the original color. It is clear, from looking at photos of engines displayed at boat shows or those depicted in catalogs, that most (not all) were painted a single color. That is, both the engine and flywheel were the same color. This should be kept in mind by those restoring old engines today. Likewise, the garish colors sometimes found on restored engines should be avoided. What to do when at a loss regarding the original color? The most prudent choices would include black, light or medium gray, medium or dark red, and dark green. As this is written, the most sophisticated collectors of engines (and automobiles) are trending away from an "over-restored" look. Instead, some are now trying to achieve a level of finish that is less perfect but more in keeping with the object as it left the factory.

What's It Worth?

Factors Increasing Value	Factors Decreasing Value	Comment
Engine is complete in all respects including builder's name tag.	Engine is missing such key components as manifold(s), brass components, timer, igniter, water pump, or name tag.	An intact engine is the most basic starting point and is always worth a premium because replacement components are rare.
Engine's manufacturer is known despite lack of nameplate.	Engine's manufacturer is unknown.	A mystery engine will always remain much less desirable than an identifiable model.
Engine is in good condition and capable of being run.	Engine is in poor condition with splits in cylinder castings.	Condition! Condition! Condition! Damaged cylinders may be repaired depending on the extent and nature of the damage, but such engines must generally be judged to be worth relatively less. Better to acquire one solid original than many clunkers.
Engine has complete provenance including no salt water exposure.	Engine's history is unknown or partly known.	As rare as a solid, complete engine in original condition, known history does much to enhance value.
Engine dates from pioneer period (pre–World War I).	Most engines built later than this period, whether marine or marinized automobile engines, are generally enclosed models that typically offer less mechanical fascination than their brass-era forebears.	The early marine engines represent the genius and handwork of those who built them and should be regarded as mechanical works of art worthy of preservation and study.
Engine is from a freshwater environment and the water jacketed parts are flawless, free of freeze or rust cracks.	Engine was operated in salt water.	Unless a saltwater engine was stored only after its water jackets were emptied and immediately filled with oil or anti-freeze, cylinder splits and damage from salt deposits are inevitable and must reduce the engine's desirability and value. Repairing cracked water jackets in salt water-cooled engines is very expensive and not always successful. Some engines with good outside appearance may have been treated with acid in an attempt to remove the salt/rust. Unfortunately, this treatment is difficult to detect and does nothing to remove salt deposits.
Engine has been restored with components and accessories that are correct to the model and brand.	Engine was restored but equipped with non-original equipment.	Often, the desire to have an operating engine results in a "cobbled" job — an engine with the wrong carburetor or timer, incorrect fittings or plumbing, and other incorrect parts that give a "homemade" look to the machine.
Engine has original paint, or its original color is known and has been duplicated.	Color is either non-original, inappropriate, or both.	Historical accuracy is always to be valued above obviously inappropriate colors or level of finish.
Engine is from a high-end builder known for quality.	Engine is a mass-produced model.	Just as a 1932 Lincoln is valued more highly than a 1932 Ford, a Scripps or Sterling must be valued more highly than a Gray or Continental. That said, certain models from important companies can be highly desirable.
Engine is recognized as especially rare.	Engine is commonly available even today.	Besides consideration of brand, rare models within a brand will command higher prices.
Engine is a racing model.	Engine is workboat-, cruiser-, or runabout-oriented.	Demand for high-performance engines makes them of special value to vintage boat racers.
Engine is local.	Engine is from another part of the country.	A locally built engine may have special appeal to collectors in the region.

Red Wing was a regular advertiser in most of the boating magazines for many years, and the brand still has a devoted following. This ad appeared in MotorBoat *in November 1924.*

Summary Collecting old marine engines is an exciting hobby that combines hands-on restoration and operating skills with the detective work involved in research and the sociability of sharing one's interest with a diverse group of like-minded people. As the engine restoration hobby matures, it is likely that demand will increase and that this will tend to increase values.

While the importance of an old marine engine is most likely to be understood by collectors and by owners of antique boats, museums have a potentially critical role to play. Currently, that role is not being uniformly fulfilled in the best possible way. It is to be hoped that museums will give much-increased attention to the conserving and interpretation of marine engines in the years ahead, sharing their importance with new generations of visitors.

Finally, the entire collecting and preservation community is best served when those restoring an old marine engine seek to achieve the highest level of quality and authenticity in their work. In this manner, the engines can best tell a true story of the way things used to be.

CHAPTER FOUR:

Researching Antique Marine Engines

Just as the collector of antique furniture seeks information about the cabinetmaker or company that built a particular piece, engine collectors seek information about manufacturers and technical information about the engines themselves. Determining the mechanical details of an old engine as they were on the day it left the factory is essential to producing a historically correct restoration. Of course, it is inevitable that questions also arise regarding the company that built the engine. For all these reasons, once an engine has been acquired, it is likely that a research phase will be needed to determine its original specifications, date of manufacture, and something about its builder.

Researching Skills Researching engine companies and their products falls generally into a category we might call "industrial archaeology." It involves digging up the past regarding people, companies, and products. Such work requires a general attitude that benefits from patience and a structured methodology. A good rule of thumb is this: Trust your own research with original source material. Original source material includes company sales literature, company ads, city directories, magazine or newspaper articles of the time, and personal reminiscences of those with first-hand knowledge of the company. City maps are useful for pinpointing addresses; insurance maps can detail size of a company's plant.

Secondary sources include articles written, by a writer with no first-hand knowledge, after a company has ceased to exist. When consulting previously published material, consider what sources the author has used. Are the author's conclusions based on evidence gleaned from original research? Is there a bibliography, or are quotes and conclusions referenced as to source? How much is conjecture, and is such material presented as honest conjecture or stated as fact? Does an article reflect more than simply a rephrasing of another previous story, which may itself be flawed? If answers to any of the previously noted questions are "no," one has no reason to trust accuracy without verifying the key points. Remember that while even the best researchers

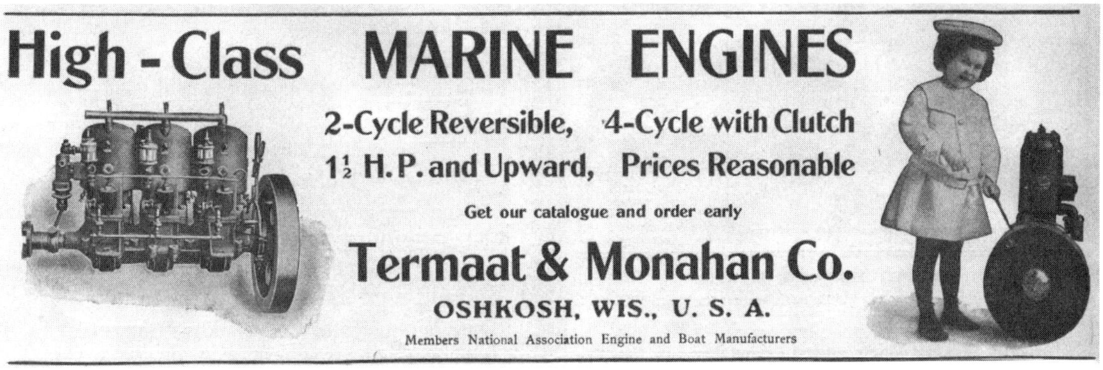

1906 T&M ad implying "so simple a child can do it"

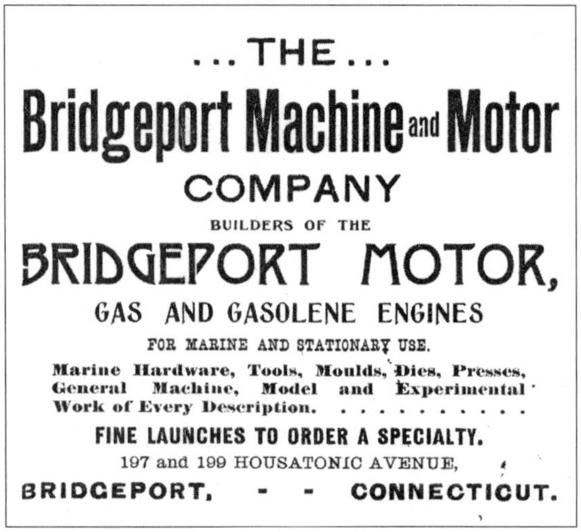

This 1901 ad for the Bridgeport Machine and Motor Company appeared in that year's Bridgeport City Directory. *Such directories are important research tools that chronicle company names, addresses, products, and personnel.*

may make errors, published material based on little or no original research or on innuendo must be taken for what it is.

Another secondary source is a descendant of a long-ago engine entrepreneur. Sometimes, descendants may have preserved primary source material about their ancestor. Often, however, descendants will know little or nothing about their forebears or have a mistaken or inflated sense of the subject.

When embarking on research into a given engine or manufacturer, it pays to be entirely open-minded, which is to say skeptical. For example, one may have a preconceived idea that suggests a connection of one company to another, but this may, in fact, be completely mistaken. One approaches the truth only by following each lead to its conclusion.

A significant research endeavor requires time, tenacity, imagination, and often cash to pay research and copying fees and postage. Some engine enthusiasts find such research so fulfilling that it becomes a hobby in itself. For those just entering the field, however, merely knowing what questions to ask may be a daunting hurdle. Rest assured that whatever joy one may find in discovering an ad for a particular engine in an old magazine, the discovery is merely the first step if one hopes to piece together a company history or get specific engine details.

Research Materials and Where to Find Them In the pursuit of knowledge about old marine engines, libraries, museums, historical societies, and even town halls are key sources. In general, we are speaking here about institutions in the same town, city, county, and state or province where the manufacturer in question was located. Here, one hopes, publications or documents relating directly to the company will be found. The advent of the Internet has provided another important tool. While they are no replacement for more traditional venues, websites may contain some very useful kernels of information: a reliable article or suggestions from fellow enthusiasts, which can be helpful. That said, unless the information presented is well organized and contains source references, it must be viewed with caution.

A word is in order regarding old boating magazines, including *The Rudder*, *MotorBoat*, *Motor Boating*, *Yachting*, *Power Boating*, and *Pacific Motor Boat*. Such magazines are among the most useful resources available to the researcher not only because they contain ads by many engine makers of the period, but also because they usually ran columns relating to engine industry personnel and company news. Often, too, the magazines published news of what engines were being installed in what boats and they published pictures of engines on display at the annual New York and Boston motor boat shows.

Local newspapers (unfortunately, rarely indexed) can sometimes reward patient study if one is seeking information about a person or event and has a specific date as reference. The same can be said regarding more general-interest

Like many engine builders, DeMooy used a racing boat to showcase its engine. This DeMooy was photographed in 1908.

magazines such as *Scientific American, American Machinist,* and *The Gas Engine* (published from about 1898 to 1929 to chronicle the growth of the internal combustion engine and its impact). *The Horseless Age,* while generally devoted to the automobile industry, also contained occasional news about marine engine developments as the two industries had so many connections.

Marine museums whose libraries own collections of prewar boating magazines and related material include The Antique Boat Museum in Clayton, New York; the Mariners' Museum in Newport News, Virginia; Mystic Seaport Museum in Mystic, Connecticut; the Chesapeake Bay Maritime Museum in St. Michaels, Maryland; the National Maritime Museum in San Francisco, California; and the Columbia River Maritime Museum in Astoria, Oregon.

Libraries with collections likely to be of use to the engine researcher include the Detroit Public Library with its invaluable Technology & Science Department (brochures) and National Automotive History Collection. The Bentley Historical Library at the University of Michigan in Ann Arbor has a collection of Michigan company catalogs. The New York Public Library, Brooklyn Public Library, the New York Historical Society, the San Francisco Public Library, and the Boston Public Library have materials that may contain information about the many engine companies that once did business in each state. State libraries such as the California State Library and other such institutions throughout the country are always worth contacting. Among the libraries at the Massachusetts Institute of Technology (MIT) is one that has a collection of *The Gas Engine* magazine.

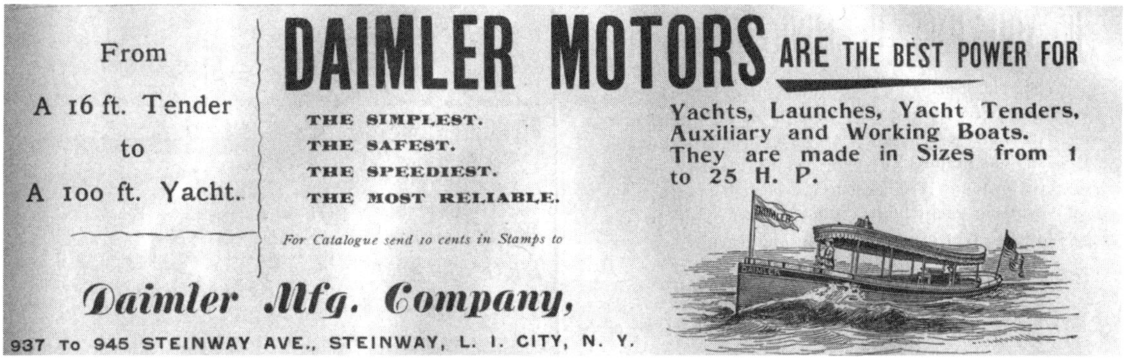

Ads like this one from Daimler in 1898 are now among the few tangible reminders that this famous brand was once a player in the American marine-engine market.

Museums or historical societies local to a long-gone engine builder, even a small one, can be excellent sources. Most public libraries have local history collections, and they may well include material that pertains to one's search. It may take the form of an entry in a local history documenting the city's industrial past, biographies of once prominent local citizens, documents relating to the incorporation of a given company, or insurance maps. Occasionally, one strikes real pay dirt and discovers a whole box of relevant material in a historical society or library reference room.

In addition to public, state, and university libraries, many state historical museums are worth contacting. One never knows what potentially valuable information may reside in such collections. For example, the Michigan Historical Museum, in Lansing, has corporate information on the state's engine builders that includes names, dates, addresses, and sometimes more. Whatever the state in which an engine was manufactured, it is worth pursuing state sources for reference. The accompanying table lists in a general way key reference materials needed and where to find them.

Finally, we should not discount the potential value of "research by walking around." Attending antique marine engine shows like those currently held by the Mystic Seaport Museum (Connecticut), the Calvert Marine Museum (Maryland), or the California Gold Expos, brings one into contact with like-minded people. One of them might well have some knowledge or a suggestion that can

Reference Materials and Where to Find Them

Key Source Materials for Engine Research	Where to Find Source Material
Local newspapers from the period during which a company existed often provide insights into the people involved and non-technical information about the engines.	• Local library • City, town, or county historical societies • Local museums
City Directories contain information about company addresses and personnel and reflect the dates during which the company was registered.	• Town/city hall • Local library • Historical societies
Boating magazines of the period, including *The Rudder, Motor Boating, MotorBoat, Power Boating,* and *Yachting* carried ads for a wide variety of companies and often published feature articles about companies and their products.	• Maritime museum libraries • Large public libraries may have some prewar boating magazines • Antiquarian book/sales literature dealers • eBay
Non-boating periodicals such as *Scientific American, American Machinist, Horseless Age,* and *The Gas Engine* often ran news stories and technical features on marine, stationary, and automotive engines.	• Engineering university libraries • Local library • Technical library
General interest magazines of 1910 and earlier sometimes included ads by the larger makers for marine engines. Advertisers in *The Saturday Evening Post, Harper's, Everybody's Magazine,* and *Country Life in America* hoped to reach an audience with increasing leisure time.	• Public libraries
Internet	• Engine-oriented websites and discussion forums, including www.oldmarineengine.com, give collectors the chance to post photos and information. Sometimes impressive expertise from participants can assist with identification and historical background, and with overcoming technical problems.

help one's own project. Of course, one may also spot an engine exactly like, or similar to, one's own and learn a great deal from it. Information about shows and dates will often be found in *Gas Engine Magazine*.

Research Step by Step Historical research is a step-by-step process with a beginning, middle, and, if one is lucky, end. When things go well, one lead suggests another. Often, one must circle back to check earlier facts and then patiently return to the last point. When things go poorly, one's search fails to uncover even a single entry in a city directory, an event that suggests the company in question was a company in name only. A sample research endeavor might proceed as follows.

1. **Ads:** Research begins with the name of the company and location as stated in a magazine ad for a specific month and year.
2. **City Directories:** Using the magazine ad's date, review the city directories to try and pinpoint the year in which the company first appears. Then work ahead until the company no longer appears. One now has a good idea of the start and end dates.
3. **Local History Collections:** Note the exact name of the company and its key personnel — usually the president, vice president, and treasurer. Research personnel in local history collections at libraries and historical societies where biographical information is likely to appear. Check the City Directory to see if a profession is attached to the name.
4. **Boating Magazines:** *The Rudder, MotorBoat, Power Boating, Motor Boating, Pacific Motor Boat,* and *Canadian MotorBoat* routinely published lists of engine companies, usually in conjunction with reports on the big motor boat show in New York. These often appeared in December and January issues. The lists are a way of verifying whether a company was still in business and of gaining insight into the features and specifications of that year's models. Engine details may also be discussed in articles about the boat show, many of which had photos.
5. **Company Catalogs:** Seek catalogs in city or state libraries which just might have some of these rare items in their collections. The Technology & Science Department at the Detroit Public Library has a good collection of brochures. Catalogs sometimes become available for purchase from dealers in antique nautical books and/or sales literature and on eBay.

The cover illustration of this collectible Acadia brochure sends a clear message: Here was an engine with the features needed for the commercial fisherman.

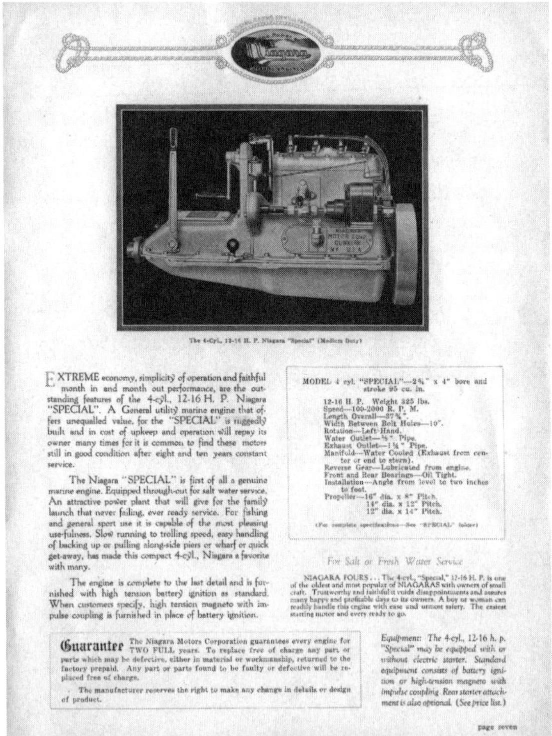

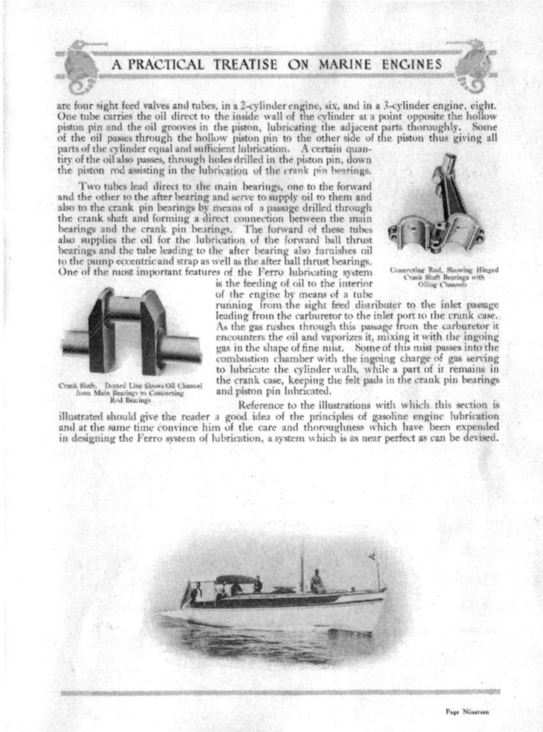

Like the best marine engine literature, this Niagara brochure guided prospective customers through all the engine's unique design aspects, metallurgy, and the features developed to ensure easy operation, reliability, and durability.

6. Newspaper Articles: With the basics now accomplished, seek more biographical or company detail in local papers of the period or in obituaries of the company's principals.

At the beginning of a project it is a good idea to prepare a manila folder for each company so that the materials gathered remain organized and in one place.

Working with Libraries and Museums One is most likely to have success when contacting libraries and museums by being well organized and knowing whom to ask for. When calling a public library to make inquiries, ask for the reference librarian first. In smaller libraries, there may be no dedicated reference librarian, but the staff is usually happy to try and answer questions. In larger libraries, there may well be a local history librarian familiar with the materials in the collection and able to look up information and respond to your query in a letter or e-mail.

Today, submitting questions to a library's "Ask a Librarian" website feature is the most efficient way of making contact. A letter remains a viable alternative. If calling, write down your questions ahead of time and make notes relating to dates and the spelling of names to help expedite the search. Similar practices are important when contacting a museum library for information. If seeking to speak with someone, remember that when talking to museum staff — even at a maritime museum — you are likely to be speaking with a professional versed in preservation methods, research, and general museum policies rather than with a fellow engine enthusiast. It is always worth asking if the museum has a curator on the staff with a background in marine engines. Be patient, and keep your eye on your goal.

Many institutions must rely on volunteers, and, volunteer or not, staff may well need very specific direction if they are to understand and dig out the information you seek. Museums in particular often seek to cull serious requests for consideration by requiring that questions be submitted in writing. Some now charge basic fees for looking up the answer to a single question. At times, a given institution may raise enough hurdles that you'll feel it is failing to meet its obligation to make information accessible to the public. Such instances do occur. Sometimes they can be overcome by patient, cordial discussion with the powers that be.

Whether calling, writing, or e-mailing a library or museum, it is important to remember that your query is almost certain to be one of many that staff members will be dealing with. When talking with a reference librarian, it is always best to take the pressure off. While some may be able to look up information while you are on the phone, be prepared to wait a few days or even several weeks. Expect to pay a fee for copying and mailing any materials you request. Finally, one should not be surprised if told that a series of questions is simply too much for the staff to handle. Often, in such cases, the institution may supply a list of local researchers willing to do the work. If not, it's worth asking for a recommendation. Expect to negotiate an hourly fee for such research.

The accompanying table lists a number of basic questions that are likely to apply to any research endeavor.

Pitfalls The idea that their work would one day be of interest to historians or engine collectors occurred to only a handful of marine engine company executives or engine pioneers. Usually, therefore, one is starting from scratch when it comes to

Key Research Questions

Research Question	Comment	Where to Look for Answers
What was the name of the company?	Don't take company names for granted. Some can be surprisingly similar yet be entirely different entities. Also, company names may have changed slightly, or dramatically, over time because of reorganizations. In such cases, more questions arise relating to why the change, and whether personnel and product changes were involved. Finally, be sure you have the correct spelling.	• Sales literature • City directories
When did the company start doing business and when did it cease?	Start at the earliest date that seems likely and work forward to the end.	• City directories • Local histories
What was the trade name of the company's engine?		• Sales literature • Ads
Where was the company located? (street address and town/city)		• City directories • Ads
Who were the company principals?	Given a name, one can often discover details such as birth/death dates, education, address, relevant personal details, or an obituary. Correct spelling and a middle initial are a must!	• City directories • Local histories • Local newspapers

reconstructing a company's story. Expect research to take time — weeks or even months. Recognize that dead ends are inevitable. Thanks in part to these realities, it is surprisingly easy to make a mistake or draw an erroneous conclusion.

Among the most basic needs is to ensure that you have properly spelled the name of the company and its product's trade name. Very often, companies took the name of their founder(s). That means you may find biographical information, addresses, even job titles in the city directory or a local history written by an amateur historian. If there is a middle initial, use it to avoid possible confusion with a relative with the same first and last name.

Engines advertised by different companies but bearing the same name always raise questions. Were they the same engine or entirely different? Did one company succeed the other and, if so, when? Was the engine simply "badge-engineered" for a boat builder or reseller? Were rights or patterns sold?

Different addresses for the same company — whether within the same city or in different cities or towns — prove a company moved but don't tell why. Often moves were caused by growth,

An engine like this 1901 Otto was used to power the U.S. Navy's first submarines including the pioneering Holland. *The cylinders were cast in pairs and the drain cocks are clearly seen in this port-side view. Three main bearings supported the crankshaft. Ignition was "hammer-break." Atop the cylinders are the exhaust valves and exhaust outlets.*

This circa 1910–1915 5-hp Corliss was pictured in one of the company's brochures. It was the smallest model in the line and was used by, among others, the Lighthouse Service in its launches. Company brochures are among the most valuable research tools in providing engine specifications and features. This Corliss had a one-piece cylinder and low-tension, "spring-hammer sparkers."

fire, or financial difficulties. It is not unusual to find that an inventor associated with a specific engine lived in one place but is referred to in press clippings in relation to another city entirely. Was he really part of the company or did he design the engine as an engineering consultant? Usually, companies were coy about such arrangements, and this camouflage can make difficult discovering who was doing what, when, and where. Proceed with caution.

Perhaps the biggest danger when conducting research is falling prey to trusting previously published material that includes errors of fact or speculation. Some sort of bibliography is needed to demonstrate what original research has actually been conducted. Likewise, researchers should pay no heed to misinformed speculation and word-of-mouth hearsay. Always seek to proceed using source material of the time. Reject or view with caution statements such as, "I heard that so-and-so was the guy who did this or that." Rely only on what can be verified. There's no shame in admitting one has uncovered only 90 percent of a company's story, or only 10 percent! The fact is that this sort of research can go on indefinitely.

Create a Timeline A year-by-year timeline can be a big help when constructing a company history, particularly in complex cases where there seem to be big gaps in events or conflicting information about when things happened. For one thing, a timeline graphically shows what one is missing and helps focus the search. For another, filling in each missing year is a satisfying way to proceed. Finally, a timeline imposes a rigid structure on the researcher and forces one to verify each year's entry or prompts one to note that a certain entry is speculative and that more digging is needed.

Summary Researching an antique marine engine and the company that built it is challenging but can also be fun. The materials necessary to research are not widely available, but they do exist in specialty museum libraries, public libraries, and historical societies. Using original source materials is the key to creating an accurate picture of a company and its products. Relying on anything for which you have not personally seen source material that appears trustworthy is almost certain to perpetuate errors made in the past. Finally, a research effort is most likely to yield success when both persistent and well organized.

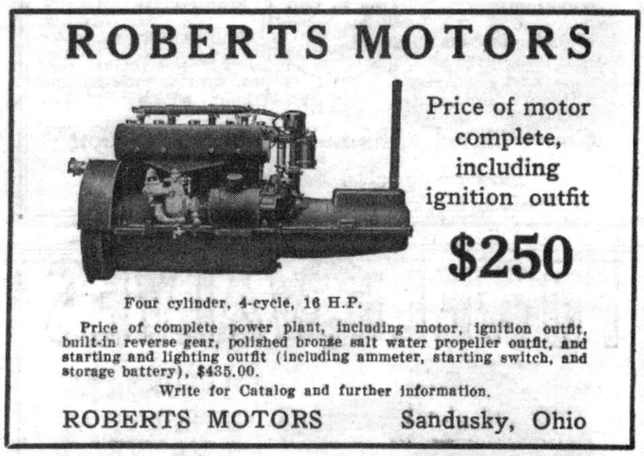

Magazine ads provide insights into how a company positioned itself in the market. They also reveal where a company was located at a given time—useful information since companies often relocated within a given city and sometimes moved to another locale.

United Engineering Works, two-cycle, hot ball, oil engine, circa 1918. Engines like this operated on comparatively cheap fuel oil or kerosene that cost about one-fourth the price of gasoline. Together with more expensive diesels, which had a higher compression ratio and needed no ignition system at all, the "hot head" spelled the end for big, gasoline engines. Ignition required to start the engine was provided by heating a ball-shaped projection in the combustion chamber with a blow torch attached to the cylinder head (On this United engine, the device was mounted on the starboard side.) or, later, electricity. The heat of combustion thereafter kept the ball at a temperature high enough for the engine to continue running. United was but one of many makers of hot-ball engines that included Bolinders, Mietz & Weiss, Kahlenberg, Venn-Severin, Dodge, and Missouri.

Part II

Portraits from a Vanished World

North American Marine Engine Companies

Introduction to the Portraits

This catalog documents American and Canadian marine engine companies and their products. Although research has uncovered significantly more information about some companies than others — usually for good reason — the format of the listings is consistent. Whenever possible, the company name, city, state (or province), and the engine trade name are included. The listings are alphabetical, but when a company name is preceded by initials or a first name, the reader is referred to the entry beginning with the company's *last name*. Canadian companies are listed separately but in a similar manner. As reference materials permitted, entries include the dates or general period during which the company did business, names of key personnel, biographical remarks, and a general description of the different engine models. Pictures reproduced from old ads or brochures are included depending on quality and availability.

The great majority of companies chronicled here were marine engine builders, many of which also offered stationary models or automotive conversions. Some companies were exclusively conversion specialists that adapted automobile or airplane engines for marine use. A few listings represent makers of automobile engines that, as a sideline, attempted to crack the marine engine trade. However, no special effort has been made to include all or even many companies that were not primarily marine engine makers as they fall outside the core subject area of this book. When a European manufacturer had an established U.S. distribution network, that maker has been included.

Those who seriously research the North American marine engine industry soon learn that there were often relationships between many companies, almost none of which were ever acknowledged. Sometimes, a company or its engine was acquired by another firm, and the engine continued in production under its previous name. Sometimes the name was changed although the engine did not. With some frequency, engines were manufactured by one company but labeled with the badge of another or that of a boat builder. Often it is possible to document such connections. In other cases, that information remains elusive.

These portraits should be considered an evolution in the scholarship regarding the industry. While comprehensive, this catalog does not pretend to include every maker or would-be maker of marine engines in North America. It includes those discovered during research that, with some exceptions, met the criteria established for inclusion — companies that were primarily marine engine makers. However, even some of the "companies" listed must be considered more as curiosities than formal business entities. Some were never officially registered as companies and so never appeared in municipal records. Often, we know about them only because of a single, long-forgotten ad or passing reference. Even some companies that seemed "official" never got beyond creation of a drawing, a prototype, an ad, or a press release. Others built only a handful of engines and folded. Additional companies that readers may discover, or further information about those included here, can be researched using the methods discussed in Chapter 4.

These Scripps V-12s powered a 57-footer built by Consolidated.

UNITED STATES

A. A. Adams & Company
See Adams, A. A. & Company

A. H. Lauson Co.
See Lauson, A. H. Co.

A. J. Houle Motor Works
See Houle, A. J. Motor Works

Able Engine Company
Peekskill, NY
Trade name: Able
As of about 1908, Able produced a modern-looking, enclosed four-cycle, 4-cylinder engine that weighed 180 lbs., produced 15 hp, and cost $125. An ad showed an intriguing $250 V-8 version as well. The sales office was located at 30 East 42nd St. in New York.

Acme Engine Company
San Francisco, CA
Trade name: Acme
The Acme was the work of designer Conrad Kriemler and a group of experienced engine men who, it was said, had financial backing from a prominent San Francisco tugboat company. Like most engine builders of the period, Acme boasted "Standardized and Interchangeable Construction" in 1918. The goal was to let potential buyers know that one's engines weren't built by hand, one at a time, but by a modern, repeatable manufacturing process that insured quality and easy replacement of parts. By 1923, Acme offered heavy-duty gasoline, benzene, kerosene, and naphtha engines of 4 to 80 hp. Make-and-break or jump-spark ignition was available.

Adams, A. A. & Company
Providence, RI
Trade name: Adams
The company offered 1- and 2-cylinder, two-cycle motors up to 22 hp by 1908. The engines had jump-spark ignition. An unusual (and ungainly) looking drive mechanism shows clearly in the company's ad. Instead of a shaft coupling on the crankshaft, the Adams had a drive shaft that was gear driven off the flywheel end of the crank. Success appears to have been elusive.

Aeromarine Plane & Motor Co.
New York, NY
Trade name: Osborne (Aeromarine by 1921)

Agnew Marine Engine Co.
Oakland, CA
Trade name: Agnew
During the '30s, Donald Agnew converted Star and Flint automobile engines for marine purposes at a shop at 1951 East 14th Street. These engines were low-cost by comparison to a purpose-built marine engine, and such conversions were not uncommon during the Depression era. The 152-cubic-inch, 18- to 30-hp Star-based W-5 sold for $175. The 268-cubic-inch six-cylinder, 68-hp Flint-based engine was labeled the E-6.

Alco Marine Vapor Engine Co.
Jersey City, NJ
Trade name: Alco
In business in 1896 building launches, Alco advertised that its engines could be started after two to five minutes of preparation. President Grover Cleveland is said to have fished from a 30-foot boat fitted with a 7-hp Alco.

Alexander & Co.
Chicago, IL
Trade name: Humming Bird
The company was an offshoot of Alexander & Crouch. It introduced a marine model circa 1906. The 1910 Humming Bird was a 1½-hp, two-cycle motor with jump-spark ignition.
See also Crouch, Chas. B. & Co.

Alexander & Crouch
Chicago, IL
Trade name: N/A
The company, founded about 1896, was dissolved in 1905 or 1906. It offered T-head engines for both automotive and marine service.
See also Crouch, Chas. B. & Co.

Algonac Machine & Boat Works (Hess Motors)
Algonac, MI
Trade name: Hess Mono-Marine
Algonac Machine & Boat Works was officially incorporated in July 1921 by Jan Matthijs Smits, Jr., a Dutch-born graduate engineer of Germany's University of Constance, and partner Joseph DeBruyn. According to the *Algonac Courier* of Dec. 9, 1921, the pair acquired the tooling and rights for Hess Motors' Mono-Marine engine. The Hess equipment was then moved from Detroit to Algonac where the engines were produced in a dedicated building. In December 1922, this Ford Model T–based engine was listed in *MotorBoat's* annual compilation of marine engine makers. The official model designation was AC. The $175 3¾" × 4" engine was rated at 5 hp at 950 rpm. A dual Bosch battery/magneto ignition system was used as was a Stromberg or Kingston carburetor.

By including not one but two company names—"Hess Motors (Algonac Machine & Boat Works)"—in its listing, *MotorBoat* and the company principals sowed the seeds for future confusion about the organization. This situation was made more complex when DeBruyn incorporated the Hess Motors *Corporation* in Algonac in 1924. Smits was not listed on the incorporation papers of that company but, meanwhile, also ran City Garage.
See also Hess Motors Corporation *and* Gierholtt Machinery Company

Allison Engineering Co.
Indianapolis, IN
Trade name: Allison
Founded in 1920 to produce improved bearings for the Liberty engine, this company was the brainchild of James A. Allison who, with his friend Carl Fisher and others, had opened the Indianapolis Motor Speedway in 1910. Soon, the Engineering Co. had developed a two-speed reduction gear for aero engines and a supercharger, both of which found a lucrative market. These and other ventures earned

1922 Allison

1911 A&B 6-cylinder

1914 Cesco by Amalgamated Engine Co.

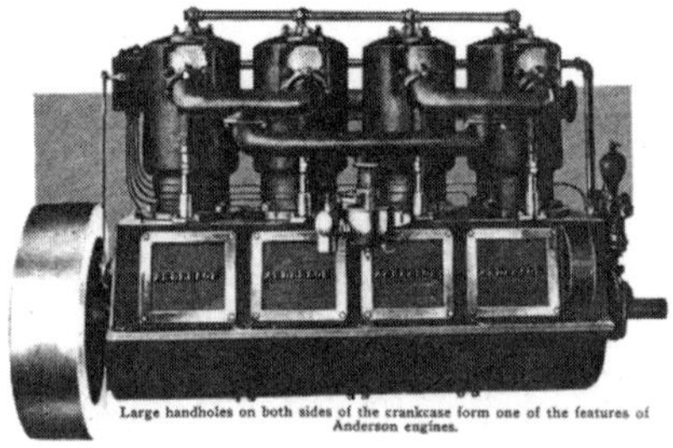

1914 Anderson

This 1929 Capitol is a Liberty V-12 converted by the Auto Engine Works.

1913 American

The 1929 220-hp Hispano-Suiza V-8 as converted by Auto Engine Works

Circa 1936 Atlas diesel built at the company's Mattoon, IL, factory

1913 Automatic

1911 American

Circa 1910 Atlas gasoline engine, make-and-break ignition, hot air feed to carburetor

1923 Automatic

Allison a fortune. The marine engine was an overhead cam V-12 introduced in 1922. It was engineered and built without regard to cost, and this was reflected in its $25,000 price. This was quite likely the most expensive marine engine ever built and perhaps the most expensive engine of any type ever built! The chief market was for immensely wealthy owners of fast motor yachts. Production was, obviously, limited — only twelve were hand-built, each with a bronze crankcase and the other main components of aluminum. This marine engine may be considered the forerunner of the Allison V-1710 that powered many fighter aircraft in World War II. Allison died of pneumonia at age 56 in 1928. The company was acquired by General Motors in 1929. None of these engines are known to survive.

Alpha Motor & Foundry Co.
Detroit, MI
Trade name: N/A
Alpha was in business during 1911 and 1912.

Amalgamated Engine Co.
Detroit, MI
Trade name: Cesco
See also Consolidated Engine Co.

American and British Mfg. Co.
Bridgeport, CT
Trade name: A and B
One of many Connecticut arms makers, this company was founded in 1903. A and B turned to engine manufacture around 1906 to 1907 and was soon advertising a 40-hp "Automobile Motor" and marine engines of 10, 20, 60, and 80 hp. The A and B was designed by Charles Frederick Herreshoff who had been born in Nice, France, in 1880 and educated (presumably as an engineer) in Glasgow before moving to the United States. Herreshoff's father was James Brown Herreshoff, older brother of yacht designers John Brown and Nathanael. Charles was chief engineer at American and British from 1904 to 1908, a time when the company was, according to *The Horseless Age* for Nov. 18, 1908, manufacturing a variety of automotive components including transmissions, frames, axles, and engines.

In an era of T-head engines, the A and B was an unusually high quality L-head, which doubtless cost less to manufacture than the former. By 1911, the engine was being touted for its favorable power-to-weight ratio. With cylinders cast in pairs and a one-piece cast-aluminum crankcase, the 5″ × 4¾″ 60-70 hp 6-cylinder A and B weighed a relatively modest 650 lbs. The 4-cylinder was rated at 40 to 50 hp. These were 1,000 rpm engines; ads claimed a favorable 1 hp for each 10 lbs. of weight.

Beginning in 1908, the engine was used by Elco in its 36-foot Express and other models. By 1915, a 100-hp model was in production and the A and B, although referred to as an "Elco" in George Crouch's report from the 1915 New York Boat Show, came in for close and admiring scrutiny. Of special note were the detachable cylinder heads sealed by copper asbestos gaskets. Pistons were lightweight, and the camshaft was a one-piece forging. The engines had electric starters and high-tension magneto ignition. As for Charles F. Herreshoff, he organized his own Herreshoff Motor Company to build an automobile in 1908, apparently even while still employed at or consulting with A and B. The company was located in Detroit in a factory formerly occupied by the Detroit branch of E. R. Thomas. The Herreshoff automobile appears to have been a high-quality car with engine models that could be linked in some design aspects to the A and B. Herreshoff left Bridgeport sometime around 1912. His car-building efforts did not meet with financial success, however, and the company was declared bankrupt in 1914. As for A and B, the company closed in 1922 and the old plant at 718 Crescent Ave. was razed in 1949. Herreshoff died in San Diego in Feb. 1954.

American Diesel Engine Co.
Oakland, CA
Trade name: Winslow Monovalve
Starting in 1932, engineer Charles Winslow, together with former Hercules chief engineer John Thysse and a tiny group of talented pattern makers and tool and die makers, began development of a unique diesel engine. The Winslow monovalve was an ingenious machine in which the intake and exhaust valves were replaced by one valve that did both jobs. Within 60 days, a single-cylinder, 117-cubic-inch test engine with a 5″ bore and a 6″ stroke was running. It was extremely advanced in its conception, relying on steel alloy rather than iron for many components and thus reducing its weight. The prototype exceeded expectations.

The American Diesel Engine Company planned to build the Monovalve in the moribund Standard Gas Engine Company plant, and soon embarked on producing a line of two-cylinder and multicylinder models producing 35 to 165 hp at up to 1,800 rpm. Some 56 of these Winslow Monovalves were turned out before an unfortunate accident in which a visiting student was injured while watching a dynamometer test. After that, the unfinished engines were sold to the Enterprise Company in San Francisco.

American Engine Co.
Detroit, MI
Trade name: American
American claimed to have been in business since 1892 although it is unlikely to have been building engines much before 1908 or 1909 when it advertised its 2-hp model for $45. The enigmatic foreword to the company's undated "Catalog No. 1" claimed, "We have made motors for years, but our entire output has been contracted for by companies who sold them under their own name as their own product. Our contracts have expired and rather than renew them we have decided to sell direct to the trade." A twist on this theme appeared in an ad claiming that while the company had previously sold through dealers, it was now selling direct.

Whatever confusion company literature and ads created for future researchers of the brand, American seems to have done a good job at engine building. Its two-port, two-cycle engines had a one-piece cylinder cast of an alloy the company called "French gray iron" that was said to be especially dense and nonporous. American warned against engines with "a loose head" that might save manufacturing costs but required a gasket that was subject to "continually blowing out." The catalog copy made a good case for the precision-fit of all components and their quality. Models included 2- to 20-hp models of 1 or 2 cylinders. Later, a 30-hp, 4-cylinder was introduced as was a 14-hp four-cycle. All were jump-spark ignition. The company was still building motors as of 1921.

American Gasolene Motor Co.
Baldwinsville, NY
Trade name: American
In the period before the marine-engine industry matured, roughly the era before about 1912, a number of companies offered engine kits for sale. The idea was that those who built their own engine would really understand it and would save money as well. The American could be had as a kit or ready-to-run. On Dec. 31, 1904, the *Post-Standard* of Syracuse, New York, reported that the American Gasolene Motor Company would soon open in a portion of the factory of the prosperous American Knife Company. The engine's designer was machinist Emery P. Van Epps of Syracuse. William Marvin and A. J. Tooley, proprietors of American Knife, were closely involved in the venture.

In 1906, the American model line was said to include 1½- to 60-hp two-cycle engines claimed to be "Lighter and more powerful than any other motor of the same bore and stroke." American also marketed its castings to would-be engine companies and offered complete 21-foot and 25-foot motor boats. American was not alone in its efforts to market engine kits but, ultimately, the idea came and went with little impact. The original hopes that American would soon outgrow its initial premises and move into its own large factory did not materialize. Instead, it is believed that the company became the Seneca Boat & Engine Co. circa 1909–1910 and that the engine, at least, was acquired by the Valley Boat & Engine Company, which moved to Saginaw, Michigan. Emery Van Epps, who had moved to Baldwinsville, returned to Syracuse and was listed as being involved in auto repairing in 1909.

American Motor Co.
Eau Claire, WI
Trade name: American
The company was producing dome-head, two-cycle engines in single- and double-cylinder models of 1¾ to 16 hp in 1908.

American Motor Co.
New York, NY
Trade name: American
This company is not to be confused with the American Motor Co. of Eau Claire. American was in business as of 1901 although the engine was not seen until the latter part of the year, and details remain unclear at the present time. In 1911, *The Gas Engine* referred to a subsequent American as "probably the first outboard motor."

Amphion Marine Engines
Milwaukee, WI
Trade name: Amphion
Amphion was the creation of Clarence J. Allen. The company was best known for its outboard motors but offered an inboard version in the early '20s. This little 3-hp two-stroke had 2½" × 2½" bore/stroke dimensions and, at 60 lbs., was 20 lbs. lighter than the outboard model.

Anderson Engine Co.
Chicago, IL
Trade name: Anderson
"Anderson engines are good engines," was the company's straightforward pitch. "Ask any user." These medium- to heavy-duty T-head engines were built at a shop at 4038 N. Rockwell Street with individual cylinders, a Schebler carburetor, and magneto ignition. The company was in full swing by the mid-teens. By 1928, the plant had been moved to 4240 Lincoln Ave. and an abbreviated model line of 1-, 2-, and 4-cylinder engines was offered under the trade name Anderson "Super."

Anderson Engine Co.
Shelbyville, IL
Trade name: Anderson
A jump-spark, 2½-hp model was advertised in *Motor Boating* in 1910. The big news, however, was a 2-cylinder, horizontally opposed engine. This four-cycle, jump-spark engine was available in 4½" × 4½" and 5" × 5" models. The valves were mounted in a chest cast into the side of the one-piece cylinders.

Anderson Motor Co.
Andersonville, IN
Trade name: Dice

Antoinette Motor Co.
E. Providence, RI
Trade name: Antoinette
An ad for this company appeared in the June, July, and August 1908 issues of *The Rudder*. The ad included a line drawing of a four-cycle, L-head V-8 that weighed less than 500 lbs. and developed 60 to 70 hp. That weight and rating was essentially the same as those for a version of the famous French Antoinette engine, which by 1906 was available in models of 40 to 400 hp. There can be no doubt that the E. Providence Antoinette Motor Co. was an attempt to import and market the French-built engine. Both engines were 45-degree V-8s, and both had offset cylinders so that two connecting rods could readily be fitted to one crankpin.

It is likely that the Antoinette was the world's first V-8. The engine was the namesake of Antoinette Gastambide, daughter of the patron of engine designer Leon Levavasseur. Although the Antoinette is now remembered primarily as an airplane en-

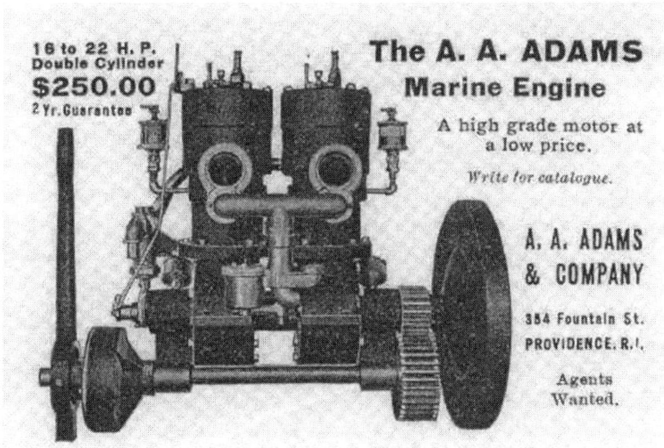

1908

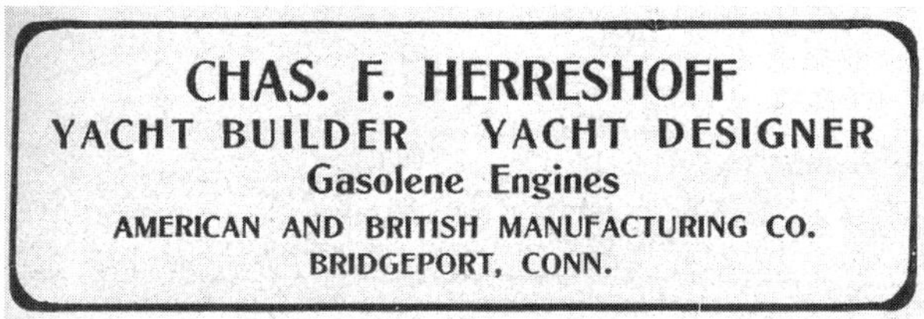

1908

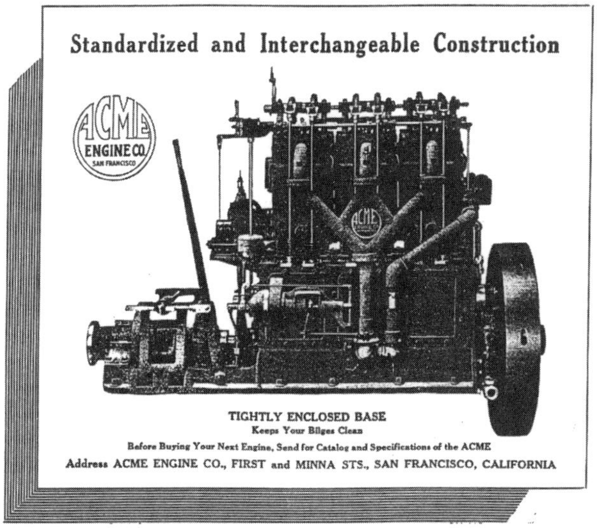

1918

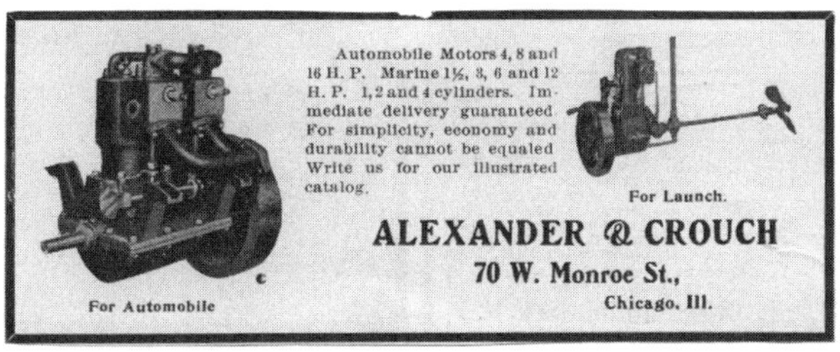

Circa 1905

1932

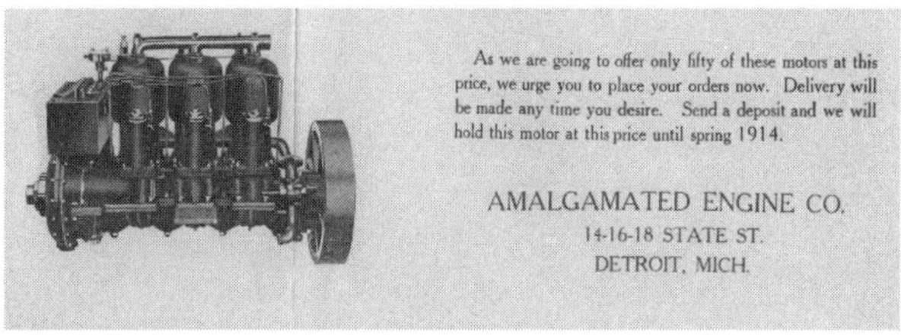

1914

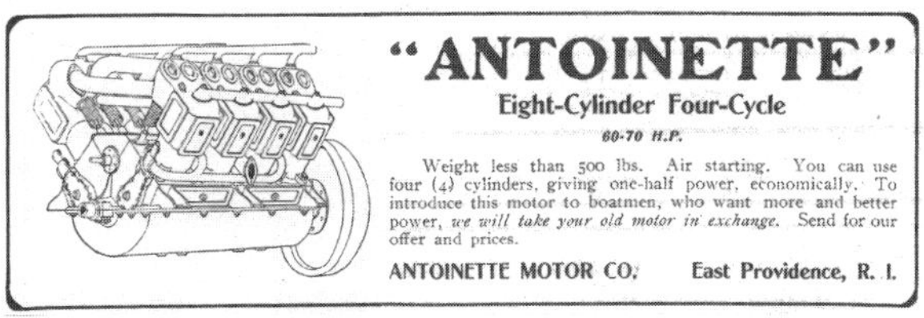

1908

gine, it achieved its first success in 1905 in a motor boat, and in 1906 a 90- to 100-hp Antoinette powered a launch named *Antoinette III* to a victory at that year's Monaco races. Those races were wildly colorful events, contested by an array of fascinating boats owned by very wealthy men and powered by fascinating motors. Among the motors, the Antoinette was recognized as being even more fascinating than most.

By 1908, the French company had focused primarily on airplane engines, and the 1909 model had lightweight machined cylinders with polished brass water jackets. It seems very likely, however, that the marine version of the engine may have had cast cylinders or else fabricated cylinders of a different design. While the French Antoinette models described in the press had automatic inlet valves, the drawing in *The Rudder* clearly shows two pushrods per cylinder leading to valve chests located on the inboard side of the vee.

The Antoinette ad in *The Rudder* claimed, "You can use four (4) cylinders, giving one-half power economically." How this was possible is suggested by a *Scientific American* article about the engine in March 1906. "The carburetion is produced by a small gasoline pump worked from the motor. It draws gasoline from the tank and sends it to eight small distributors placed at the top of each motor on the Inlet. These regulate the supply of liquid causing the spray action which is needed. The output of the pump is variable at will...." What's more, the 1906 Antoinette, at least, was direct reversing, thus dispensing with the need for a "friction clutch or a reversing mechanism."

Despite the company's growing focus on aviation, fast boats powered by the Antoinette continued to be built. In the summer of 1908, a Fauber hydroplane powered by a 50-hp Antoinette made news when it achieved a 38-mph speed on a quiet section of the river Seine. That same summer, a reference to the Rhode Island Antoinette appeared in June. A 200-hp Antoinette was installed in a boat named *Autowin II* owned by a wealthy MIT graduate, engineer Edwin S. Webster of Boston. Unfortunately, no details were provided. (Webster was a founder of what would become one of the world's largest engineering firms, Stone & Webster.)

The Antoinette Motor Company of E. Providence, Rhode Island, appears to have been a casual attempt to market a version of the French Antoinette in the U.S. There is not a single record that indicates the company was ever incorporated or otherwise registered in Rhode Island.

Arrow Machine Works
Detroit, MI
Trade name: Arrow

Arrow Motor & Machine Co.
See Pausin Engineering Co.

Art Machine Co.
See Fulton Engine Co.

Aster Motors
New York, NY
Trade name: Aster
The company advertised "Motors for Automobiles and Boats" in 1905.

Astoria Iron Works
Astoria, OR
Trade name: Troger-Fox

Atkin-Wheeler Co.
Huntington, NY
Trade name: Achilles; AWC
This company resulted from the enthusiasm and ambition of boat designer William Atkin and his wealthy friend Cottrell C. Wheeler. In 1906, the pair opened the Red Boat Shop with the backing of the immensely wealthy August Heckscher who, among other things, owned a bank in then pastoral Huntington on Long Island. In 1914, according to Atkin's old friend, designer Weston Farmer, the pair built *Cabrilla*, a 115-foot yacht, for Heckscher. It was powered by V-8 motors of Atkin's own design. However, the boat was repowered with Van Blerck V-12s in 1916. Atkin-Wheeler was essentially a boat-design and building company, not an engine company. Possibly, the Achilles was a re-badged two-cycle motor.

Atlantic Company
Amesbury, MA
Trade name: Atlantic
By 1910, if not earlier, the company was offering a line of 1- to 4-cylinder, two-cycle, three-port engines in ratings of 3 to 40 hp. They were equipped with jump-spark ignition and a Schebler carburetor. Paragon reverse gears were used. In 1911, it was reported that service departments had been established in Boston and New York. The company also built a highly regarded line of "sensibly priced" boats, a "semi-speed clipper launch," and the Poyen-designed, "seagoing 'Gurnet' dory." The latter looked nothing like a traditional or round-sided dory, but more like a launch. In 1912, Atlantic introduced a pair of 4-cylinder models. These T-head engines developed 20 and 40 hp. They were equipped with a Schebler Model L carburetor and a Paragon reverse gear.

Atlas Gas Engine Company
San Francisco/Oakland
Trade name: Atlas
Founded in 1904 by former employees at San Francisco's early engine maker, Hercules, Atlas had been in production for only a short while on stationary, portable, and hoisting engines when the earthquake struck in 1906. After an infusion of capital, the company moved to Oakland where a new factory was built at 11th Avenue and East 9th Street. Soon, it would cover 15 acres. Marine engines were added to the model line, and with sales agents located around the Pacific, revenues reached a million dollars by 1909.

One-, 2-, 3-, and 4-cylinder marine engines were offered of from 4½ to 200 hp. These heavy-duty T-head engines with make-and-break ignition were principally used in commercial applications. An example of their durability was a 150-hp model that covered 35,000 nautical miles with the sum total of repairs being merely nine dollars. Atlas executives began development of a diesel engine in 1914 when it was evident that rising gasoline prices were simply having too negative an impact on commercial operators and the threat of other companies then at work on diesels became obvious. The company's diesel development was well advanced by 1916 when Atlas merged with the Imperial Gas Engine Company to become Atlas-Imperial.

See also Atlas-Imperial *and* Imperial Gas Engine Company

Atlas-Imperial
Oakland, CA
Trade name: Atlas-Imperial

The first Atlas diesel, an experimental, 6-cylinder 4,580-cubic-inch model that required an elaborate air system to inject fuel, was installed in 1916 in the Seattle ferry *Vashon Island*. It was still powering the boat after World War II. In 1919, however, the engine was fitted with the breakthrough technology that would finally make marine diesels truly practical. This was the "common rail" or "solid injection" fuel system, in which fuel pressure was developed and maintained by mechanical pumps that fed injector nozzles into the cylinders. Atlas was said to be "the first company west of the Mississippi River to develop a commercially successful solid-injection fuel system."

The classic, apple-green–painted Atlas-Imperial diesel engine was a slow-turning (350 rpm), heavy-duty machine available in from 2- to 6-cylinder models. Essentially indestructible, these gentle but powerful engines could do everything except lose weight and size.

During the mid-'30s, Atlas introduced more modern-looking engines with the inner works enclosed and speed ratings increased to as high as 900 rpm. A line of compact, 950 rpm models, with a new constant pressure ATIMCO SYSTEM injector was developed for small work boats. These and the larger engines were built in a factory erected in Mattoon, Illinois. None of this ensured survival much beyond the end of World War II. Engines like the classic Atlas-Imperial were entirely supplanted by the high-speed, lightweight diesels developed by General Motors. Around 1950, White-Superior bought the company to get its name and the right to make spare parts, the latter itself not exactly a growth industry. A number of Atlas Imperial diesels survive today, some still at work powering the tugs in which they'd been installed during the 1920s.

Auger Engine Co.
Minneapolis, MN
Trade name: Auger

Auto Engine Works
St. Paul, MN
Trade name: The Capitol

By 1912, this company offered an extensive line of four-cycle engines that were available with iron or aluminum crankcases. The latter engines were about 150 lbs. lighter than their less expensive counterparts. The headless, side-valve cylinders were, on multicylinder models, cast in pairs. Horsepower ratings covered a wide part of the market, from 5 to 100 hp. Boat builder Ramaley was an important customer. By the mid-'20s, the company was offering marinized aircraft engines under "The Capitol" label. These included the Liberty V-12, the 220-hp, Hispano-Suiza V-8, Hall Scott, and the 1,947-cubic-inch 525–600-hp Wright Tornado V-12. Capitol was still preaching the light weight gospel in 1936, noting its products were particularly well adapted to fast boats.

Automatic Machine Co.
Bridgeport, CT
Trade name: The Automatic

One of several engine companies in the manufacturing center that once was Bridgeport, Automatic was begun in 1899 as a general machine shop and maker of specialty machinery. The address was 145 Stratford Ave. In 1903, Frank I. Hitchcock joined the company as vice president. Hitchcock was interested in the potential of marine engines but was likely frustrated in his efforts to spur Automatic's production of such machines because he soon left to found his own company. In 1910, Automatic moved to 113 E. Washington and then added "iron founders" to its list of specialties. The date when Automatic introduced its first engines, two-cycle models of 4 and 7 hp, is currently unknown. However, by 1910, if not earlier, these were joined by four-cycle models. In 1911, it was announced that a 3-cylinder, 37½-hp heavy-duty Automatic would power a 50-footer across the Atlantic. Apparently, that voyage did not happen.

By 1912, the company divided its product line into two models, Launch or Heavy Duty. The Launch models (5½″ × 7″) were available in 1-, 2-, 3-, or 4-cylinder engines of from 6 to 24 hp. All were rated at 425 rpm. The Heavy Duty models (7½″ × 9″ up to 60 hp and 8½″ × 11″ for the 65- and 90-hp models) were offered in from 1 to 6 cylinders and 12½ to 150 hp. They were rated at 300 or 425 rpm. A 2-cylinder 25-hp Heavy-Duty Automatic weighed 3,100 lbs. compared to a 24-hp, 4-cylinder that weighed 2,000 lbs. These engines all had separately cast cylinders and were available with jump-spark or make-and-break ignition. All these models, it is believed, were open-base designs with gravity lubrication. In 1916, Automatic introduced a new line of enclosed-base, pressure-lubricated engines intended for cruising yachts. These L-head engines had mechanically operated valves, and were offered in 4- and 6-cylinder models of from 30 to 150 hp. The engines were protected from over-revving by an enclosed governor.

The post–World War I Automatics were available in models up to 200 and 250 hp, these being T-head machines. Automatic continued production of an extensive line of open-base engines of 3 to 150 hp, still with make-and-break or jump-spark ignition, and enclosed base models of 16 to 240 hp with American Bosch battery and jump-spark ignition. These were hefty, slow-turning engines aimed at commercial applications. For example, the open base

150-hp engine was rated at 275 rpm and weighed 15,340 lbs. while the enclosed 150-hp model was rated at 250 rpm and weighed 20,000 lbs. Although adapted to run on either gasoline or kerosene, the Automatic must have encountered increased competition from diesels. It is not clear when engine production ceased, but the company remained in business through 1943.

Auto Specialty Company
Detroit, MI
Trade name: Liberty Dwarf
As the name implied, the Dwarf was a little 2½″ x 2¼″ motor that weighed just 30 lbs. The motor developed 1 hp at 900 rpm. It was a jump-spark, two-port engine distinguished by a polished, cast bronze water jacket. Such an engine turned a 7″ 3-blade propeller. When bored out to 3″, the motor produced an additional ½ hp. Motors like this were of a popular genre—"canoe motors."

Ayers Gasoline Engine Works
Saginaw, MI
Trade name: Ayers
Machinist William Ayers was a San Franciscan who moved to Michigan to be part of the burgeoning engine and automobile industries. Together with Seymour Ayers, presumably his brother, William incorporated the Engine Works in Saginaw 1901. The business was located at the corner of Hamilton and Bristol. The initial product was a moderately priced automobile. Marine engines were announced in 1904. The first model appears to have been a two-cycle with a vertically mounted check valve mounted in a casting at the rear of the cylinder. The Ayers's limited automobile making efforts ceased in 1905 and they left Saginaw for Rochester, Michigan, in 1907. What was then known as the Ayers Engine Co. filed for dissolution in either May of 1909 or August of 1910, depending on which source one accepts. Despite that, a 1911 lineup was published by yet another entity, the Ayers Engine & Motor Co. in Trenton. The advertised models ranged from a 1-cylinder, 800-rpm, 1½-hp, presumably two-cycle model that weighed just 43 lbs. to a four-cycle, 6-cylinder, 30-hp engine of 525 lbs. One must assume that these motors were primarily, if not entirely, sourced from other makers and there is no evidence that any were sold.

B. F. Brown Gas Engine Co.
See Brown, B. F. Gas Engine Co.

B. F. Sturtevant Co.
See Sturtevant, B. F. Co.

Baby Engine Co.
Stamford, CT
Trade name: N/A

Backus Water Motor Co.
Newark, NJ
Trade name: Backus Valveless/ Backus

Badger Motor Co.
Milwaukee, WI
Trade name: Badger Junior
In 1912, this company succeeded A. H. Lauson Co., but Andrew H. Lauson remained involved as manager. The *City Directory* listed the address as 917–19 3rd. The company continued to be listed in the 1913 edition of the *City Directory*. Although documentation has yet failed to surface, it is probable that Lauson had developed a compact L-head, 4-cylinder engine that attracted the attention of J. D. Termaat, L. J. Monahan, and investor E. H. Fahrney, all of whom departed Termaat & Monahan, probably in 1914. The trio acquired Badger and promptly relocated it to Oshkosh.
See also Universal Motor Company

Baird & Henselwood Co.
See Baird Machine & Mfg. Co.

Baird Machine & Mfg. Co.
Detroit, MI
Trade name: Yale
William and Andrew Baird opened a machine shop at 51–53 Fort Street East around 1908 or earlier. A subsequent address was the 26 Goebel Building. The Yale was a two-cycle, dome-head, 9-hp motor offered as a kit. Another business entity, Baird & Henselwood, was apparently formed to market the Yale. As others learned, the Bairds discovered that a "build-it-yourself" engine proved to have quite limited market appeal, given that few men had well-equipped machine shops in which to do the necessary finishing and assembly work. The Bairds also owned W. J. Baird & Co., a firm that built or sold machinery of now-unknown type.

Baldwin Motor Co.
Baldwin, NY
Trade name: Baldwin
The company was in business in 1903.

Ball Mfg. Co.
Stamford, CT
Trade name: Ball

Baltimore Oil Engine Co.
Baltimore, MD
Trade name: B.O.E.C.

Barber Brothers
Syracuse, NY
Trade name: Barber
"Built for business by practical men," is how *MotorBoat*'s William Nutting described the Barber engine in Oct. 1914. Charles Barber got his start in the engine business by building two-port engines for D. M. Tuttle in Canastota, 25 miles east of Syracuse. When Tuttle was able to take over his own manufacturing, Barber began building a line of two-cycle engines at his factory on Grape Street next door to the Brennan Motor Company. Barber's engine was a three-port design, and Barber claimed to be the "First Marine Engine to Use the Jump-Spark" in 1904. While such claims are always risky, Charley Barber did recognize early that to get the best from the three-port's comparatively high (600 to 800 in his case) rpm potential, jump-spark was preferable to make-and-break. (Barber was among those who paid Joseph Day a license fee on the 3-port patent.) The Barber timer was easily adapted to multicylinder

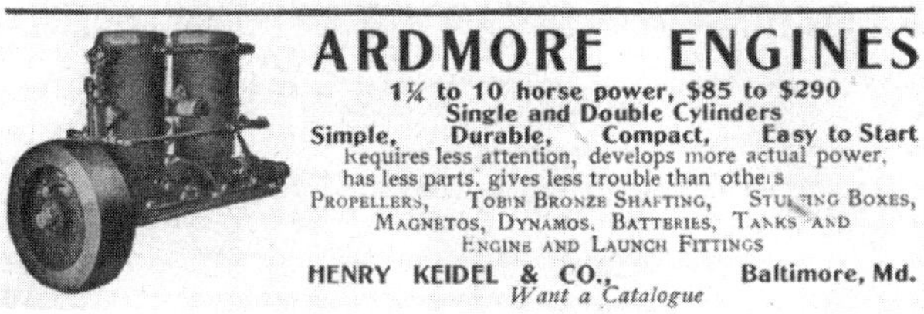

1906

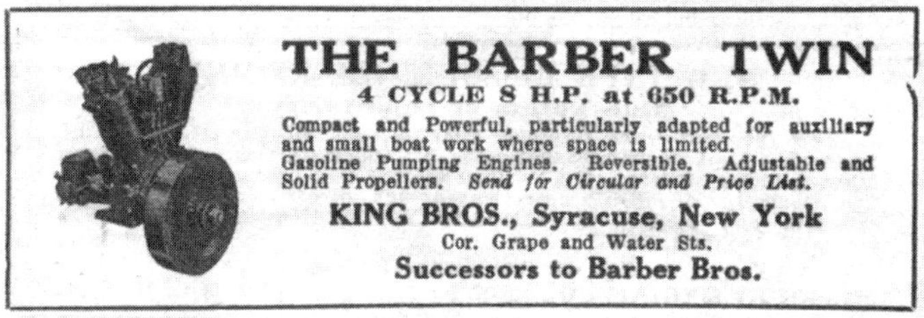

1923 ad by King Bros. for the Barber V-twin

The Wright Tornado aircraft engine converted by the Auto Engine Works took center stage in this 1931 ad.

1913

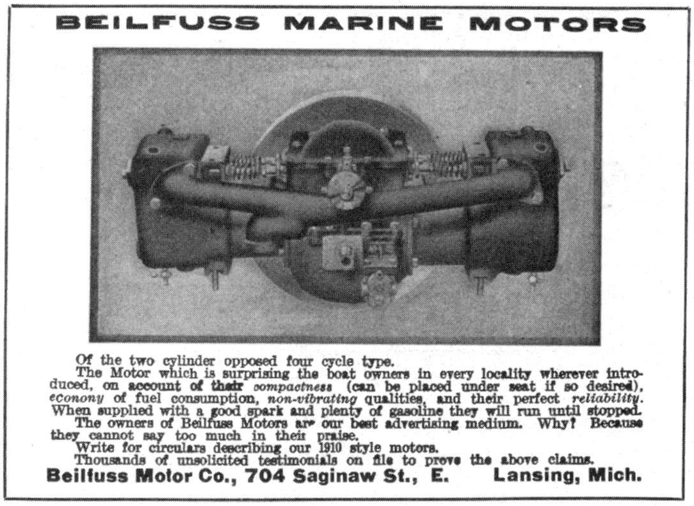

1911

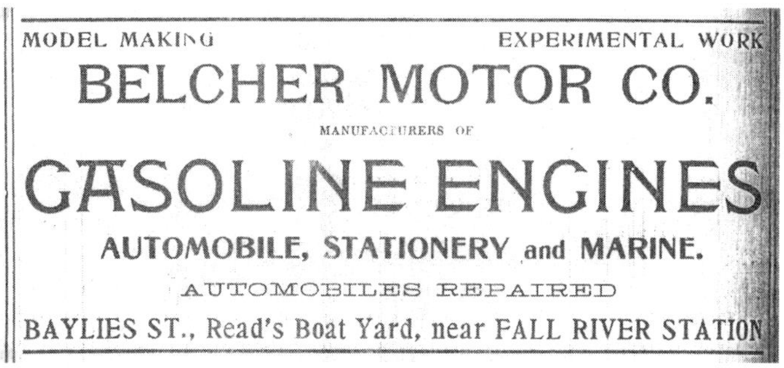

1906

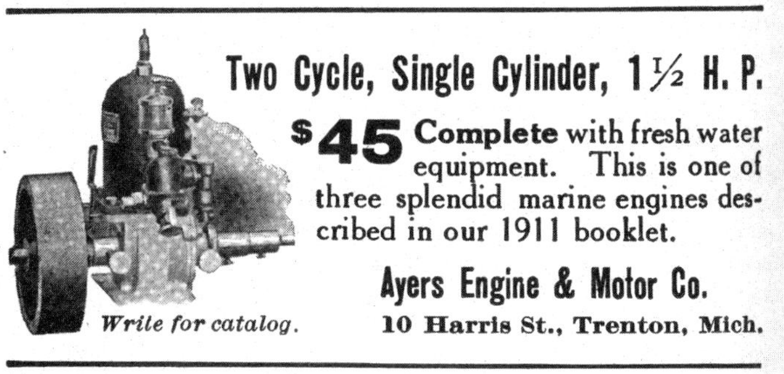

1911

engines. What's more, Barber designed a carburetor specifically adapted to a three-port engine, in which the intake portion of the stroke was necessarily shorter than that of a two-port. Barber's "Straight Line" carburetor directed air through a straight passage in which the throttle was located. The float chamber was a separate unit. Barber also designed and built reversible propellers of from 10 to 22 inches in diameter.

Barber gained much publicity after a 30-foot launch named *P.D.Q.*, owned by wealthy enthusiast George Boldt (address: the Waldorf-Astoria in New York City), achieved a speed of 18 mph with his 24-hp Barber in 1906. The Barber was available in sizes ranging from a 27-cubic-inch one-lunger of 2½ hp to 2-, 3-, 4-, and 6-cylinder engines. The big 8″ × 8″ doubles and triples (1,206 cubic inches) found homes in the engine rooms of tugboats working on the Erie and Oswego Canals. In 1922, an ad of the King Brothers of Syracuse claimed that they were Barber's successor. The ad depicted an 8-hp V-twin and a single-cylinder engine with overhead valves.

See King Brothers *and* Wonder Manufacturing Company

Barker Motor Co.
Norwalk, CT
Trade name: Barker

C. L. Barker commenced producing a line of two-cycle engines as early as 1909. By 1911, the lineup included 1-, 2-, and 4-cylinder engines of from 1¼ to 30 hp. Make-and-break ignition was standard, jump-spark optional. The engines were fitted with a carburetor of Barker's own design. The company was still in business as The Barker Factory in the mid '20s with a line of 1¼- to 8-hp two-cycle motors. In 1929, Barker offered a marinized version of Ford's Model A engine rated at 40 hp at its maximum 2,200 rpm. Barker also supplied Henry Keidel & Co. with engines re-badged as Ardmores.

See Keidel, Henry & Co.

Bath Marine Construction Co.
Bath, ME
Trade name: N/A

Bauers Brothers
Baltimore, MD
Trade name: Bauers

The brothers were in business by 1905 advertising a four-cycle engine that could be put together "with a monkey wrench and screw driver."

Bay State Motor Works
Holyoke, MA
Trade name: N/A

Bayonne Launch Co.
Bayonne, NJ
Trade name: Ackers

Beaver Mfg. Co.
Milwaukee, WI
Trade name: Beaver

The 1908 Beaver was available in an opposed configuration of 12 and 18 hp, and a vertical of 28 and 30 hp. However, the company, begun in 1905, achieved more success with a line of four-cycle, en bloc engines for farm tractors and trucks.

Beilfuss Motor Co.
Lansing, MI
Trade name: Beilfuss

Based on patents granted to Richard Beilfuss, this horizontal, four-cycle engine was produced beginning in 1904 at a shop at 701 E. Saginaw. Originally available as a single-cylinder motor, by 1911, 2- and 4-cylinder Beilfuss engines were listed. These were, by the standards of the day, high-speed engines. The 7½- and 12-hp twins ran at 900 and 800 rpm, respectively, while the 18-hp twin was rated at 1,000 rpm. Apparently, Beilfuss achieved some measure of success although presumably on a local basis. The company roster in 1912 listed Beilfuss as superintendent, Dr. C. M. Watson as president, and Edward M. Reitz as secretary/treasurer.

Beith Motor Co.
Providence, RI
Trade name: Beith

The Beith was a 2-hp, three-port, two-cycle engine. The idea at Beith was to focus on a single model and produce it with dedicated tooling so that quality and pricing would be favorable. Although the engine was equipped with make-and-break ignition, the company claimed it could operate at 1,000 rpm. A ¾″ Kingston carburetor was used. Beith production occurred during 1914.

Beiver Motor Co.
Kalamazoo, MI
Trade name: Beiver

The company was in business in 1911.

Belcher, George E. Machine Shop
Stoughton, MA
Trade name: Vanguard

Regarding the Vanguard, *MotorBoat* reported in Feb. 1915, "The motor is of the open-base type, the cylinders being carried on turned-steel columns. In place of the base compression that is common to ordinary two-cycle motors, the mixture…is compressed in the closed space beneath the piston. In order to carry out this design, the connecting rod cannot be attached to the wrist pin and carried directly in the piston, but a piston rod attached to the piston and running to a cross head must be interposed between the piston and the connecting rod…. The advantages claimed for this open-base compression are wonderful accessibility, the impossibility of flooding the cylinders with a rich mixture, no leakage from cylinder to cylinder or out the main bearings, as is the case with crankcase compression, no uneven wear on the cylinders because the side thrust of the connecting rod is on the guide." A single-cylinder 4- to 5-hp and 8- to 10-hp twin were available, both rated at 700 rpm. Three- and 4-cylinder models were also offered.

The Vanguard had previously been built in Boston (beginning about 1910–1911), and it is now unclear what the relationship was between the

1911 Barber

1905 Bauers Bros.

1915 Blakely single with integral tank

1915 Blakely twin

Circa 1916 Bolinders Model M for the schooner Sierra, San Francisco

1913 Bridgeport double

1915 Bridgeport triple

1915 Bridgeport four-cycle, 3-cylinder

1915 Bridgeport single

1906 Brown-Talbot

1905 Brown-Cochran

earlier company and Belcher. Open-base engines like this, whatever their theoretical benefits, never challenged such conventional stalwarts as Palmer, Lathrop, Mianus, or other conventional two-cycle models, and the decline of the two-cycle marine engine certainly ended what success, if any, Vanguard or Belcher could have had.

See Vanguard Engine Co.

Belcher Motor Co.
Fall River, MA
Trade name: Belcher
In 1904, Gilbert and Chester Belcher claimed to have had 10 years of experience building and experimenting, but their company ad was not specific regarding details of what they had built. In fact, Gilbert H. Belcher was listed in the City Directory as a piano tuner. The Belcher shop was located in Read's Boat Yard on Baylies Street and, from 1908 to 1913, the Motor Company did appear in the City Directory. Gilbert died in 1913; thereafter, the company no longer appeared in the City Directory. Chester S. Belcher reappeared in 1913 as a machinist working on Davol Street, which was the location of the Mechanics Foundry.

See Mechanics Foundry

Belle Isle Boat & Engine Co.
Detroit, MI
Trade name: Skidoo, Little Hummer
During the years from 1900 until a time after World War I, Belle Isle was just one of the denizens of a Detroit neighborhood known generally as "Motor Boat Lane." The company was located at Jefferson and Park View Avenues. The focus here was on small, two-cycle machines. The first products in 1907 were the 2-hp, $23 Skidoo and the 5-hp, $46 Little Hummer. These motors were a dome-head design with jump-spark ignition. Like so many engine companies, this one underwent name changes — sometimes dramatic — as its business activity and financing evolved. In 1910, the company was known as the New Belle Isle Motor Company and the engine was a $40, 4- to 5-hp one-lunger with jump-spark ignition rated at 700 rpm. In 1911, the firm was merged with the Collapsible Steel Form Company to become known as the Concrete Form & Engine Company, not a promising moniker in the crowded and difficult marine engine business. The company filed for dissolution on August 29, 1912. The Belle Isle name, however, survived. By the mid-teens, the Belle Isle Boat & Engine Co. was offering stock 25- and 36-foot V-bottom runabouts, almost certainly powered by motors from other makers.

Benton, Thomas P. & Son
La Crosse, WI
Trade name: Benton
Benton claimed to have built his first engine in 1898. By 1909, he was advertising 2- and 4-cylinder marine models of 2 to 60 hp that were "correct for any service."

Bessemer Gas Engine Co.
Grove City, PA
Trade name: Bessemer
The company was begun in 1899 by a mechanically oriented physician named E. J. Fithian and a machinist named John Carruthers in Grove City, Pennsylvania, some 60 miles south of where Edwin Drake's first oil well had been drilled in Titusville in 1859. While the company soon dominated a good portion of the oil-field market for gasoline engines, Fithian was realistic in his assessment that Bessemer's future lay with the diesel. When John Carruthers died in the early '20s, Fithian purchased Carruthers's share in the company. He promptly acquired a license from Atlas-Imperial in Oakland, California, and the company began development of what would be introduced in 1923 as the Bessemer Atlas Diesel. These were similar to the California engines but carried the legend "Bessemer Atlas Diesel" on the inspection covers. Three-, 4-, and 6-cylinder models were available in ratings of 50 to 1,000 horsepower. "For thirty years they have been Pacific Coast favorites," said a company ad in December 1923, "and are now brought to the East Coast, the Great Lakes, Gulf and Eastern rivers through the huge plant of the Bessemer Gas Engine Co., one of the largest manufacturers of gas and oil engines in the world."

The first customer for one of these engines may well have been the Borough of Grove City itself, which purchased a 360-horsepower stationary model on November 21, 1923. Only four engines were sold in 1923, but the following year, more than two dozen were built. A number of these were bought for use in fishing schooners in Boston and Gloucester, and another, a 230-horsepower 6 cylinder was installed in the yacht *Speejacks II*, successor to the *Speejacks*, which, in 1923, had become the first gasoline-powered yacht to circumnavigate the world. In 1925, Packard president Henry B. Joy ordered a Bessemer for his new yacht, *Spray III*.

While it found ready customers among wealthy yachtsmen and New England schooner owners, Bessemer did not make inroads into the river towboat fleet until 1926. That June, the Iron City Sand & Gravel Company in Pittsburgh ordered a 3-cylinder, 60-horsepower Bessemer for its new *Poppy* and a similar engine went into another towboat, the *Virginia*. If the towboat was to be a stern-wheeler rather than screw-driven, a stationary version of the Bessemer was ordered. For example, in January 1927, Charles Zubik, among the most colorful and independent of Pittsburgh's towboat operators, purchased a 110-horsepower, 3-cylinder Bessemer for his 71-foot *Tornado* and a 235-horsepower, 4 cylinder for his 73-foot *Smoky City*.

Wonderful as it was, the Atlas-Imperial was recognized by Bessemer engineers as representing World War I–era technology. After the war, the great cast iron marine gasoline engines with individual cylinders and exposed pushrods quickly gave way to ever more streamlined models as marine engine companies looked to automakers as role models for such things. In December 1927, Bessemer introduced its R-line of engines with en bloc cylinders "strongly reminiscent of automotive practice."

The 6- or 8-cylinder R was a world removed from the Atlas-Imperial, and it was of fabulous specification by any measure. It had what Bessemer called an overhead camshaft (although the cam was actually mounted alongside the top of the cylinder head rather than directly above the valves), replaceable cylinder liners, and an enclosed flywheel. Mounted in removable cages in the cylinder heads was a pair of forged-steel exhaust valves and a pair of inlet valves — 4 valves per cylinder! The 10-inch diameter crankshaft was built to Lloyd's specifications. It was unusual in that it was forged in two sections, which were bolted together. Each crank section served one block of cylinders; the crankshaft itself was carried in main bearings 12¾″ in diameter. A pinion mounted at the coupling drove the camshafts through a 2:1 reduction gear. This feature contributed to the engine's comparatively compact dimensions and to its smooth operation. When running at 300 rpm, the engine was said to be as "quiet and relatively free from vibration as the most delicately balanced 8 cylinder automobile engine."

The 6- and 8-cylinder R series engines shared 14″ × 18″ bore-stroke dimensions, the six producing up to 675 horsepower, the eight developing up to 900 horsepower, both at a maximum 375 rpm. They were, of course, direct reversing, and they had a motor-driven lubrication system that included an oil cooler. An engine-driven fuel pump delivered fuel under pressure through a common rail to the injector valves at each cylinder. These, as was typical of common rail systems, were operated by the camshaft. These R models found their primary market in commercial vessels and big (170- to 180-foot) commuter yachts designed by Henry J. Gielow in New York. The Atlas-based Bessemer, which continued to be manufactured, and which must have been considerably less expensive, was the model ordered by towboat and fishing vessel owners.

Bessemer recognized that its magnificent R was not destined for volume production. Although the R may well have been the basis for the company's reputation as builder of "the Cadillac of diesel engines," Bessemer introduced a line of smaller displacement models in 1929. The EP-6 and EP-8 shared 7½″ × 10″ bore/stroke dimensions and developed their 150- to 200- horsepower and 200- to 275-horsepower ratings at a comparatively high 500 to 670 rpm. Three- and 4-cylinder versions were also available. A larger 6- and 8-cylinder version, the FP, had a 9″ bore and 12″ stroke. The company now spoke of having improved power-to-weight ratios to 50:1 or 70:1, a progressive move. These engines were different in their architecture from the R. Most notably, the camshaft was now located in the block, operating the overhead valves — two per cylinder — by lifters and pushrods. The EP was fully enclosed, and its one-piece block, like the R models, carried a replaceable liner for each cylinder. The crankshaft, still built to Lloyd's specs, was now a one-piece forging. Full-pressure lubrication remained. In December of 1929, Bessemer announced that 41 of its diesels were scheduled for installation in vessels for the upcoming season. In Grove City, the factory now covered 12 acres and contained 734 machine tools and two dozen overhead cranes; it employed 1,400 people.

In the early '30s, Bessemer, under the Cooper-Bessemer name, introduced another new series, the N. Now, the cam was moved higher in the block, operating a short pushrod via a roller follower. A patented Cooper-Bessemer injection pump contained twin plungers, each replaceable and each capable of supplying the camshaft-driven injectors with adequate fuel pressure. As on the P models, single intake and exhaust valves were used, and the cylinder heads were individually removable for servicing, weighing 435 lbs. apiece. In 1930, a larger displacement version of the N, the J model, was introduced. Offered in 6- and 8-cylinder models, the J had all its iron parts cast of Meehanite, although specific engines that would run continuously at higher rpm had alloy pistons to save weight. The JS had 13″ × 16″ bore/stroke dimensions; the JTB 12″ × 15″. These engines and others, including the larger displacement L models, were steadily developed by Cooper-Bessemer, each one carefully built to each customer's specific application needs.

A Bessemer (or Cooper-Bessemer) engine, properly displayed and interpreted, would be a marvelous acquisition for any museum local to the company or any marine museum, especially one dedicated to inland river commerce.

See also Cooper-Bessemer

Bicknell Mfg. & Supply Co.
Janesville, WI
Trade name: Bicknell

Black Rock Machine Co.
Bridgeport, CT
Trade name: Aristos

In Dec. 1910 Black Rock emerged as the third and final producer of Clarence Van Auken's ingenious horizontal (two pistons per cylinder), two-stroke engine. "The explosion occurs in the center of the machine and drives both systems outward. The crankshafts are so arranged that propellers turn in opposite direction. Two reverse gears are fitted so that a boat can be very easily controlled without the use of the rudder."

Under Black Rock, the three-port engine was offered in 2-, 3-, and 4-cylinder models ranging from 8 to 65 hp. In the company's entry in *Power Boating* in 1912, a special note was included: "Note that although a definite stroke length is given, figures should be doubled to obtain total stroke and gas expansion." Bore/strokes were: 3³⁄16″ × 3½″ on the 8- and 12-hp models, 4″ × 4¼″ on the 18-, 27-, and 36-hp models, and 4¾″ × 5″ on the 48- and 65-hp models. Cylinders were cast separately. A Bosch magneto supplied current to the spark plugs. A Schebler carburetor was standard as was a reverse gear mounted integrally. Clever in conception and impressive to examine, the engine was no more a success for Black

Rock than it had been for its predecessors. Subsequent to 1911, the company never again engaged in engine production but produced specialized rubber-cutting machines and other equipment. The company name changed to Black Rock Mfg. Co. in 1917. Black Rock remained in business until 1960.

See also Van Auken Motor & Machine Works

Blakely Engine Co.
Muskegon, MI
Trade name: Blakely
Introduced in 1914, the Blakely was a 40-pound 2½-hp motor designed for canoes. The $35 price included a fuel tank and battery.

Blanchard Machine Co.
Cambridge, MA
Trade name: Blanchard
The company produced an oil engine designed by Wolcott Remington. It was available in 1- to 4-cylinder models of 8 to 100 hp. The Blanchard was advertised in 1912 as "The Cheapest and Safest Power." The company stressed its engine could be started in three minutes and lacked "troublesome electric ignition devices."

Bliss, F.A. Motor Co.
Brooklyn, NY
Trade name: N/A

Blomstrom, C. H. Motor Co.
Detroit, MI
Trade name: Blomstrom
Although C. H. Blomstrom's primary activity involved automobiles, he advertised a pair of 600-rpm marine engines, a ¾-hp and a 1½-hp model. These were listed from 1904 to 1908 when the company was closed. Launches were also offered.

Blount & Lovell Motor
Boston, MA
Trade name: B & L
The engines shown by the company at the 1911 Boston Boat Show were T-heads of 1-, 2-, 3-, 4-, and 6-cylinder configuration. The cylinders were individual castings. The 1912 line, referred to as the Model "E," was rated at up to 800 rpm and ranged from 10 to 65 hp. The company then was located at 102 High Street.

Blount Engineering Co.
Boston, MA
Trade name: Blount
"Clean, Quiet and Reliable," claimed Blount of its engines in 1916. "Any Size. Any Service." The Blount was offered in models of 2 to 6 cylinders, 12 to 70 hp.

B.O.E.C.
See Baltimore Oil Engine Co.

Bolinders Company, Inc.
New York, NY
Trade name: Bolinders
Jean and Carl Gerhard Bolinder began producing four-cycle engines in their Stockholm, Sweden, shop in 1893 as J. & C. G. Bolinders. A hot-bulb, two-cycle oil engine followed in 1902 and was soon refined with direct reversing. The company then went on to become one of the important engine builders in Europe. In 1921, a new model, the NE, was introduced, which featured an improved fuel-spray pattern across the hot bulb. Air starting and a new electric-starting system based on a glow plug were available. These engines were imported into the U.S. by Bolinders Company in New York beginning, it is believed, in 1922 when 13 models of from 20 to 500 hp were listed. The engines were represented on the west coast by Henry Lund & Co., which had offices in the Marine Building in San Francisco.

The Bolinders was famous for its precision machining: "There are parts machined to half-hundredth of a millimeter, which is to say one five-thousandth part of an inch, and accuracy of similar nature is carried out all over the engine, because its design calls for such care to insure proper and reliable working."

While the market focus was always commercial, as the range expanded, smaller horsepower models would have been suitable both for smaller workboats and pleasure boats. By 1928, the lineup included 28 engines, the smallest of which was a single-cylinder 4¹³⁄₃₂" × 4¹³⁄₃₂" that developed 7 hp at 900 rpm at a weight of 330 lbs. In 1932, Bolinders joined with agricultural equipment maker Munktell to become Bolinders-Munktell. Production was then centered in Eskilstuna, Sweden. The Bolinders line remained available into the late '30s in the U.S., making it one of the most successful European engines on the American market. The company is still in business today as Volvo Construction Equipment, and the antique marine engines are highly prized by collectors. Some are still in use powering "narrow boats" along England's many charming canals, and in other vessels.

Boothbay Engine Co.
Boothbay, ME
Trade name: Boothbay
Boothbay was a subsidiary of the Rice Brothers Shipyard established in 1892. Of the three brothers involved, it was Henry Rice, the eldest, who developed an interest in engines. The first engine, a two-cycle, 2-hp one-lunger with make-and-break ignition, was completed in 1902. It led to a 450-rpm, 6-hp (5½" × 5¼") version and then a 2-cylinder, 850-rpm, 12-hp (4½" × 5") motor. It is estimated that Rice built about 15 engines a year until 1911.

Often, at one of the big boat shows, Rice would bring a handsome, varnished, launch-style hull fitted with one of its own engines. As demand increased after 1911, a four-cycle line was introduced. A new foundry and assembly facility were built to permit increased production. That year, 100 engines were produced. Thereafter, volume increased each year until a fire destroyed the engine company in 1917. Surviving equipment was bought by a local machinist named Thurston who continued a modest production until 1924. It is said that the Rice-built engines were painted red but that some later models were gray. The Thurston engines were gray.

1915

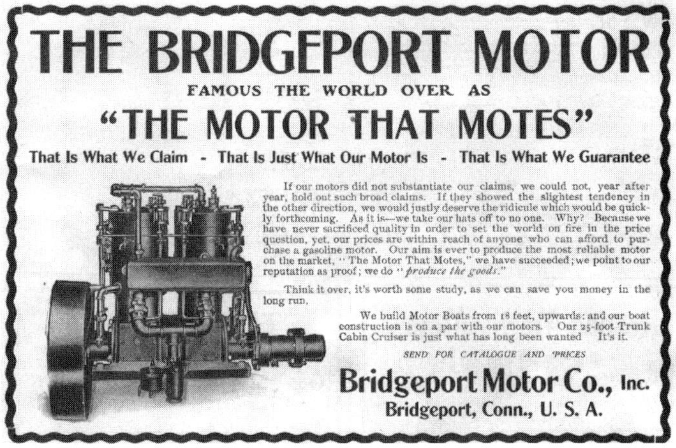

1906

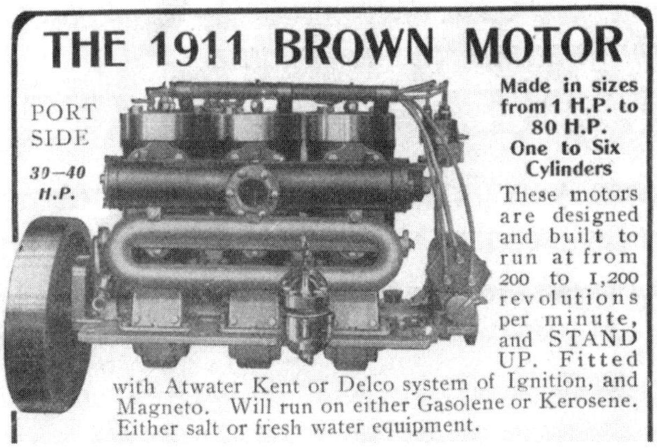

1911

1906

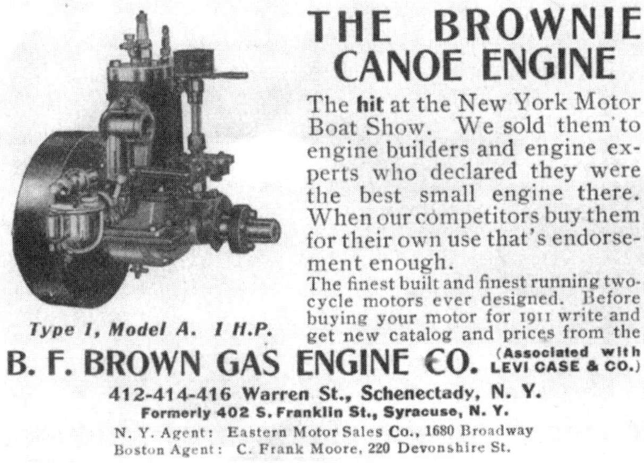

1911

Borden & Selleck Co.
Chicago, IL
Trade name: Borden & Selleck
The company offered a variety of hardware and machinery under its own name and claimed that marine engines, as well as stationary models, were among them.

Bosserdet Yacht & Engine Co.
Detroit, MI
Trade Name: N/A
Another denizen of Detroit's Motor Boat Lane neighborhood, Bosserdet was located at Jefferson Ave. and Parkview Ave. from 1910 until 1917. No information about the company's engines or boats has yet surfaced, and it is not clear whether engines of its own make (unlikely) were installed in Bosserdet boats.

Boston Gasoline Engine Co.
Boston, MA
Jump-spark or make-and-break engines were offered in 1904.

Bowen, George B. Co.
Buffalo, NY
Trade name: Bowen

Boyer
Oakland, CA
Trade name: Boyer

Brennan Motor Mfg. Co.
Syracuse, NY
Trade name: Brennan Standard
Patrick Brennan established his company in 1897 and began selling engines in 1900. Addressing a primarily local clientele, Brennan learned what customers were looking for and focused on three models, the four-cylinder 20-hp "M," the top-seller four-cylinder 25-hp "4B," and the six-cylinder 36-hp "6B." Brennan's engines found homes in automobiles, trucks, tractors, and boats. One happy owner of a Brennan-powered Thomas automobile wrote the company in Nov. 1904 to share his satisfaction with an engine that could produce 48 mph and "climb almost any hill on the second speed." The model line than ranged from 6 hp to 16 hp.

By 1911, Brennan offered 11 models, a trio of doubles rated at from 10 to 20 hp, six 4-cylinder models of 20 to 70 hp, and pair of sixes (7" × 8") of 80 and 100 hp. Brennan kept prices reasonable by purchasing castings and forgings in quantity and doing the final boring and finishing work in his own shop. All this made for a long-lived company. In 1928, the line had matured to include an impressive array of medium speed (600 to 900 rpm) and high-speed (1,000 to 2,800 rpm) models. Among them was the 6-cylinder, 250-hp Gold Cup and the 200-hp Mater Six. All these engines had Stromberg carburetors and force-feed lubrication. Brennan did not survive the Depression.

Bridgeport Machine and Motor Co.
See Bridgeport Motor Co.

Bridgeport Motor Co.
Bridgeport, CT
Trade name: Bridgeport
This important company, long a key employer in Bridgeport, was incorporated on Jan. 1, 1900, under the direction of F. H. Stillman as the Bridgeport Machine and Motor Co. The initial location was at 197 Housatonic Ave. In 1904, the company changed its name to the Bridgeport Motor Co., Inc., and moved across the Pequonnock River to 105 Kossuth St. There, at a location close to what is now Rt. 95 south, Bridgeport would remain for over three decades. Eventually, the plant would grow to a 20,000-square-foot facility stretching from 91 to 111 Kossuth Street.

The first ads gave a good indication of Bridgeport's marketing direction: "cruising launches, common sense launches, yacht tenders." In fact, the engines also became favorites of commercial fishermen. The Bridgeports were robust two- and three-port medium- and heavy-duty two-cycle motors with removable cylinder heads. One- and 2-cylinder models were offered, the largest having a 6½" bore and 7" stroke and a rating of 400 rpm. It is worth noting that Bridgeport was, from the beginning, a diversified company. Besides marine and stationary engines, it offered marine hardware, tools, dies, and presses, and performed general machine work. Before World War I, the company claimed its export business was larger than its domestic sales.

In 1915, when *The Rudder*'s editor, Thomas Fleming Day, replaced the old Knox in his famous Seabird, he did so with a single-cylinder 5½" × 5" Bridgeport that developed 4 hp at 500 rpm. It was enough to push the 27-foot, V-bottom boat at 7 mph. In 1916, the line expanded to include 24-, 36-, and 48-hp four-cycle engines. These were heavy duty 7½" × 9" engines rated at 375 rpm. Bridgeport cited several advantages for their selection of an L-head configuration. These included: good cooling of the exhaust valve and uniform temperature in the combustion chamber, elimination of the second camshaft needed in a T-head, and an easily enclosed valve train for quiet operation.

By 1916, the reliability of its engines prompted Bridgeport to adopt its famous slogan as maker of "The Motor that Motes." By the early '20s, the two-cycle models ranged from 4½ to 22 hp, the four-cycles from 7 hp to 60 hp. The 5½" × 6½" 4-cylinder Bridgeport 304 was introduced in 1924, aimed at the market for cruisers and auxiliaries. The engine remained based on individual cylinders rather than adopting en bloc design. By 1928, the four-cycle models predominated although four single-cylinder two-cycle models were still listed in 1928. These included 4½- and 6-hp models rated at 500 rpm and presumably two-port engines, and 8- and 11-hp engines rated at 800 rpm and presumably three-port designs. Every engine had a Schebler carburetor and Joe's gear.

The company remained under the direction of the Stillman family through 1937. In 1938, the firm relocated to 386 Mt. Grove St. and was listed under the direction of Henry Brautigam. The company ceased to appear in the city directory after 1940.

Brigham
Trade name: Brigham

Brockway
Trade name: Brockway

Brooks Mfg. Co.
Saginaw, MI
Trade name: Brooks
Beginning about 1906, boatbuilder Brooks re-badged engines from companies that may have included Gray and DuBrie (who designed the first Gray engine). Within a few years, Brooks literature was presenting "Engines to Rush the Smaller Boat." These included 2½-, 4-, and 6-hp models whose origin remains, at present, unknown.

Brown, B. F. Gas Engine Co.
Schenectady, NY
Trade name: Brown, Brownie
Prior to 1911, the company was located in Syracuse. It offered engines of 1 hp to 80 hp, 1 to 6 cylinders. By 1912, the line included 17 models. The Brownie was a 1-hp "Canoe Engine" that, with its aluminum base, weighed only 35 lbs. A 2-cylinder, 2-hp, 50-lb. Canoe Engine model was also offered. These motors were described in *The Rudder* in 1910 as "stout little machines as finely finished as a sewing machine." Perhaps the most unusual specification of these little 2¾″ × 2¼″ engines was their rpm rating. The engines were said to operate smoothly at between 200 and 1,200 rpm, but in 1912 that was increased to 1,500 rpm for the Canoe models.

The larger model, the Brown, was also noted for its fine finish. The rpm rating on the 3- to 6-cylinder Browns (and a hefty 5¼″ × 4½″ single) was listed as 750 rpm. All the engines were fitted with a Kingston carburetor and jump-spark ignition. In 1910, George Crouch covered the Boston Show and took special note of Brown's display. He reported that the 3-cylinder 30- to 40-hp engine "was fitted with bronze heads, manifolds and base … The manifold containing both inlet, exhaust and water jacket … is in one piece." At the 1911 Boston Boat Show, the company displayed a big 6-cylinder engine that had two carburetors. Each was mounted on a peculiar oval-shaped intake manifold that does not appear to be integral with the exhaust manifold mounted directly above it. Perhaps the inlet manifold was intended to equalize delivery of the mixture to all 3 cylinders. This was presumably a three-port engine. A 3-cylinder version with a similar set-up was shown in a company ad. Another engine that was pictured, however, showed the single carburetor feeding into a short manifold leading to check valves in the base. It seems quite likely that B. F. Brown was among that number of companies that hedged its bet by offering customers the choice of a two- or three-port configuration. Whatever the hoped-for result, these seemingly interesting engines did not achieve any lasting market success, and there is no record of the company surviving after 1912.

Brown-Cochran Co.
Lorain, OH
Trade name: Lacy
The company advertised both two- and four-cycle models in 1907. These engines may have been developed earlier by D. E. Lacy.
See also Rathbun-Lacy

Brown-Collins Engine Co.
Hartford, CT
Trade name: Brown-Collins
The company offered two-cycle 1- to 4-cylinder engines of 1½ to 20 hp circa 1910.

Brown-Talbot Co.
Salem, MA
Trade name: Brown-Talbot
This company was in business from 1904 through 1916 when its lineup consisted of six models, three singles and a double, ranging from 4 to 20 hp. Brown-Talbot advertised itself as a many-faceted company whose products included automobile engines, launch motors, and power dories. Surviving ads for the "launch motors" reveal a conventional-looking two-port, two-cycle engine. A feature of the Brown-Talbot was its very light, large diameter, low-lift check valve, a type intended to minimize the typically poor throttling of a two-port engine. Brown-Talbot claimed the check valve would permit its engines to achieve the speed and controllability of a three-port engine but, at the same time, said it was "prepared to furnish [its engines] in the three-port type if desired." Ignition was make-and-break with jump-spark an option. The company's principal business office was at 53 Washington St. in Salem but a Boston office also was opened in the Old South Building. It is believed that machinist Auguste Talbot was one of the company principals.

Brownback, H. L. Co.
Norristown, PA
Trade name: Brownback, Shark
A successful engineer with a background in airplane engine development, Henry Brownback also offered marine engines or marine conversions. In 1924, the company was known as H. L. Brownback & Associates and was located in New York City. Ads showed a four-cycle engine with individual cast-iron cylinders. Equipped with a Joe's gear, the Shark had Bosch starting and ignition equipment and listed at $1,350.

Brownell, F. A. Motor Co.
Rochester, NY
Trade name: Brownell
Incorporated in Oct. 1907 by Frank Brownell after he left Brownell-Trebert, the company produced a line of 3-, 4-, and 6-cylinder overhead-camshaft engines for marine, automotive, and industrial applications. The crankcase included an integral oil reservoir from which the oil was drawn by a pump to feed cylinder walls and bearings. In 1908, the camshaft was relocated to the crankcase, operating the overhead valves by pushrods. Specifications for the 1910 50- to 60-hp 4-cylinder included 5½″ × 6½″ bore/stroke dimen-

sions, a 900-rpm rated speed, jump-spark ignition, and a weight of 1,090 lbs. Reverse gears, clutches, and automobile and truck transmissions were also built. Records indicate the firm ceased operation in 1913.

Brownell-Trebert
Rochester, NY
Trade name: Brownell-Trebert
Four-cylinder, 327-cubic-inch engines developed by the partnership of Frank Brownell and Henry Trebert were powering several of the fastest "speed launches" in the 1906 races for the Gold Challenge Cup. It is possible that the engines were an overhead cam design. Some stationary engines were also offered. The company was in business in 1906 and 1907. Each partner then went on to start his own engine works.
See also Brownell, F. A. Motor Co. *and* Trebert, H.L.F. Motors

Bruns, Kimble & Co.
Jersey City, NJ
Trade name: American Boy
The company produced a 2-hp engine with make-and-break ignition.

Bucket
Trade name: Bucket
The company was in business in 1912.

Bucksport Launch & Engine Works
Bucksport, ME
Trade name: Soper
In 1911 Bucksport listed a 10-model lineup that included two series of Soper engines of 1, 2, 3, 4, and 6 cylinders although it seems unlikely that any, or many, of the multicylinder models were actually built. One line had 3¼″ × 3¼″ dimensions and ratings of 3 to 12 hp. The other had 4½″ × 4¼″ dimensions and ratings of 5 to 30 hp. Ignition was jump-spark. Maine collector Hugh Durgin reported that the engines did not carry the usual nameplate but that instead, the word Soper was stamped into the top of the removable cylinder head. These engines were built in the Soper foundry on the Penobscot River in Bucksport. The Soper was a two-port engine but could be run as a three-port by blocking off the appropriate port and unscrewing the casting covering the other. Jump-spark ignition was used.

Buda Co.
Harvey, IL
Trade name: Buda
Established in Buda, Illinois, in 1881, this company specialized initially in the manufacture of railroad equipment. Buda's first gasoline engines were developed to power its own track-inspection cars but, by 1915, were also available in marine versions. Four-cylinder Budas of 197 and 318 cubic inches were then listed as suitable for restricted class racing. By the '20s, Buda had moved to Harvey, Illinois, and was producing a line of fine, 4-cylinder, 80- to 130-hp gasoline engines with Schebler carburetors, full-pressure lubrication, and Paragon or Joe's reverse gears. In 1928, the company introduced a 30-hp L-head six displacing 638 cubic inches with dry sump lubrication and a crankshaft that was 3 inches in diameter.

Given the rising cost of gasoline, Buda acquired the U.S. rights to the diesel-engine patents of Germany's Maschinenfabrik Augsburg-Nürnberg (M.A.N.) in 1926. The company promptly introduced a 4-cylinder (6″ × 9″), 100-hp diesel that weighed 6,500 lbs. This and subsequent models used "solid" injection (The method of injecting fuel was mechanical and didn't require cumbersome air pumps and related equipment.) based on an American Bosch fuel injection pump. From that beginning, Buda went on to become one of the leading and earliest makers in the U.S. of what were then considered to be high-speed, lightweight diesel engines.

Buda acquired the Lanova (a combination of the name of its German inventor Fritz Lang and the Latin word for new, *Novum*) patent for a comparatively lower pressure (625 lbs. per square inch) combustion chamber. This resulted in remarkably smooth, quiet operation. The chamber's shape meant that "no flame or fuel strikes the piston or cylinder wall. Consequently, piston temperatures are no higher than in common gasoline engines." The compression ratio was also a relatively low 12:1, which required glow plugs for starting. Valves were overhead. Two injection nozzles per cylinder were said to be as easily removed for cleaning as a spark plug.

In 1929, a 2,000-rpm model appeared that pointed the way to some of the most compact diesels built before World War II. In 1934, Buda unveiled its latest DM models. The line included a 4-cylinder 186-cubic-inch, 40 hp engine intended for smaller yachts and workboats. It was fitted with a 2:1 reduction gear and electric starting, and commentators noted they believed the engine to be "the smallest diesel ever exhibited at any boat show." It was these Buda diesels that were acquired by the Navy, fitted with American-made Ex-Cell-O fuel injection, and became the basis within the service for what was called "dieselization." These DM diesels were high-grade engines. A 100-hp six, the DM-468, was introduced in 1935. It shared the DM series' 4¼″ × 5½″ bore/stroke dimensions and had a 3-inch diameter crankshaft supported by 7 main bearings. All oil passages were internal.

Like all companies, Buda saw sales drop terribly during the Depression, but it continued its engine development. In 1936, the company unveiled a new line of 4- and 6-cylinder Silver Crown marine engines together with refinements to the diesels. Sales manager G. B. Wright went on record as saying, "We are looking forward to a greatly increased volume of business for 1936…" Buda emerged from the Depression and made it through World War II. The company's assets were purchased by Allis-Chalmers in 1953.

Buffalo Auto Truck and Motor Co.
Buffalo, NY
Trade name: Buffalo Arrow
"…[W]e have achieved a triumph that marks the greatest advance in the construction of gasoline motors since

The Brownie canoe motor of the B. F. Brown Co.

1911 Brownie twin

Model DM-6-17, furnished with or without reduction gear; electrically started

1932 Buda DM-6-17, 125 hp @ 1,500 rpm

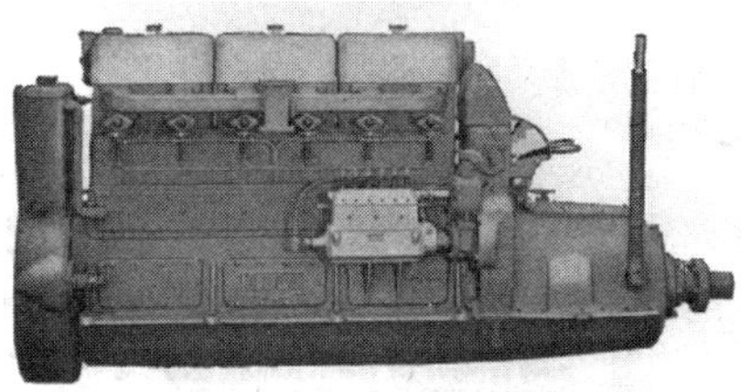
1932 Buda DM, fuel-pump side

SINGLE-CYLINDER, HEAVY-DUTY

1911 Buffalo

1911 Buffalo six

1913 Buffalo

1923 Buffalo

1901 Buffalo

1923 Buffalo 14-30 hp

4-cylinder Heavy Duty Type. Also built in 2 and 6 cylinder.

1923 Buffalo

1911 Cady

1923 Cady

they have became such a necessity not only for pleasure but for commercial purposes," claimed the company catalog. This advance referred to the company's method of avoiding explosions in the base of a two-cycle engine. It simply involved a typical deflector plate on the intake-port side of the piston that directed the intake mixture to the top of the cylinder, where it helped drive exhaust gas out of the exhaust port. Since many engines used a similar piston, one must take this particular claim for what it's worth.

More unusual, however, was that the engine had a horizontally mounted "throttle valve" in the intake port that "instantly governs the speed of the motor at the will of the operator" and a vertically mounted valve that positively sealed the opening of the transfer port to the combustion chamber. The engine's operating range was said to be from 100 to 600 rpm. Both jump-spark and "hammer-spark" ignition were available; the company recommended the former as "more reliable."

Two-cycle Buffalo Arrows were available in 1- to 6-cylinder models, and some models were adapted for cars and trucks, the plan being to sell them directly to manufacturers. The company also offered T-head four-cycle engines, a 2-cylinder 18-hp model, and a 4-cylinder, 36-hp version.

Buffalo Gasolene Motor Co.
Buffalo, NY
Trade name: Buffalo
Founded in 1899, this company had a gloomy beginning. Its efforts to build engines and chassis for the emerging automobile market were plagued by a Selden patent-related lawsuit (Rochester, New York, patent attorney George Selden tried with some success to block would-be auto makers from selling their cars unless they paid him a royalty.) and by customers who couldn't get their new cars to operate. "Nothing but complaints from morning 'till night," is how those days were remembered in a *MotorBoat* article of April 10, 1924. Then, one day, a man whom Louis Fisher, the company president, would always remember as "the nut," visited the factory. Despite being warned against his plan, this customer purchased a 4-cylinder engine for installation in his boat. A few months later, Fisher received a letter quite unlike those he'd gotten from other customers. This one was glowing in its praise of the engine. The company promptly entered the marine engine business, became a stalwart producer of high-quality marine engines, and managed to survive through World War II.

The early Buffalo engine, circa 1907, was an open-base T-head seemingly covered with levers, rods, cams, pushrods, and manifolds. By 1911, signature models were medium-duty, four-cycle engines that ranged from 2 to 225 hp. By 1917, the Buffalo emerged with cylinders cast in en bloc pairs, a neatly enclosed crankcase and transmission, and jump-spark ignition. In 1919 a 44-foot cruiser with a 4-cylinder, 665-cubic-inch Buffalo won the Scripps Reliability Cruise, a nautical version of the Glidden Tour for automobiles, designed to highlight reliability.

By 1916, the Buffalo model range included engines of 2 to 150 hp. Target markets were work boats, speed boats, launches, and cruisers. The company introduced a higher rpm model — available with aluminum or iron base — for express cruiser/pleasure use in 1924. In 1929, the company introduced the Buffalo-Knight, a 6-cylinder, 80-hp sleeve-valve engine that drew on the resources of car maker Willys-Knight. Its sales appear to have been nonexistent. However, Buffalo still had its then-popular 6-cylinder models that included the Clipper (3⅛" × 3⅞"), the 100-hp Navigator (4½" × 5"), and the 200-hp Dreadnought (5⅞" × 7"). A company magazine named *The Herd* published news and testimonials. The rugged Buffalo had success in at least one other application: Many were adapted for use with drilling rigs in Pennsylvania's oil fields.

Buick Auto-Vim and Power Co.
See Buick Manufacturing Co.

Buick Manufacturing Co.
Flint, MI
Trade name: Auto Vim, Buick
Often, marine engine companies were the result of a combination of desire and chance meetings, and that seems to be what led David Dunbar Buick into the business. Like so many of the men who would pioneer Detroit's automobile industry, Buick had an enthusiasm for both boats and motors. In 1899 Buick sold his interest in Buick & Sherwood, a maker of plumbing equipment — Buick had devised a way to coat a cast-iron bathtub with white porcelain, among other things — and founded the Buick Auto-Vim and Power Company in 1900 or 1901 to build gasoline engines for marine and stationary use. Buick was then commodore of the Detroit Yacht Club, and he had happened to meet engineer Walter L. Marr when the latter was repairing a motor boat. The result of the ensuing collaboration was a single-cylinder, four-cycle motor. Marr apparently worked for Buick for 13 months, at which point he became dissatisfied and left to focus on developing an automobile. Production of the Auto Vim motors is believed to have been rather limited. Late in 1902, Buick reorganized his company as the Buick Manufacturing Co. When Marr returned to work with Buick and engineer Eugene Richard, the trio developed a new overhead valve engine and launched the Buick automobile in Flint, Michigan, in 1904. According to 1904 ads, both two- and four-cycle marine engines were then offered, but production ceased when the focus became automobiles.

Burdick Gas Engine Co.
St. Paul, MN
Trade name: Little Jap
Burdick built two-cycle motors with a dome-shaped base, "which insures higher base compression" that presumably led to more speed. The engines were on

the market in 1906. Ratings were from 1½ to 16 hp.

Burdick-Putnam Gasoline Engine Co.
St. Paul, MN
Trade name: Burdick-Putnam
The company was presumably a predecessor or successor to the Burdick Gas Engine Co.

Burnoil Engine Co.
South Bend, IN
Trade name: Burnoil
This firm built oil engines under license from the R. M. Hvid Company, which had its headquarters in Battle Creek, Michigan, and, in 1916, a sales office in Chicago. Rasmus Martin Hvid (pronounced Veed) was a Danish-born, naturalized U.S. citizen who had developed a combustion chamber designed to improve vaporization of diesel oil without the aid of an air injection system. The latter was the bane of early diesel engines.

Burpee and Letson Ltd.
South Bellingham, WA
Trade name: Burpee and Letson
This extensive foundry, it is believed, produced an engine of its own.

Burtt Mfg. Co.
Kalamazoo, MI
Trade name: Burtt
Frank Burtt was in business from 1912 to 1915.

Busch-Sulzer Bros. Diesel Engine Co.
St. Louis, MO
Trade name: Diesel
The brewer Augustus Busch had met Rudolph Diesel, and Busch became the original U.S. licensee of Maschinenfabrik Augsburg-Nürnberg (M.A.N.), which controlled diesel's patents and built his engine. The St. Louis-built marine engines were intended only for commercial applications. Despite his large investment in plant and staff, Busch never saw the rewards that he'd hoped for.

C. C. Riotte Co.
See Riotte, C. C. Co. *and* Standard Motor Construction Co.

C. F. Sparks Machine Co.
See Sparks, C. F. Machine Co,

C. H. Blomstrom Motor Co.
See Blomstrom, C. H. Motor Co.

C. M. Giddings
See Giddings, C. M.

C. N. Cady Co., Inc.
See Cady, C. N. Co., Inc.

C. T. Wright Engine Co.
See Wright Engine Co.

Cady, C. N. Co., Inc.
Canastota, NY
Trade name: Cady
C. N. Cady built his first experimental engine in 1894, and, crude as it was, it was still in service 17 years later. Cady commenced producing his two-cycle engines, it is believed, around 1905. The company was reorganized as C. N. Cady, Inc., in 1911. The motors were sold to, among others, the U.S. Navy and to the Light House Service. By 1923, the line ranged from 1½ to 8 hp in two-cycle models plus a 12- to 16-hp four-cycle, the Ford-based Cadyford—"Ford Replacements in Stock Everywhere."

Caille Perfection Motor Co.
Detroit, MI
Trade name: Perfection
An early maker, Caille was begun by Arthur Caille (pronounced "kale") and his brother Adolph. The company listed ten engines in the 1911 model line, and half of them were 1-cylinder models. This resulted in what now seems like overkill considering that the ratings included a 2, 2½, and 3 hp as well as a 6 hp and an 8-hp version. Still, there was a big jump in weight from the 2-hp Caille at 90 lbs. to the 2½ at 140 lbs. The 8-hp twin at 200 lbs. weighed 100 lbs. less than the 8-hp one-lunger, but the latter was a heavy-duty engine. In 1911, Caille introduced a new heavy-duty model, an 18- to 20-hp twin. The most powerful Caille then was a 4-cylinder (4½" × 4½") that produced 30 hp at 600 rpm. Caille expanded the factory in 1912 to meet demand, and claimed to be the world's largest builder of two-cycle motors in 1916. This may be accurate. By 1919, the company had 201 employees, an impressive number for any marine engine maker of the era. A 4-cylinder, four-cycle followed, the 3½" × 3½" 14-hp Aristocrat in 1917. The two-cycle models were by then dubbed the Perfection, Unit, and Heavy Duty. Today, Caille is equally well known for its outboards, which first were offered in 1913.

California Gas Engine Co.
San Francisco, CA
Trade name: California

Camden Anchor-Rockland Machine Co.
Rockland, ME
Trade name: Knox
In January 1901 *The Gas Engine* reported that the Rockland Machine Co. had acquired the business of Portland's Lamb Gas Engine and Power Co and would manufacture two-cycle marine and stationary engines. It was not until some years later, however, after further company development, that the Knox engine emerged. Without doubt the most successful of Maine's marine engine builders, Camden advertised in all the boating magazines of the day, displayed its wares at the Boston and New York shows, signed sales agents, and found a market well beyond its home state.

Knox engines were used in recreational boats (the company offered 19- to 30-foot launches of its own design and construction), work boats, and government craft. In fact, Camden was an innovative firm that built on its initial success. That success was founded on make-and-break or jump-spark two-cycle engines of three-port design, "far superior to the ordinary two port in *economy, simplicity* and therefore *reliability*."

The first Knox motors were likely built in 1906, but in 1911 the motors underwent some apparently significant upgrades. Among these were enlarged ports to increase rpm, and the addition of an auxiliary air port. Essentially, this was simply an automatic check valve (patented) that allowed air into the top of the transfer port "ahead of the explosive gas" to help clean the cylinder and eliminate backfiring. The "plug" of the make-and-break igniter was changed from bronze to iron after Knox learned iron could better withstand the intense heat of the combustion process. While Knox's multicylinder models were available with make-and-break ignition, the company's sales literature recommended jump-spark as more reliable. When Knox upgraded its engines' port arrangement, it also much enlarged the water jacket space to improve cooling. Also introduced was a new carburetor that Knox called the "D." It's likely that any confusion with the Schebler Model "D" was deliberate, but it's worth noting that Knox had developed a succession of carburetor models and that these were the only parts of the engine covered by a patent.

Despite the success of engines lubricated with an oil/gas mixture, Knox was a persistent advocate of its mechanically driven oiler. All the engines had removable heads. A "combination" stationary and marine motor was also available that included a governor and permitted the motor to be removed from the boat and used for other purposes in the off-season.

In 1912 a 3-hp Knox provided auxiliary power for the yawl *Sea Bird*, famous for its transatlantic passage with *The Rudder* editor Thomas Fleming Day and crew. The Knox consumed 0.15 gallons per mile over 967 miles and used six gallons of oil and three lbs. of grease. That engine was one of a lineup that included double- and triple-cylinder models. One of the battleship *Michigan*'s launches had a 15-hp Knox twin. In fact, Knox motors could be found in all manner of craft including those of its own design and construction. In 1911, for example, Knox built a 38′ × 9′ raised-deck cruiser powered by a 3-cylinder, 22-hp Knox. That motor would drive the boat at cruise speed of 10 mph.

Whatever virtues it espoused for two-cycle motors, Knox offered four-cycle models by 1911. That year, a 20-hp 2-cylinder and a 40-hp four were listed. T-head cylinders were cast in en bloc pairs, and the base was neatly enclosed by plates with the word "Knox" cast on them. The engine featured water-cooled valve guides and seats. The later Model "G" "workboat engine" was introduced in 1920. This was a "medium-weight" (the single weighed 400 lbs, the double 900 lbs) valve-in-head motor available in two single-cylinder models of 5 to 6 hp (5″ × 5¼″) or 7 to 8 hp (5½″ × 6¼″) and two doubles, a 10- to 12-hp (5″ × 5¼″) and a 16 hp (5½″ × 6¼″). Double-ignition was standard.

"Boys! She's a Humdinger" promised a *MotorBoat* ad in Feb. 1923. By then, Knox models were available in ratings up to 75 hp. The original color of the Knox is believed to have been dark green although, by 1919, colors mentioned were vermilion (red) with black trim. Engine production ceased around 1925 by which time it is estimated that some 5,000 Knox engines had been built.

Cameron Marine Car Co.
Beverly, MA
Trade name: Cameron

At the Boston Show in 1910 the Cameron was described as a light-weight, high-speed, four-cycle engine. It had horizontal valves mounted on each side of the cylinder head where they and the actuating gear were readily accessible. Everett and Forrest Cameron had established a car company in 1903 that made cars in a bewildering variety of locations in Massachusetts, Rhode Island, Connecticut, and elsewhere. The effort to crack the marine engine trade seems to have met with no success. The company ceased operating in 1920.

Campbell Marine Motor Co.
Minneapolis (Wayzata), MN
Trade name: Campbell

This four-cycle engine had an open base design to reduce heat on bearings — "a current of air passes through the engine" — and permit easy servicing. That said, the motors came equipped with nickel-aluminum side plates to protect "the parts from the elements and dirt, and prevent the disagreeable spattering of oil." The headless (no "packing" to "leak") cylinders had side valves easily accessed for cleaning and grinding. Each cylinder had twin spark plugs and Atwater-Kent ignition equipment. Schebler carburetors were used.

In 1910 Campbell's was an extensive line that ranged from 1 to 6 cylinders and 5 to 60 hp. Testimonials cited the engine's reliability and fuel efficiency. In the parlance of 1910, good economy meant the two gallons per hour turned in by a 21-hp triple (5″ × 5″) installed in a heavy keel sailboat. Campbell remained in production at least through 1922.

Carl Engine Works
Philadelphia, PA
Trade name: N/A

Carlisle & Finch Co.
Cincinnati, OH
Trade name: Carlisle & Finch

This four-cycle engine was the work of Robert Finch and Morten Carlisle, two former General Electric employees who started a business to repair and manufacture electrical equipment in 1894. After the Carlisle & Finch searchlight became a big success, the pair became intrigued with the promise of internal combustion. They determined that a uniform pressure on the crankshaft could be had in a 4-cylinder engine, which would mimic the pressure exerted on the crank of a double-acting, single-cylinder steam engine. The engine had bore/stroke dimensions of 4″ × 4″ and "wipe break spark" with igniters. As was typical in those pre-carburetor days, a vaporizer was developed. The engine was a decidedly ungainly looking affair to modern eyes, with an open crank-

1908 Caille

1935 Chris-Craft A-120 V-8, 250 hp

1913 Caille

1915 Caille

Chrysler six

1905 Clifton

1915 Clay

1913 Knox by Camden Anchor

1906 Campbell, 70 hp

1905 Clifton

case. A chain from a sprocket on the side-mounted camshaft drove the shaft that worked the igniters. The engine was built starting in 1898 and was said to have been marketed until the United States entered World War I when the demand for searchlights prompted the company to concentrate on their manufacture only.

Carlson & Russ, Inc.
Richmond, NY
Trade name: Rich, Scavenging

Carlyle-Johnson Machine Co.
Manchester, CT
Trade name: Bud-E; Carlyle-Johnson
These en bloc 2-cylinder motors were available by 1913 and were designed to be "as compact and complete as it is possible to imagine." The Bud-E was a 3″ × 3″ 5-hp motor rated at 1,200 rpm. With a total weight of just 118 lbs., the Bud-E was intended for use in canoes, lightweight tenders, rowboats, and little hydroplanes. In 1915 a 15-hp model (4½″ × 4″) was added to the line, the intended application being light runabouts. Referred to by the company as "unit motors," these engines had cylinder castings that included the inlet and exhaust manifolds.

Carson Motor Co.
Detroit, MI
Trade name: Aristocrat
Located at 654 Franklin St., Carson seems to have succeeded the Herrmann Engineering Co. as builders of a 9- to 12-hp, 4-cylinder, four-cycle engine. A Sept. 1916 ad in *Motor Boating* touted a patented starter.

Chadwick Engineering Works
Pottstown, PA
Trade name: N/A

Chalmers
Detroit, MI
Trade name: Chalmers
This engine was almost certainly the same 4-cylinder used in Chalmers automobiles. It is unlikely that it proved popular in marine applications. Hugh Chalmers, head of the company, was a motorboat enthusiast who owned, among others, an expensive runabout powered by a Van Blerck engine.

Champaign Bros.
Ithaca, NY
Trade name: Ithaca
As of 1912, Ithaca gasoline engines were used in Ithaca motor boats advertised by Champaign Brothers.

Chandler Dunlop Co.
Seattle, WA
Trade name: Standard Kid; Clift
Like a number of other marine engines, these were based on Ford Model-T parts — a 1-cylinder 3 hp and 2-cylinder 6 hp.

Chapman's, L.M. Sons
Old Town, ME
Trade name: Chapman

Charles A. Scheppele
See Scheppele, Charles A.

Charles B. King Co.
See King, Charles B. Co.

Charles L. Seabury & Co.
See Seabury, Charles L. *and* Consolidated

Charles P. Willard & Co.
See Willard, Charles P. & Co.

Charles White Engine Co.
See White, Charles Engine Co.

Chas. P. Crouch & Co.
See Crouch, Chas. P. & Co.

Chase Machine Co.
Cleveland, OH
Trade name: N/A
The firm was listed in the 1904 *Marine Buyers' Guide.*

Chase Yacht & Engine Co.
Providence, RI
Trade name: Chase
In 1906, the company advertised four-stroke, T-head engines in 1-, 2-, and 4-cylinder models of 7½, 15, and 30 hp.

Chesapeake Engine Co.
Oxford, MD
Trade name: Chesapeake

Chicago Boat & Engine Co.
Chicago, IL
Trade name: Chicago

Chicago Caloric Engine Co.
Chicago, IL
Trade name: Caloric
An ad in *The Rudder* in May 1906 is now one of the only indications that this company existed. The term "caloric" was applied to what are commonly better known as hot-air engines. Since the engine pictured in Caloric's ad appears to have some sort of heating device atop the cylinders rather than an ignition system, it is likely that the engine was, indeed, a hot-air device. Engineer John Ericsson, most famous as designer of the *Monitor*, had tried without success to build a hot-air engine for big ships. It seems unlikely that the little Caloric enjoyed any more success.

Chris Smith & Sons Boat Co.
Algonac, MI
Trade name: Chris-Craft
With a successful background in designing and building racing boats and runabouts beginning around 1910 — Van Blerck and Sterling were the most typical engines used — Christopher Columbus Smith founded his company to build production motor boats in 1921. Soon, Chris-Craft runabouts and cruisers became American icons. A separate marine engine division was established in 1927. The Chris-Craft A-120 V-8 was introduced in 1928 after extensive testing. Its design and production were guided by Jay W. Smith when he became president of the company begun by his father, Chris. The 825-cubic-inch A-120 was the firm's first in-house marine engine. Approximately 249 A-120s had been built when production ended in 1941. The A-120 had a 5″ bore

and 5¼″ stroke and, with a downdraft Holley carburetor, developed 250 hp at 2,200 rpm. It weighed 2,200 lbs. Some of the engines were fitted with copper, rather than aluminum, cylinder heads to further improve cooling performance under extended high-speed operation. The A-120 has since become a popular engine for use in restored boats of its era.

While the A-120 was exclusively a Chris-Craft product, the company purchased engines from a variety of manufacturers for use in its own boats. Among those engines were Kermath and Chrysler, but Chris-Craft assigned its own model names to the engines it bought. The Chris-Craft A, B, K, M, and W models were Hercules blocks equipped with Chris-Craft manifolds and a marine transmission. These engines were painted the familiar Chris-Craft blue.

Christie Machine Works
San Francisco, CA
Trade name: Christie
In 1911, the Christie Machine Works was located at 191 Fremont. Robert M. Christie was listed as the company's manager, and he was assisted by Harry Christie. It is likely that the company had been founded by Robert and Harry's father. Christie advertised a V-type four-cycle motor in *Motor Boating* in Sept. 1912. The company remained in business at least through 1915.

Chrysler Corporation
Detroit, MI
Trade name: Chrysler
Chrysler was the only one of Detroit's Big Three to enter the marine engine business. Development was spurred by speedboat enthusiast Walter Chrysler, and the engines were developed under the direction of H. E. Fromm, an experienced marine engine man. The first engine was introduced in 1927, the 100-hp, L-head Imperial. It was supplied by Chrysler's marine engine division known as Amplex to Chris-Craft, Dodge, Dart, and other production boat builders. Many other models followed: 6-cylinder models included the 190-cubic-inch Ace, the 224-cubic-inch Crown, the 309-cubic-inch Imperial, the 8-cylinder, 323.5-cubic-inch Royal Eight, and the 384-cubic-inch Majestic. In 1931 the 4-cylinder, 35- to 55-hp Regal was introduced. Its rubber-cushioned mounts absorbed vibration, and these were adopted on the other models as well. In 1936 a new Royal Eight (3¼″ × 3⅞″) was introduced that replaced the Imperial. It was equipped with a 2½:1 reduction gear and two updraft carburetors. By 1936, too, deep or shallow oil pans were offered, enabling engines to be installed at up to a 15-degree angle without the need for an additional oil pump. Chrysler also claimed it was a leader in adapting reduction gears to smaller-bore marine engines.

Such engines, which had benefited from intense automotive development and, particularly, the large volume production involved in the automobile industry, meant that marine adaptations could be sold at favorable prices compared to a lower-production, exclusively marine engine. These Chryslers met the everyday needs of boaters and were popular in craft as diverse as a 16-foot runabout and a twin-screw, 70-foot cruiser. To cite just one example of a production boat with a Chrysler, the 38-foot Wheeler Playmate of 1932 had a 105-hp Chrysler Imperial that drove it at 12 to 14 knots. Engines like the Chryslers, together with the gradual introduction of the compact diesel engine by other companies, spelled the end for the purely marine engine. Chrysler production continued well into the post–World War II era before ceasing in the late 1970s.

Clarence J. Allen
See Amphion

Clark Manufacturing Co.
Fond du Lac, WI
Trade name: N/A

Clarke Gas Engine Co.
Evansville, IN
Trade name: Clarke
Founded in 1898 by Clarence and John Clarke, the Clarke Gas Engine Co. was one of the many small shops throughout the country that produced engines for a local clientele. One didn't expect to find such companies included on magazines' national lists of marine engine builders. According to Clarke enthusiast Keith Kinney, the Clarkes found customers who were operators of ferries on the Ohio and Mississippi. These operators typically purchased 2-cylinder engines. One-, 2-, 4-, and 8-cylinder engines were offered. During their years in business, the brothers built approximately 300 engines, most before the onset of the Depression. The business closed in 1941. The focus here was four-cycle, medium-duty engines. The first models were L-heads, but overhead-valve engines followed.

Clark's Multiple Engine Co.
New York, NY
Trade name: N/A

Claude Sintz
See Sintz Gas Engine Co.; Sintz, Claude. *See also* Sintz, Guy L.

Clay Engine Mfg. Co.
Cleveland, OH
Trade name: Honest Clay
These sturdy engines owed their design and manufacture to an inventive machinist named Edward Henry Clay of Cleveland, Ohio. In 1898, Clay, with the help of his brother, Cassius, constructed a four-cycle, overhead valve engine for his own use. As sometimes happened in those early days, several of Clay's neighbors were impressed enough by the engine's performance to ask him to build engines for them. This was the foundation for a company, whose focus was always the fisherman and whose motto, "The Honest Clay," expressed a message of dependability and economy.

At some point in his company's development, Edward Henry Clay began using the image of the American patriot, Henry Clay, as a symbol of the firm's integrity. In ads, the company referred to "the statesman whose integrity proved a symbol for public life and endeared him to the hearts of the people. We have not chosen the name HONEST CLAY for our engines without full realization that we must measure up to the ideal incorporated in the words."

The Clays were all heavy-duty types with crankshaft and connecting rods machined from carbon steel forgings, cams made of cold-rolled steel, and valves whose cast-iron heads were threaded and welded to cold-rolled steel stems. By about 1917, the engines had progressed so that they were not only overhead valve, but overhead cam as well. This valve mechanism was arranged so that the cylinder heads could be removed without affecting it. A pressure lubrication system was standard, and, as a practical matter, "All nuts, bolts, studs, and fittings are standard and can be purchased at any hardware store." This entire approach to design and manufacture was apparently quite effective, for when Edward Clay made a trip to the Gulf Coast in the 1950s, he quite accidentally stumbled across one of the machines still performing daily service in a fishing boat. One-, 2-, and 4-cylinder models were offered of from 4 to 100 hp.

The Clays sold their company in 1917. The name was changed from the Clay Engine Company to the Clay Engine Manufacturing Company whose president, Horace Newsome, reported a capitalization of $100,000. The model line grew steadily until, by 1919, it included models ranging from 4 to 100 horsepower.

Cleveland Ice Mach. & Mfg. Co.
Lorain, OH
Trade name: Lacy
An ad for this particular Lacy appeared in *Motor Boating* on Dec. 12, 1912.

Cleveland Marine Motor
Cleveland, OH
Trade name: All-in-One
Cleveland was in business in 1910.

Clifford, H.F.
Concord, NH
Trade name: Granite State, Yale

Clift Motor Co.
Bellingham, WA
Trade name: Clift
Con Clift was an experienced marine engineer before he opened a marine engine business on Chestnut Street in 1912. His first engine was a big (10" × 12") 4-cylinder four-cycle that developed 100 hp and was promptly bought by the Pacific American Fisheries which ran a fleet of fishing and cannery vessels. The popularity of this and subsequent engines prompted Clift to move to a new building on Whatcom Creek Waterway in 1915; this facility was doubled in size three years later. In 1919, there were 15 men working at the company, and production reached 60 engines per month. In 1930, Clift introduced new models including single-cylinder engines (3¾" × 4½" and 5" × 6"), doubles (5" × 7" and 5¾" × 7"). By 1931 Clift had built more than 4,000 motors, most for the commercial trade. Castings for the engines were made in San Francisco at the Union Iron Works.

Clifton Motor Works
Cincinnati, OH
Trade name: Clifton
Founded in 1891 and located at 229 and 231 E. Clifton Ave., the company by 1903 was advertising two-cycle stationary and marine engines up to 16 hp. By 1910 Clifton was building heavy-duty, four-cycle engines of from 10 to 90 hp. These engines had individually cast cylinders, jump-spark ignition (battery or magneto), and an open or closed base depending on customer preference. Bore/stroke dimensions ranged from 5" × 6" on the double to 6½" × 7" on the triple and fours to 8½" × 11" on the largest displacement four and six. Most were rated at 400 rpm. Clifton found a good customer in the U.S. Navy after a 6-cylinder engine fulfilled all requirements of a day-long test session. "So great has been the demand for these engines this year that they have had to add a good many machine tools, make special jigs and fixtures and go into the manufacture on a larger scale than before," reported *The Gas Engine* in January 1910. Clifton was in business through the end of 1918.

Climax Engineering Co.
Clinton, IA
Trade name: Climax
See Lamb

Clipper Lawn Mower Co.
Dixon, IL
Trade name: Clipper

Coleman, Walter & Sons
Providence, RI
Trade name: Alpha

Columbia Engine Co.
Detroit, MI
Trade name: Columbia
See Detroit Motor Car Supply Co.

Columbia Machine Works
Brooklyn, NY
Trade name: Columbia
"It Runs," claimed Columbia's ads in 1901. The engine was a single-cylinder, four-cycle with a valve chest on one side of the cylinder. The exhaust valve was mechanically driven, but the operation of the inlet valve was presumably automatic on the piston's down-stroke.

Columbus
Trade name: Columbus
A four-cycle 20- to 25-hp motor was said to be available in *Motor Boating* in Feb. 1912.

Comack
Trade name: Comack
A two-cycle Comack twin was advertised in *Motor Boating* in Feb. 1912.

Comet Motor Works
Chicago, IL
Trade name: Comet

Commonwealth Motor Co.
Chicago, IL
Trade name: Quayle

1906

1911

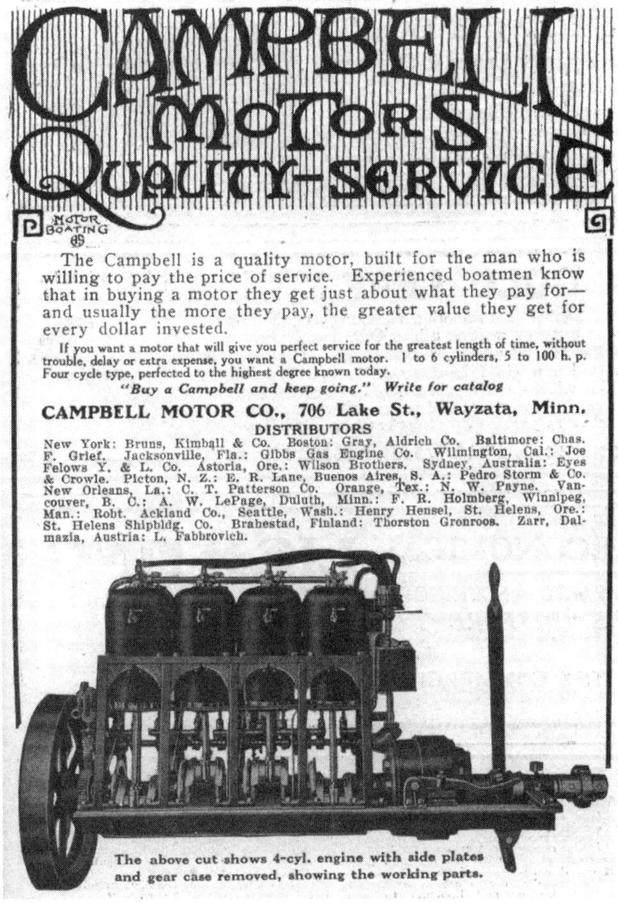

1913

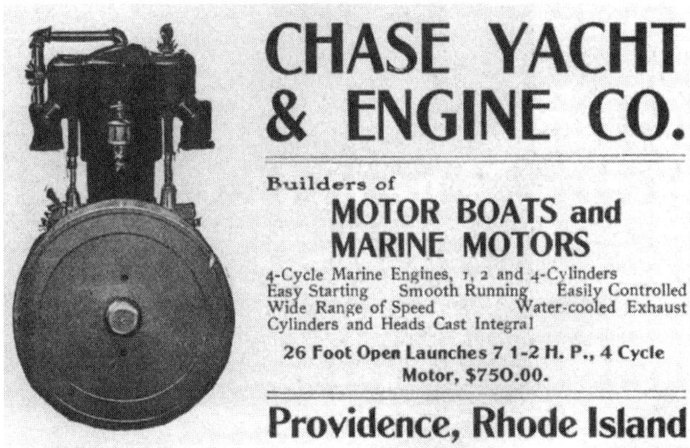

1905

The Quayle was a heavy-duty oil engine.

Concrete Form & Engine Co.
Detroit, MI
Trade name: Belle Isle
See Belle Isle

Connerton
Trade name: Connerton
A 2-cylinder, 12-hp, four-cycle Connerton was advertised in *The Rudder* in Dec. 1912.

Consolidated C.S.M. Corp.
Bridgeport, CT
Trade name: Park City, Bull Pup, Fairfield
C.S.M. stood for Consolidated Sewing Machine Co., which was located at 1047 Broad St. The company entered the marine engine business in 1908 with a two-port, two-cycle engine of now unknown origin but perhaps the Bull Pup built by the Fairfield Motor Co. However, in Feb. 1910, *MotorBoat* shed some light on the company. "Park City Automarine motors, as well as Fairfield motors were exhibited by the Consolidated C.S.M. Co. ... The Park City motors are of the high-speed type... One improvement is the use of a water-jacketed manifold, another is a new type check valve, with a vertical lift, which is fitted in a valve chamber in the manifold."

None of the engines, presumably built for the company by one or more vendors, seems to have made a significant impression — certainly there was plenty of local competition — and production may not have lasted very long. In 1916, the company relocated to Fairfield Ave. and became a distributor for Velie and Interstate automobiles and Stewart trucks. It ceased to appear in city directories after 1920.

Consolidated Engine Co.
Detroit, MI
Trade name: Cesco
The Cesco (acronym for Consolidated Engine Company) was advertised in 1913. Consolidated was likely a successor to the Amalgamated Engine Company.

Consolidated Shipbuilding Corp.
New York, NY
Trade name: Speedway
This company emerged in 1919 when the Gas Engine & Power Co. and Charles L. Seabury & Co. simplified its name. Consolidated could trace its roots to 1885 and the building of marine machinery and yachts. The history of Consolidated, together with those of its predecessor companies, would fill a sizable book. A wide range of engines was offered, all of top quality; by 1917 the company's factories stretched for 1,800 feet along the Harlem River. The plant was then engaged in defense work, building minesweepers.

Yacht designer L. Francis Herreshoff wrote that the Speedway was "noted for its long life, quietness, and reliability." In 1912, ads claimed a Speedway had covered 15,500 miles in five years without a repair. By then, aside from a 5-hp, two-cycle one-lunger, all Speedways were four-cycle engines. The 6-cylinder 100-hp model and its less powerful siblings were closed-base designs. However, open base ("stanchion") Speedways of 130 to 275 hp were also available, the biggest a 6-cylinder with an 11″ bore and 12″ stroke. These big sixes were fascinating engines with "perfect" make-and-break ignition. Starting was by compressed air, the tank fed by an engine-driven compressor.

The clientele was both diverse and wealthy, so much so that Consolidated developed, in the mid-teens, a line of engines designed expressly for use in 16- to 35-foot Speedway yacht tenders. The engines had aluminum bases to reduce weight. At the 1916 New York boat show, one of these tenders, an enclosed "Coupe," was shown together with a 43-foot cruiser powered by a 6-cylinder 6″ × 6″ engine and a 30-foot sedan runabout. That year also saw the introduction of an L-head 6-cylinder (5¾″ × 7″) that produced 130 to 150 hp. Smooth and quiet in operation, this gorgeous engine had cylinders cast in pairs and an enclosed flywheel.

The 1919 model lineup reflects the quality and range offered by Consolidated. General characteristics included a L-type cylinder with the cylinders cast in pairs integral with the head and water jackets. The bed plate was a single casting and carried a nickel-steel crankshaft supported by large main bearings. Of course, the valve gear, which operated intake and exhaust valves of identical sizes, was enclosed. These basic features were applied to an extensive lineup, which included the following:

- High Duty (1,000 to 1,200 rpm) models including the 4-cylinder K (22 to 28 hp) and Z (35 to 44 hp); 6-cylinder Model N (50 to 66 hp) and M (130 to 150 hp), and the 8-cylinder M (175 to 200 hp); These engines were the lightest in the range and had an aluminum frame and base. They were primarily intended for fast launches, runabouts, and day cruisers.
- Medium Duty (600 to 1,000 rpm) Model M 4-cylinder (48 to 75 hp), 6-cylinder (75 to 130 hp), and 8-cylinder (100 to 175 hp). These engines had a cast-iron frame and base and were intended for a variety of work boats, small passenger boats, and cargo boats. They had electric starting.
- Heavy Duty (400 to 600 rpm) Model L 6-cylinder engines of 80 to 115 hp, 120 to 160 hp, and 200 to 250 hp). These were heavily constructed motors designed for cruisers and houseboats, tow boats, cargo boats, and passenger boats.

By the early '20s, Consolidated's most popular yachts were 60- to 65-footers. In 1922 and 1923 the company introduced new engines designed for such boats. These included the 185 hp Model MR (5¾″ × 7″) with removable heads and single intake and exhaust valves. The 4-valve 300 hp Model R (6¾″ × 8″) was a 6-cylinder that produced 185 hp at 1,300 rpm.

Consolidated always refined and upgraded its engines, staying contemporary with en bloc cylinders and overhead valves. The company made big news at the 1929 New York Boat Show with the introduction of a 300-hp, solid-injection diesel engine. The 700-rpm engine was described by Penn State's dean of engineering, R. L. Sackett, as being "attractive in appearance, of refined design, and looks like a race horse." Consolidated ceased boatbuilding in 1952 (engines had long since been out of production), but remained in the storage business into the 1980s.

Continental Engine Co.
Chicago, IL
Trade name: Continental
This company existed circa 1898 to 1913. The plant was located in Dallas City, Illinois. A marine engine was introduced circa 1904.

Continental Motor Manufacturing Co.
Detroit, MI
Trade name: Continental
Although never a purely marine engine maker, Continental was among the longest-lived of any American engine company, and marine conversions of Continental blocks were not uncommon. Continental had its origins in 1902 when engineer Ross Judson and his brother-in-law Arthur Tobin recognized the growing automobile industry as a ready market for internal combustion engines. A four-cycle, 2-cylinder engine was on display by 1903 and Continental — the name chosen to create an association with "European quality" — soon enjoyed rapid growth thanks to an era of "assembled cars." At the time, companies that couldn't afford to develop their own engines, including Studebaker, Peerless, Hudson, and Willys, all used Continentals. In addition to its Detroit factory, Continental built a second plant in Muskegon in 1905.

By 1911, Continental was listed occasionally along with marine engine makers in the bigger boating magazines. The line then consisted of seven 4-cylinder, 1,000-rpm motors ranging from 20 to 55 hp. Each engine had cylinders cast in pairs, Bosch high-tension magneto ignition, splash lubrication, and a Schebler carburetor. The engines were also converted for marine application by a variety of companies. Joe Van Blerck's Continental-Van Blercks were built at his plant in Red Bank, New Jersey, during the late 1920s, but Gray was certainly among the largest customers.

By 1928, Continental claimed to be supplying 128 automobile companies, and it also began making engines for light airplanes. After struggling through the Depression, Continental became a major engine supplier for the war effort and after the war took over Gray Marine. That was the start of ongoing corporate ups and downs and Continental itself became a part of larger corporations. Today it still makes engines as part of Teledyne.

Cook Engine Co.
Cambridge, IL
Trade name: Cook
In 1914 Cook offered a line of 2½- to 20-hp motors of 1 to 4 cylinders.

Cooley Manufacturing Co.
Waterbury, VT
Trade name: Cooley
William Cooley established his company in 1882 and began making engines, it is thought, during the late 1890s. Marine models were offered in 1- and 2-cylinder versions. The Cooley was a two-cycle motor with make-and-break ignition. Commenting on the engines displayed at the Boston Show in *MotorBoat* in Feb. 1910, George F. Crouch noted that a single-cylinder 5-hp and a 2-cylinder 10-hp Cooley were on display. He reported that customers could choose either a detachable or integral head. He also noted that "centrifugal ring oilers for crank pins will be used on all Cooley motors, in place of the hollow connecting rod which has been previously used as an oil reservoir." Ads for the Cooley appeared in *Motor Boating* in Dec. 1911 and *MotorBoat* in Dec. 1913 with a Boston address, which was that of the sales agent. It is believed that engine production ceased after 1913. No Cooley is known to have survived.

Cooper-Bessemer
Mt. Vernon, OH
Trade name: Cooper-Bessemer
The C. & G. Cooper Co. was begun as an iron foundry in 1833 by Charles and Elias Cooper. The company built locomotives during the Civil War and not long thereafter acquired the license to the steam engines developed in Providence, Rhode Island, by George H. Corliss. The Cooper-Corliss was, as the company liked to say, "the first Corliss Engine built in the West." Some of these machines were 200-ton giants, with pistons seven feet in diameter and flywheels that weighed more than one of Cooper's locomotives. The company also began manufacture of portable steam engines for use on farms and in 1875 introduced a highly successful steam traction engine.

None of this would be enough, the company recognized, for it to remain viable long into the 1900s. Development of steam turbines by Westinghouse and the apparently limitless potential of gasoline and natural-gas engines meant that Cooper would need to head in a new direction. The first step was taken with the introduction in 1909 of a huge, natural-gas engine and compressor for use on a pipeline in West Virginia. The manufacture and installation of such machines after World War I, together with smaller 80- to 170-horsepower models, made Cooper the leader in pipeline compression engines. Its field engineers formed a rugged and romantic group known as "erection men" who somehow managed to get the largest of engines put together and operating in the remotest of places.

To advance its diesel prospects, meanwhile, Cooper in 1927 acquired the rights from Maschinenfabrik Augsburg-Nürnberg (M.A.N.) to that company's four-cycle diesel engine and began investigating its potential development as the Cooper-M.A.N. But the company still felt vulnerable, and in

April 1929 it merged with the Bessemer Gas Engine Co.

The quality of the Cooper-Bessemer engines was reflected in their price. In the mid-'30s, a 4-cylinder EP cost some $8,000 at a time when wages were often some 65 cents an hour. A few years later, an 8-cylinder JS sold for nearly $27,000 in 1942 and nearly $32,000 in 1946. But the engines remained in production throughout the harsh years following the stock market crash, and Cooper-Bessemer sales even increased to nearly a million dollars in 1930 before dropping to under $300,000 in 1931 and $66,000 in 1932. After that, things slowly began to improve. In his *Inland River Record* compiled in 1945, Fred Way, Jr., listed a total of 31 towboats powered by Bessemer or Cooper-Bessemer engines, about eight percent of the total. The majority of these boats were built between 1939 and 1943. Production began to dwindle during the '50s, however, and the last marine engine was built, as nearly as anyone can now remember, in the early 1960s.

See also Bessemer Gas Engine Co.

Corliss Gas Engine Co.

Petaluma, CA (plant); San Francisco, CA (sales office)

Trade name: Corliss

This builder of heavy-duty gasoline engines—"The Most Modern Motor Made"—had no connection to the Corliss of steam engine fame. The company was named for owner Edward M. Corliss who, according to those with local knowledge, arrived in Petaluma from San Francisco in 1908. The 1915 Corliss catalog claimed the engines were the outcome "of over *twenty* years experience." The wording suggests that, if Corliss itself did not date to 1895, then a predecessor company did. That firm may have been Marcellus Graham's, builder of the Oriental engine. Corliss's newly built plant—erected at First and H Streets, giving the company access to newly dredged Petaluma Creek—reputedly cost $40,000.

The 1915 catalog listed stationary and marine engines of from 1 to 6 cylinders and 7 to 340 hp. The weight of these machines is one measure of what was really meant by a "heavy-duty" engine. A 7-hp Corliss marine engine weighed about 1,170 lbs.; a 180-hp, 4-cylinder weighed 27,550 lbs., and a 6-cylinder, 300-hp model weighed in at nearly 42,000 lbs. Each cylinder configuration was available in several bore/stroke dimensions so that a variety of horsepower ratings could be selected. The six, for example, was available in 50-, 65-, 85-, and 100-hp models.

These were T-head machines with "spark hammer" (make-and-break) low tension ignition. Despite its size, the gray iron Corliss cylinder and cylinder head always remained a one-piece casting, "a vast improvement over the removable cylinder-head construction, inasmuch as it eliminates entirely all possibility of trouble from leaky packed joints." Crankshafts were machined from solid forgings. The engines were painted green.

Corliss focused on the commercial and government market for marine engines. Its 5-hp, single-cylinder engine was installed in Lighthouse Service launches while the engine rooms of tugs and survey vessels became home to larger models. The extent of the line was extensive, and 31 models were listed for 1916. These included a half-dozen 2-cylinder engines ranging from 12 to 65 hp and from 900 lbs. to 9,600 lbs. The biggest Corliss was a 40,000 lb. giant of a six. This 14½" × 17" engine developed 350 hp at 250 rpm. Neither this extensive line of marine engines nor Corliss's stationary engines, used for water pumping on farms and in homes, generated revenues adequate for the company's survival. As gasoline prices rose in the period from 1910 to 1915, the day of the big, slow, turning gasoline marine engine ended. Atlas and Imperial survived the downturn by merging and developing a practical diesel. Corliss was absorbed by The Standard Gas Engine Company in 1916, a move that doubled Standard's capacity and boosted its workforce to 300 people.

Cornwell, R.M. Co.

Syracuse, NY

Trade name: Wonder

See Wonder Manufacturing Co.

Corson Motor Co.

Detroit, MI

Trade name: Aristocrat

This Aristocrat was a 4-cylinder en bloc with a removable head. (The Caille Aristocrat had a nonremovable head.).

Cottrell & Sons, C.B. Co.

Westerly, RI

Trade name: Cottrell-Holmes

This was a 12-cylinder, 350-hp engine listed as of 1918. Almost certainly, it was a Liberty engine converted for marine use.

Cragg Motor Mfg. Co.

Detroit, MI

Trade name: Cragg

See Gilmore-Cragg

Craig, James Engine and Machine Works

Jersey City, NJ

Trade name: Craig

James Craig, Jr., was building multi-cylinder (producing about 5 hp per cylinder) four-cycle marine engines in a shop at 556 West 34th Street in New York City by 1902. At that time, the company was known as James Craig Marine Gasolene Engines. The camshaft was described as being "mounted on the cylinder casings parallel to the crankshaft." From that position on the engine's port side, it operated big, L-shaped rocker arm castings that opened the overhead valves. The individual cylinder castings were mounted well above the "crankpit" and the innards were covered by a flat plate with a handhold. In 1902, Craig offered a single-cylinder 5½-hp model in addition to 3- and 4-cylinder, 400- to 500-rpm engines of up to 45 hp.

By 1906 the U.S. Navy had ordered 250-hp Craigs for installation in four submarines then being built at the Fore River shipyard in Massachusetts. Craig gained significant public exposure

1902

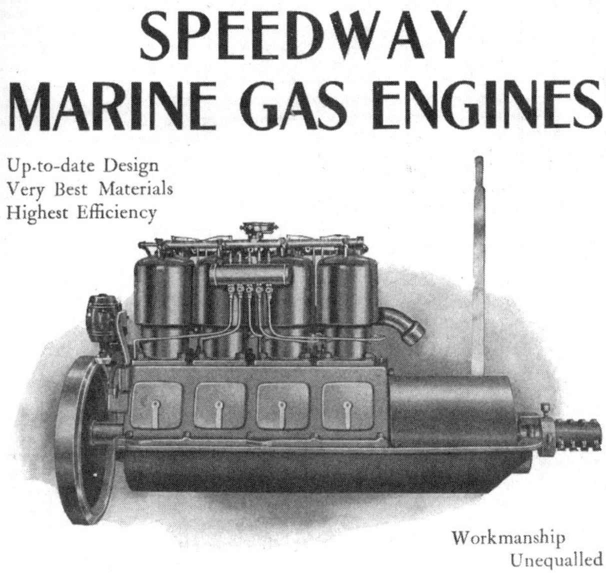

1906 Speedway by Consolidated Shipbuilding ad

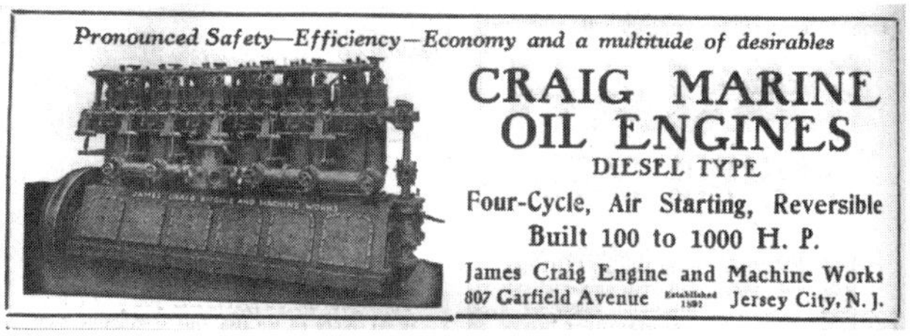

1915

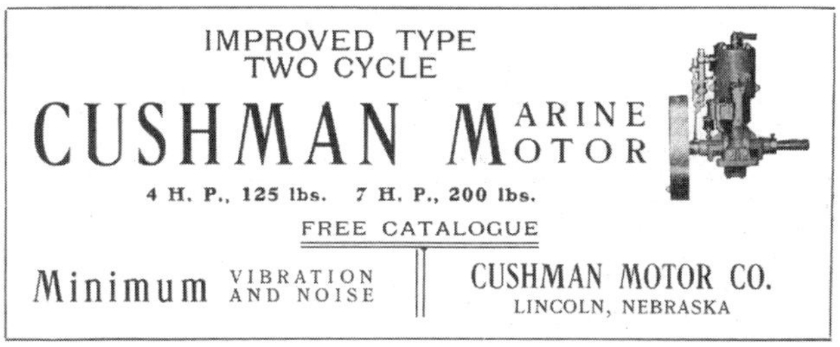

1905

1906

1911

in 1908 after a Craig-powered motor boat, Ailsa-Craig, won a race from New York to Bermuda. After further refinement of his engine, Craig won another race from Huntington, New York, to Marblehead, Massachusetts, in 1911. By the mid-teens, the company had a plant on Garfield Ave. in Jersey City where the model line included both gasoline and big (9½" x 11") oil engines. At the 1916 New York Boat Show, a show featuring a variety of terrific engines, a big 6-cylinder (9½" x 12") Craig Diesel was judged by many to be the "greatest engine at the show." Unfortunately, a two-stage air compressor was needed for injection. Always problematic because of leaks and other difficulties, such systems became obsolete with the development of "common rail" injection based on a mechanical fuel injection pump.

Craig Marine Gasolene Engines
New York, NY
Trade name: Craig
See Craig, James Engine and Machine Works

Crescent Marine Motor Co.
St. Louis, MO
Trade name: Crescent
Located at 116 Ferry Street, this company was in business from 1907 to 1911 and claimed to offer motors of 3 to 80 hp. These were two-cycle engines.

Cross Gear & Engine Co.
Detroit, MI
Trade name: Cross-Kysor
The 220-hp Cross-Kysor V-8 was a marine conversion of the elegant Hispano-Suiza V-8 aero engine of World War I. When sleeved or de-stroked to the Gold Cup maximum 625-cubic-inch limit, the motors were successful in Gold Cup racing. Today, wealthy collectors continue to seek and restore the "Hisso" for use in restored or recreated boats harking back to the Roaring Twenties.

Cross, M.O. Engine Co.
Detroit, MI
Trade name: Cross
The 1906 Cross was a four-cycle motor said to have a potential of 1,800 rpm. The cylinder was cast with a nonremovable head to avoid gasket leaks; the valves, mounted in "in cages quickly removed for regrinding," appear to have been overhead. The 2-hp engine sold for $65, and the company advertised models up to 85 hp.

Crouch, Chas. P. & Co.
Chicago, IL
Trade name: Junior
Located at 472–488 Carroll Ave., Crouch was likely the successor company to Alexander & Co. The official transfer of patterns and the incorporation dates have proved difficult to pinpoint but probably occurred in 1911. Nine models were offered. These included the single-cylinder Models C (1¼ hp), F (2 hp), and G (3 hp); the 2-cylinder Models H (2½ hp), T (9 hp), and E (6 hp); and the 4-cylinder A (5½ hp), T (9 hp), and U (14 hp). Ads claimed a two-year guarantee.

Cummins Engine Co.
Columbus, IN
Trade name: Cummins
"In 1937," Clessie Cummins wrote, "after eighteen years of existence, the company operated with black ink on the books. My long battle had finally led to victory." The battle had begun in 1919 with Cummins's quest to build and market a compact diesel engine for marine and automotive applications.

In 1919, the company entered into an arrangement to supply Sears, Roebuck & Company with 4,500 1½-, 3-, and 6-hp diesel engines. Most were returned to Sears by customers who either took advantage of the Sears 60-day guarantee or else because of problems caused by defects in the parts Cummins received from its suppliers. The contract was terminated in 1922 after some 3,000 engines had been built.

This episode was followed by problems with Cummins 1-cylinder, Hvid-type diesels on the Gulf Coast in 1922. These difficulties included both carbon filling the holes in the injector cups and, even worse, mechanical failures. "In the next three years," wrote Clessie Cummins in his memoirs, "we designed, built and tested nearly three thousand separate fuel injection systems . . . always our approach returned to the two basic ideas of eliminating the old separate fuel cup and using a plunger to drive the fuel charge out into the combustion chamber. Eventually, these two basics were synthesized in the Cummins injector which, along with the exclusive Cummins fuel metering system, has made Cummins engines outstanding in the high-speed diesel field."

With bad memories still fresh in the minds of Gulf Coast fishermen, Cummins took his next engine, the Model F, to Seattle in 1924. Despite its promise and even some later success with government agencies, the initial engines sold ran into problems that involved some poorly manufactured pistons, rings, and valves. What's more, when run at the low rpm needed for trolling, the engine's combustion chamber temperatures were too low to burn off the overly high asphalt content in the then-available fuel. Development of light-grade diesel fuel would be necessary for the ultimate success of the high-speed diesel. By 1939, with General Motors well along in its development of high-speed diesels, "light diesel fuel" became widely available.

Finally, in 1928, the model "U" was introduced, an engine that Cummins called the first completely enclosed diesel ever produced, its reliability demonstrated in a Packard sedan and in trucks. The Cummins Model "U" produced 10 horsepower per cylinder out of a 4½" bore and 6" stroke. Its compact size and enclosed design attracted great attention at the 1928 Boat Show in New York. Although the first "U" engines produced failed because of problems with poorly cast cylinder heads, the engine was successful after this problem was cured. The new engine had an electric starter, a major advance in making the diesel practical for smaller craft, especially pleasure boats. It also had the latest version of the fuel injection system that Cummins engineers had been working on for so many years. Fuel was delivered by a plunger pump at a low pressure (30 to 50 lbs.) to an injector in each cylinder. Suction created as the piston was on its down-stroke drew fuel into a vaporizing chamber where it simply pooled above nozzles that led to the combustion chamber itself. In a 1928 *Motor Boating* article, Cummins described what happened next:

"On the compression stroke, the full force of the charge of combustion air in cylinder, under maximum pressure of 500 lbs. compression, which heats it up to 800 to 1000 degrees Fahrenheit, is forced through the small nozzles directly through the little pool of fuel, thus breaking it up into a hot, rich gas. This mixture is too rich to burn in the plunger chamber but, at top center, this plunger is suddenly depressed through a suitable cam mechanism, which drives this rich gaseous mixture out through the distributing nozzles into the red hot air in main cylinder combustion chamber. Here it mixes readily with the oxygen contained in the air and combustion takes place at once.

"Because the charge is fully prepared beforehand, having been preheated and atomized through 720-degree crank travel, combustion takes place immediately on contact of the gas with the combustion air and, as a consequence, no advance in timing is necessary."

Cummins and his engineers had made practical the intent of the Hvid patents that it had long since abandoned, and it produced an engine that offered control of operating speed that rivaled a gasoline engine. As introduced in 1928, the Model "U" was rated at 1,800 rpm. It was available in 1- through 6-cylinder models in ratings of 8 to 48 horsepower. The latter weighed 2,750 lbs with its Joe's reverse gear, a power-to-weight ratio of 67:1. By contrast, when Cummins showed its Model HMR in 1936, the 2,100 rpm, 55-horsepower version weighed 2,100 lbs, a 38:1 ratio. By then, Cummins was at last on the verge of financial stability.

By the time of the Boat Show in New York in 1934, Cummins was listing six models for marine application. The 6-cylinder Model HMR developed 100 horsepower at 1,500 rpm and weighed 2,500 lbs, including its Joe's reverse gear, a power-to-weight ratio of 25:1. Obviously, this was an engine of an entirely different conception from the big West Coast diesels. A 100-horsepower Washington-Estep weighed 18,500 lbs, a 185:1 ratio; a 100-horsepower Union diesel weighed 16,075 lbs. In 1936 Cummins introduced its most compact engine yet, the AMR-6 (3½" × 5"), which produced 65 hp at 1,900 rpm and a weight of 1,600 lbs. It was specifically designed for smaller cruisers and as an auxiliary in larger sailboats. By then, too, some production boat builders were offering the Cummins diesel as an option to the standard gasoline engines.

Cunningham, James & Son Co.
Rochester, NY
Trade name: Cunningham
A 19th-century builder of dog carts, sleighs, and carriages, this company began building automobiles in 1907. The engines were supplied by Continental and Buffalo. In 1911, the company introduced its own 4-cylinder engine and, in 1916, a 442-cubic-inch V-8. Efforts were made to introduce the V-8 to the marine trade, but there is little evidence of any success.

Curtiss Aeroplane (Motor) Co.
Hammondsport, NY
Trade name: Curtiss
Curtiss V-8, V-12, and 6-cylinder aero engines were occasionally adapted for use in racing boats. The 1914 V-8 weighed less than 700 lbs. and developed 200 hp.

Cushman Motor Works
Lincoln, NE
Trade name: Cushman
Born to farming families who moved to Nebraska, Everett Cushman and his cousin Clinton were more interested in engines than agriculture. Based on patents relating to seals that would stop loss of compression from the base of a two-cycle engine, the two young men incorporated the Cushman Motor Company on Sept. 18, 1902. The first engines were intended for stationary applications on farms. However, the cousins were interested in boats, and in response to a contest to see who could build the fastest boat powered by a single-cylinder engine, they entered and won.

The Rudder magazine article about the event brought attention to the Cushman motor and a marine-engine line was added to the farm engines. "The valveless, cussless motor," said a 1905 ad. Business difficulties forced a reorganization in 1906 and the hiring in 1909 of an experienced manager. Thereafter, the focus was on the farm engine market, which far exceeded the marine business in terms of potential volume. A 4-hp, 4-cycle, 4-cylinder motor was introduced in 1910 and became an immediate success. It, too, was advertised initially for boat applications. "The Motor of Merit" said an ad in 1911. Still, the direction to the farm market was clear, and Cushman Motor Works left marine engines behind. The Cushmans themselves left the company after World War I, and it was later purchased and took a new course that made it famous as a maker of small engines and scooters.

Cushnoc Motor Co.
Augusta, ME
Trade name: Cushnoc
Cushnoc existed from 1909 to 1913. The nature of its engine is currently unknown.

Daimler Motor Co.
New York, NY
Trade name: Daimler
When he was asked about the early days of his engine's development, Gottlieb Daimler always referred to the motor boat as the vehicle that did the most to quash prejudice against using gasoline as fuel. Thanks to interest in the motor boat, marine engines made

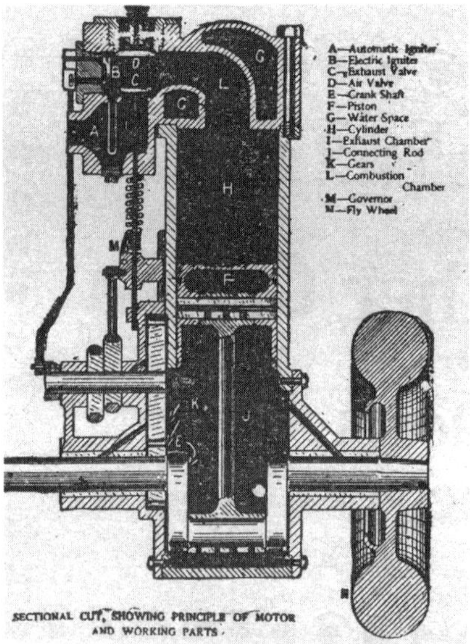

1901 Columbia

Corliss, circa 1915

Cooley, circa 1899

1905 Speedway by Consolidated Shipbuilding

1911 Fox

1915 Detroit fitted for kerosene

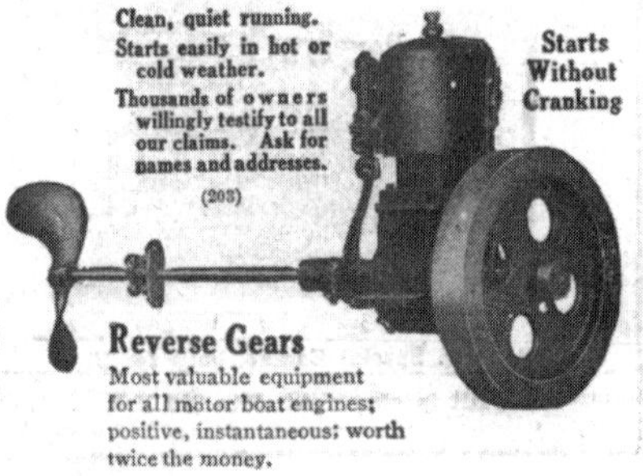

1915 Detroit Engine Works

1906 Doman

1911 Doman

1913 Doman

1923 Doman

up a significant portion of Daimler's early production. Healthy, if wealthy, markets developed for the boats in the U.K. and the U.S. where piano maker William Steinway paid Daimler a hefty $10,000 license fee for manufacturing rights in 1888.

What Steinway received for his money, besides the rights, was a set of plans with metric dimensions that needed to be converted into inches so a prototype could be built. This task was performed in Hartford, Connecticut, by a 20-year-old former apprentice at Pratt & Whitney named George Long. Long was then entrusted with building the prototype engine for the National Machine Company, which was formed in Hartford to produce the Daimler. The company was credible in advertising the light weight of its motors and in 1900 was among the very rare firms to try and quantify fuel economy: "less than a pint of gasoline per H.P. per hour." One- and 2-cylinder models were offered.

It is a measure of just how competitive the American market was — with its soon-to-burgeon homegrown engines — that the Daimler had negligible impact. One of those who did purchase a Daimler single-cylinder for a launch was railroad magnate and Florida developer Henry Flagler. Flagler's 1891 Daimler has been restored and now resides in the Connecticut State Museum.

Danford, O.
Fall River, MA
Trade Name: Danford
Two- and four-cycle engines were made.

Dantzenbecker, H. E.
Philadelphia, PA
Trade name: Special, Crown
According to William Deed, reporting in *The Rudder* in May 1930, the Crown was an engine "in which weights were reduced to a minimum, their little 2½ horsepower yacht tender engine having lignum vitae bearings and aluminum alloy base. They used Parsons' white brass for bearings of their larger engines."

Davenport Motor Co.
Davenport, IA
Trade name: Davenport
This two-cycle motor was advertised as being available to "responsible parties" for a 3-day trial period in 1901.

Davis Manufacturing Co.
Milwaukee, WI
Trade name: Davis

De Moore
Trade name: De Moore
A company ad appeared in *Motor Boating* in Dec. 1925 for a two-cycle, 14-hp twin.

DeMooy Brothers Co. (DeMooy Machine Co.)
Cleveland, OH
Trade name: DeMooy
Established in 1889, DeMooy offered two-port, jump-spark engines of 1½ to 60 hp by 1906. Racing in the Country Club Regatta on Lake St. Clair in August 1907, the DeMooy brothers finished second in a launch that, with a stock engine, achieved a then-very-impressive 20 mph. The engines were built at a plant at 1833 East 55th St. "The feature which distinguishes the DeMooy from other two-stroke engines is the use of an auxiliary air valve, which is guaranteed to give at least 10 percent increase of power without increasing the fuel consumption," reported the company in the January 1908 *Gas Engine*.

By 1912 the line had altered, split between two-cycle motors of 8 to 40 hp and a range of heavy-duty (275 rpm) four-cycle engines of 1 to 6 cylinders and 10 to 40 hp. The two-cycle engines had jump-spark battery ignition. The four-cycles had high-tension magnetos. The two-cycle models were distinguished by having one carburetor for each cylinder, whether 2, 3, or 4 cylinders. The four-cycle engine was served by a single carburetor. Engine production may have ceased in 1916 but the Palmers & DeMooy Foundry was in business in the '20s.

Dean Manufacturing Co.
Newport, KY
Trade name: Fox
The Dean Manufacturing Company in Newport seems likely to have been the successor company to the Fox Reversible Engine Works of nearby Covington. In 1905, Fox introduced a 5″ × 5″ two-cycle motor with a one-piece cylinder/head casting. Speed was regulated by advancing or retarding the spark and a reverse lever did "away with a reversing clutch or propeller." Jump-spark ignition was used. The hope was that by standardizing on a single model, the moderately priced engine would find a wide market. A stationary version was offered in kit form.

Whatever success Fox experienced is now unknown. However, by 1909, Dean's lineup of Fox two-cycle engines was as comprehensive as any in the United States. It included 1-, 2-, and 3-cylinder motors that ranged from 3½ to 30 horsepower. In 1910, the range included 4-, 6-, and 8-cylinder models, the latter rated at up to 200 hp at 800 rpm. It became a renowned engine thanks to the performance of a racing boat named the *Br'er Fox II*. This was a sophisticated lineup: There were 3½- and 7-horsepower, three-port motors intended for 14- to 22-foot pleasure boats as well as heavy-duty, two-port models for commercial use. Another series of pleasure-boat motors, billed as "four-port" types, appeared in 1910. These were three-port engines with an additional "accelerator port" in the intake manifold that could be adjusted to furnish auxiliary air to the cylinder independent of the carburetor. Ignition was jump-spark. Carburetors were Fox or Schebler depending on the specific engine.

Dean molders (those who made the castings) went on strike in October 1913 and 25 company workers filed attachments for unpaid wages. While none of the employees sought more than $15 (perhaps a week's pay) each, Dean doubtless had additional obligations, and the company was closed.

DeCourse Dickenson Mfg. Co.
Los Angeles, CA
Trade name: Western Reliance
The company was in business in 1915.

Defoe Boat & Motor Works
Bay City, MI
Trade name: Stork
Boat builder Defoe was advertising its cruisers as either kits or completed boats by 1911.
See also Stork

Delaware Machine Works
Wilmington, DE
Trade name: Harris

Delaware Marine Motor Co.
Wilmington, DE
Trade name: duPont
Founded in 1916 by E. Paul duPont, the company offered valve-in-head, 20-, 40-, and 60-hp models of from 2 to 6 cylinders.
See also duPont

Delaware River Motor Co.
Chester, PA
Trade name: Keystone
In 1904, the company listed 2- and 4-cycle models of 1 to 25 hp.

DeLong Engine Co.
Webster, NY
Trade name: DeLong
G. Edwin DeLong founded this company after he left Elbridge. A $39.50 2½-hp two-cycle was advertised in 1913 and 1914.
See also Elbridge

Delray Motor Co.
Detroit, MI
Trade name: Delray
Its 1908 ad claimed Delray offered a four-cycle, 20-hp, 4-cylinder motor and a three-port two-stroke of 1½ hp.

DeMooy
See also DeMooy Brothers Co. (DeMooy Machine Co.)

Denison Electric Engr. Co.
New Haven, CT
Trade name: Yale
The company was one of several that marketed a complete motor or a kit with the name Yale. In this case, at least, the company was in the same city as Yale University. Ads appeared in 1900. Whether Denison had a connection to the Baird Machine & Mfg. Co. in Detroit, manufacturers of a kit for a Yale engine, is unknown.

Desberon Motor Co.
New York, NY
Trade name: Desberon
This company was in business in 1904.

Detroit Auto Engine Specialty Co.
Detroit, MI
Trade name: Liberty, Liberty Dwarf
Incorporated in September of 1905, this company was well capitalized at $75,000. The product here was a two-port, two-cycle engine with a polished bronze water jacket. Models ranged from a 50 lb. 1½-hp one-lunger to a 300-lb., 14-hp 2-cylinder. After competing through the heyday of the two-cycle engine, the company was dissolved in August 1919.

Detroit Auto-Marine Co.
Detroit, MI
Trade name: Auto-Marine
Detroit Auto-Marine became the successor to the Detroit Lackawanna Co. in 1905. Presumably, the acquisition of Detroit Lackawanna brought with it rights to Joseph Day's patent on a three-port, two-cycle engine. The Briton's rights to patent protection did not expire until Oct. 1906, and the ornate cast nameplate on each Auto-Marine engine showed renderings of both the U.S. flag and the Union Jack. Offerings from this firm by 1906 were claimed to range from a 1½-hp single to a 4-cylinder, 20-hp version. These were all jump-spark engines. The smallest models, the 1½- and a 3-hp versions, were most widely advertised and perhaps the only ones actually built. Their prices were $33.15 and $44.00, respectively. That did not include shaft, coil, or batteries. The company, which had addresses at 66 Congress St. and at 82 (and/ or 81) E. Congress, claimed that it would build 10,000 engines in 1906. That anticipated volume was projected, in part, because Detroit Auto-Marine for a time supplied the W. H. Mullins Co. with motors that carried a Mullins Motor Co. label. Whatever the arrangement was, it did not continue as expected. Detroit Auto-Marine filed for dissolution in Feb. 1907. It reemerged that year as the Detroit Gas Engine & Machinery Company at the 66 Congress St. address.
See also Detroit Gas Engine & Machinery Company

Detroit Engine Works
Detroit, MI
Trade name: Detroit
From an 8½-acre factory on the Detroit River, this firm was shipping its two-cycle engines to customers in the U.S. and abroad — India, Turkey, China, Egypt, and Panama, to name a few — by about 1907. A distributor was also established in London. In addition to marine gasoline engines, Detroit offered a variety of stationary engines set up for farm applications and as generating and pumping outfits, kerosene-fueled stationary and marine engines, marine reverse gears, tractors, and machinery.

The company became a conglomerate. It also produced motors under the labels of the Columbia Engine Co. and the Detroit Motor Car Supply Co. (Sandow); those engines were identical to the Detroit-badged models. Other businesses were the Michigan Steel Boat Co. and the Detroit Boat Co. Detroit Engine Works principals included president Hugo Scherer and secretary/treasurer Frederick Wadsworth. Both were very wealthy men who engaged in business interests as diverse as real estate development, automobile bodies and components, internal combustion engines, and even "Wadsworth Light Kerosene Tractors." The latter were manufactured under the auspices of the Detroit Engine Works.

Like many in the industry, Detroit stressed the quality of its tooling and, like many, claimed to have "the best equipped factory in the United States." That said, these Detroit motors, with their single-piece cylinder castings (to avoid potential head gasket leaks), seem to have been as well made as any of the Michigan-built two-cycle engines. The basic castings were made of what Detroit called "French Grey Iron," the exact alloy said to be proprietary. "Ordinary iron will not flow readily enough to make thin castings of the grade we use; and many manufacturers are compelled to use thick heavy castings because they do not have the proper foundry facilities." A counter-balanced crankshaft was said to reduce vibration. High-tension, jump-spark ignition was fitted. These engines were, it is believed, finished in black enamel ("beautiful heat proof enamel") highlighted by polished bronze fittings.

Detroit claimed it spent two years investigating whether a two- or three-port design was preferable only to abandon the "three-port type of engine as too impractical to offer to the owner of a launch." This decision could have been influenced by Joseph Day's demand for payment for a license on his three-port patent during the period before the rights expired after 1906. The company claimed better throttle control for its two-port motor, better economy, and better efficiency, yet it still claimed its engine could achieve 1,000 rpm. All this may be a matter of semantics, and the brochure copy is not without aspects that are confusing. The 1910 brochure described the virtue of a check-valve–equipped two-port motor versus a three-port with a carburetor, yet then went on to describe the Detroit float-feed carburetor and presented a cutaway drawing that revealed no check valve. What's more, carburetors from a variety of suppliers including all the major brands — Schebler, Lunkenheimer, Krice, and Essex — seem to have been used at one time or another. By 1911 the model line included 11 engines. These 1- and 2-cylinder motors ranged from 2 to 20 hp. Ignition was jump-spark.

Whatever the merits of its engines, it does not seem that the Detroit Engine Works was well-positioned to survive the downturn in the market for two-cycle engines that struck beginning about 1910 and continued for the next several years. The company introduced a four-cycle, 4-cylinder engine in 1911 that, with bore/stroke dimensions of 5″ × 5¼″, developed 40 hp at 650 rpm. Forty-three people were said to be working at the Detroit Engine Works in 1914, about the same as at Kermath. However, while Kermath was regularly advertising and displaying at shows, the Detroit Engine Works appeared on no lists of marine engine companies published in the boating press after 1911. The extent of marine engine production was likely in steep decline by then, when only the occasional tiny ad appeared. In 1917, the company became the Standard Motor Parts Company, which ceased operation in about 1920.

Detroit Gas Engine & Machinery Co.
Detroit, MI
Trade name: Major

The Major was a two-cycle engine that had previously been known as the Detroit Auto-Marine and was said to be available in a similar, if not identical, range of models — 1½ to 24 hp. It is plausible that Detroit Gas Engine & Machinery was responsible for production of what had been the Detroit Auto-Marine even before the latter filed for dissolution in early 1907. The company was located at 66 E. Congress St., the same address as one of those listed for the Detroit Auto-Marine Company. Like Detroit Auto-Marine, this company was short-lived. It is believed that its assets were acquired by another Detroit firm, the Charles A. Strelinger Co., in late 1907.

See also Detroit Auto-Marine Company

Detroit Lackawanna Co.
Detroit, MI
Trade Name: Auto-Marine

This company is believed to have been in operation from 1904 into 1905, producing a little 1-hp, two-cycle engine. Evidence has yet to surface indicating possible or probable connection to the Lackawanna Mfg. Co. of Newburgh, NY.

See also Detroit Auto-Marine Co.

Detroit Motor Car Supply Co.
Detroit, MI
Trade name: Sandow

The company operated under the umbrella corporation Detroit Engine Works. The Sandow marine engine appears to have been a Detroit with a different badge.

See Detroit Engine Works

Detroit River Iron Works
Detroit, MI
Trade name: N/A

This company first appeared in 1901. It was located at 667 Atwater Street East. As of 1909–1910, the Works advertised stationary and marine gasoline engines.

Dingfelder Motor Co.
Detroit, MI
Trade name: Dingfelder

After a failed attempt at producing a 500-lb "motor buggy" powered by a 3½-hp engine in 1903–04, Max Dingfelder regrouped and began building marine engines a few years later. In 1906–07, he produced a 4-cylinder 5¾″ × 6″ four-cycle engine that developed 34.5 hp, enough to drive a 40′ launch. He exhibited a 60-hp six at the first Detroit Motor Boat Show in 1908.

See Maximotor Makers

Dirigo Engine & Machine Works, Inc.
Portland, ME
Trade name: Dirigo

Dirigo was managed by George H. Scanlan, who, prior to his engine-building venture, had been involved in various executive capacities with several Portland businesses including his own George H. Scanlan, a maker of machinery, in 1890. In 1895, he was treasurer of the family's Scanlan Co., makers of bicycles and machinery. He managed

1902

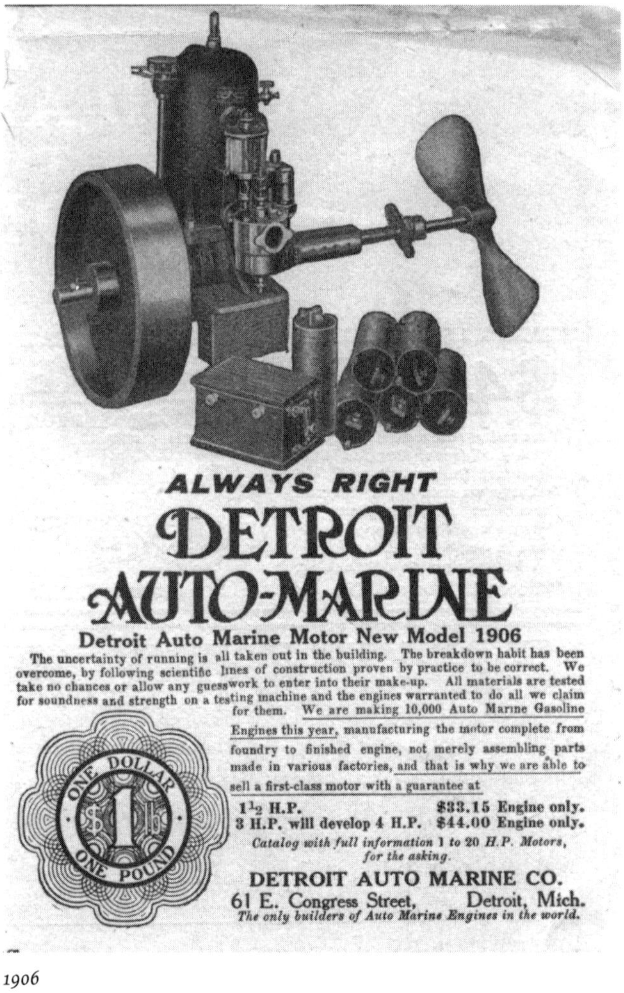

1906

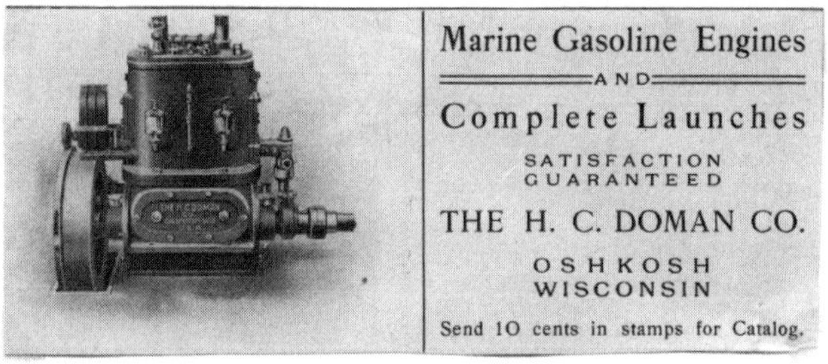

1906

1906

1913

the Puritan Cycle Mfg. Co. in 1898 before starting Dirigo in 1899. Dirigo built two-cycle engines until 1903. Scanlan left Portland for New York City in 1909.

Disbrow, W. C. Jr.
New York, NY
Trade name: Disbrow
In 1920, the company offered a four-cycle, 37-hp, 4-cylinder motor.

Doak Gas Engine Co.
San Francisco/Oakland, CA
Trade name: Doak
Primarily known for its horizontal-cylinder stationary engines intended for operation on gasoline or "California engine distillate," Doak also built vertical marine engines. The circa 1915 Doak was an open-base, four-cycle engine with overhead valves driven by push-rods off a gear-driven camshaft on the engine's port side.

Dodge Engine Co.
Boston, MA
Trade name: Dodge
The 1910 Dodge was a two-cycle, three-port engine with an open crankcase and removable head. Serviceability seems to have been a key motivation in the engine's design: "The cylinder head may be readily taken off at any time for the purpose of removing burnt carbon or making repairs, without moving other connections, thus eliminating all danger of leaky joints. After removing the cylinder head, the piston may be taken out without disturbing the cylinder in any way." Sales, if any, seem to have been few.

Dodge Mfg. Co. (formerly Burnoil)
Mishawaka, IN
Trade name: Burnoil
See Burnoil

Dodge Sales and Engineering Co.
Mishawaka, IN
Trade name: Dodge Heavy Oil Engine

Doman, H. C. Co.
Oshkosh, WI
Trade name: Doman
"Responds Like 'A Thing of Life,'" said Doman ads of the 1910 period. By then, sales had surpassed $50,000. H. C. Doman began his career building steam engines in 1890 but immediately saw the potential of internal combustion. He built two-cycle one-lungers, added cylinders, and went through the process again with four-cycle engines. By 1910, Doman offered medium- and heavy-duty, T-head, four-cycle models of 5 to 55 hp. The largest was a 450 rpm 6-cylinder (7″ x 9″) that produced 90 hp at a weight of 3,723 lbs. These hefty Domans had separately cast cylinders and were available with either jump-spark or make-and-break ignition. Lubrication was by mechanical oiler and splash. In 1915, Doman introduced the en bloc Junior, a thoroughly modern little 4-cylinder 3½″ x 3½″ motor that weighed just 225 lbs. All cooling and oil passages were in the head and cylinder castings, eliminating external piping. The Junior was a side-valve motor with a detachable head and was perfect for tenders and small runabouts.

During World War I, Doman converted his factory to defense work, and he retired at war's end. The addition of machinery to build war materiel prompted a postwar expansion into production of electric generating plants. These were marketed as products of the Universal Products Company while the engines were built under the auspices of the Doman Engine Division. According to R. M. Schriber who became its vice president, the company changed its name in 1919 to Universal Products Co. (not to be confused with the Universal Motor Co. also in Oshkosh) of which the Doman Engine Division was a part. The 4-hp Doman Bull Dog was designed to be the least expensive engine of its class. Next up the line was the 5- to 7-hp Standard Fish Boat Engine. The 15- to 18-hp LM-4 was a fully enclosed engine designed for runabouts and speedboats. The 40- to 50-hp TM-4 was the largest (4¾″ x 6″) motor then built by the company. In 1927, with further expansion, the company was renamed the United States Motors Corporation.
See Universal Products Co. *and* United States Motors Corporation

Domestic Eng. & Pump Co.
Shippensburg, PA
Trade name: Domestic

Dow Pump and Diesel Engine Co.
Alameda, CA
Trade name: Dow

Du Brie Motor Co.
Detroit, MI
Trade name: DuBrie
Born in Sandusky, Ohio, in 1874, Stanley R. Du Brie received his degree in mechanical engineering at the University of Ohio in 1896. Along the way, he'd also become a machinist, having worked at the F. M. Underwood Gas Engine Company in Sandusky. That background prepared Du Brie to become both a prolific and successful engine and automobile designer. The peripatetic nature of Du Brie's career and the fact that his name showed up on engines and cars built by companies in which he had a design role but no active management role tends to make deciphering his career something of a puzzle.

Du Brie's first engine was a two cycle with hot-tube ignition that was probably built around 1895 and was said to be working perfectly 15 years later although it had by then been converted to electric ignition. Upon graduation from college, Du Brie established the Du Brie Motor Works in Fremont and sold his first two-cycle marine engine in the fall of 1896. This was a 2-hp two-cycle motor that sold for $375. He continued in production until 1900 when he sold the business on Jan. 1, 1900, to the Clauss Shear Works in Fremont. Whether the engine's name was then changed to the Sandusky or whether that was a wholly new engine is unclear.

Du Brie himself moved to Penetang, Ontario, where he designed a two-passenger runabout for the Charles Beck Mfg. Co., the four-cycle DuBriemobile built in Toronto, and a line of two-cycle marine engines of 1 to 10 hp for Peter Payette & Co. in Penetang.

In July 1904 Du Brie moved to Detroit and designed the two-cycle, 2-cylinder (7″ × 7″) De Loach automobile and the two-cycle Du Brie-Caille automobile, the latter for the Caille Brothers. Within the first year, more than 1,000 examples were sold. Two-cycle Caille marine engines followed.

Other engines designed by Du Brie included the Bense marine and stationary models, and the Waterloo. Du Brie designed the first Gray marine motor in 1906. At about the same time, the Du Brie Motor Company began building a line of two-cycle engines of from 2½ to 30 hp, at a plant on Griswold and State Streets circa 1908 to 1911. In 1911 Du Brie was named chief engineer for the Peerless Marine Engine Company in Detroit, builder of a 9-hp two-cycle one-lunger. The engine could turn a 17″ × 23″ three-blade propeller at 550 rpm and was said to be suitable for 18- to 30-foot launch-style boats.

Despite his title in the Peerless catalog, it seems unlikely that Du Brie would have left his own company. Instead, he appears to have kept up his consulting/design work as a sideline, lending his name to entrepreneurs eager to associate themselves with a recognized industry pioneer. Du Brie kept his own business — the Du Brie Motor Company — going at the corner of Griswold and State Streets. "We are pioneer engine builders and our engines are in every quarter of the globe," said a 1911 ad. It was a small operation, employing 14 people in 1912.

In February 1923 the company was either reorganized as, or replaced by, DuBrie Marine Motors (the name appeared both as DuBrie and Du Brie) at 7450 Jefferson Ave. The company focus was on a single-cylinder, four-cycle, Model T–based motor. It developed 5 hp and was equipped with a Bosch magneto and Holley carburetor. The price, with propeller, was $97.50. Said the company: "The motor is designed so that 42 parts may be replaced or serviced with similar parts now being in Ford engines. These parts include pistons and rings, piston pin, connecting rod, all bearings, timing gear, valves, valve spring and tappets, spark plug, carburetor and nuts and bolts." A year later, the company reported that many orders had to be canceled by "unexpected demand."

In 1924 the company's address changed to 5626 McGraw Ave., and the company claimed it had "taken over the engine production of the Kermath-Whitcomb Company of Detroit, and will be in a position this year to fill orders promptly." Whether or not that was the case, the company was sold to one F. T. Smith and relocated to 567 Beacon St. in Detroit. Its motto: "The motor that is showing thousands the way to happiness." The engine was still in production in the late '20s. As if to perpetuate the confusion relating to a man who designed many motors for many companies, a magazine article about the firm was titled, "The Smiths of DuBrie."

Ducro Mfg. Co.
Buffalo, NY
Trade name: Niagara
See Niagara

Dunn Motor Works
Ogdensburg, NY
Trade name: Dunn
This company, one of many in the upstate New York region, offered a line of four-cycle gas and kerosene models. All were the handiwork of Walter E. Dunn. In 1911, the lineup included models of 1 to 4 cylinders producing up to 16 hp. That year, the company introduced the "Dunn 4-cycle Reversible Outfit." This was a 1¾-hp one-lunger and was said to be reversible by adjustment of the camshaft and timer.

duPont Motors, Inc.
Wilmington, DE
Trade name: duPont
Organized shortly before U.S. entry into World War I (as the successor to the Delaware Motor Co.), duPont turned its attention to building exceptionally beautiful, limited-production automobiles after the Armistice. However, marine engines were still being advertised in 1925.

Duryea Motor Co.
Saginaw, MI
Trade name: Duryea
Charles Duryea, who had organized America's first automobile company in Springfield, Massachusetts, in 1895, established another horseless carriage company in Saginaw in 1911. In Dec. 1912 a Duryea engine intended for marine use was advertised. Production, if any, was probably very small.

Dutcher Machine Co.
Fulton, NY
Trade name: New Parker
See J. J. Parker Co.

Dutton Engineering Co.
Perth Amboy, NJ
Trade name: Dutton
The company was in production of 1½- to 24-hp engines by 1902. An ad showed an en bloc twin with jump-spark ignition and a solid flywheel.

E. E. Winkley
See Winkley, E. E.

E. H. Godshalk & Co.
See Godshalk, E. H. & Co.

E. M. White & Co.
See White, E. M. & Co.

E. Louvet & Son
See Louvet, E. & Son

E. W. Graef
See Graef, E. W. *and also* Reeves-Graef

Eagle Co.
Newark, NJ
Trade name: Eagle
In 1910 the Progressive Manufacturing Company changed its name to The Eagle Co. Production moved from Torrington, Connecticut, to a

1906 Essex

1911 Dunn

1911 Dunn

1902 Dutton

1905 Eclipse

1905 Eagle

The Eagle three cylinder model.

1911 Eagle

1913 Model "1K" 3½ H. P. "Eagle" Engine

1913 Eagle single

1913 Model "2K" 7 H. P. Eagle Engine

1913 Eagle twin

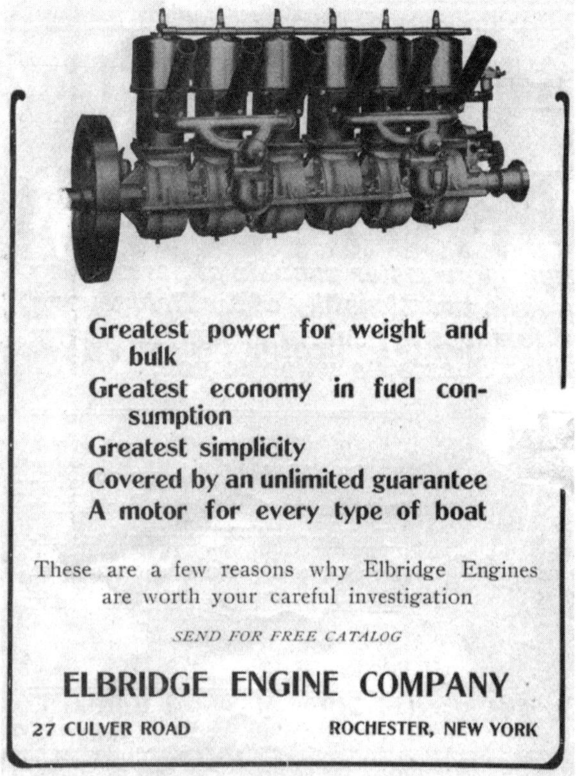

1911

1911

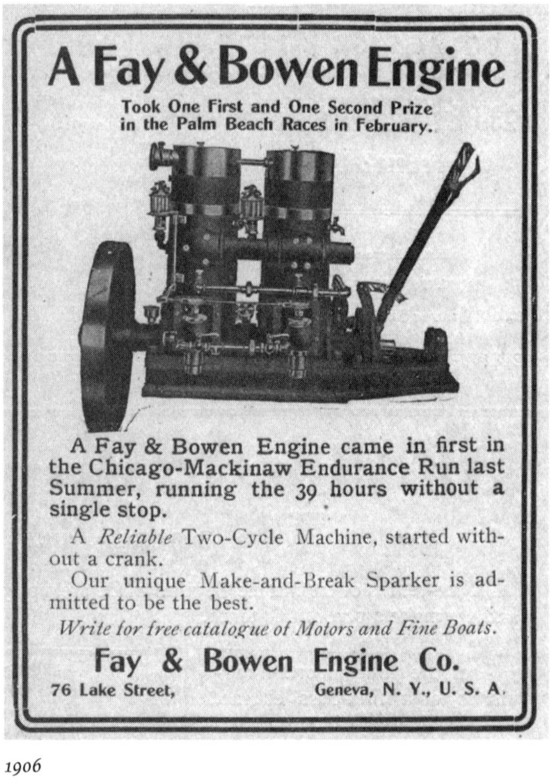

1906

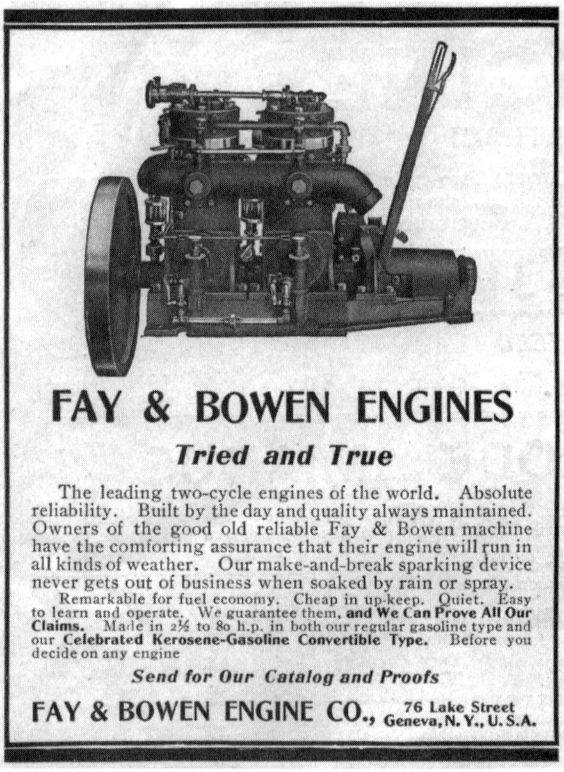

1911

100,000-sq.-ft. Newark, New Jersey, plant where 15 models of 1½ to 25 hp were produced. These engines were available in an updated heavy-duty model and in newly introduced "semi-heavy-duty" and medium-speed models. The slower turning (400 to 450 rpm) heavy-duty models (D and E) and the single-cylinder A (1½ hp) and F (2½ hp) were equipped only with make-and-break ignition. Other models were available with jump-spark or make-and-break ignition. Perfex magnetos and K-W battery-ignition components were used as were Paragon clutches. Press reports in 1910 noted the company had improved its make-and-break igniter. Eagle kept all its drawings and patterns in a fireproof room. Hope of 3,500 engines per year wasn't realized, and production returned to Connecticut in 1913 amidst the broad decline of the two-cycle marine engine.

See also Progressive *and* Standard Co.

East Davenport Mach. & Novelty
Davenport, IA
Trade name: Kuhner
See Kuhner Engine & Machine Co.

Easternhouse Engine Co.
Taunton, MA
Trade name: Pearl

The Pearl was offered in 6- to 100-hp models by 1908.

Eclipse Motor Co.
Mancelona, MI
Trade name: Eclipse

In existence prior to 1901, Eclipse probably got its start as the Eclipse Motor & Boat Company in Traverse City. The proprietor, Brownslow Starbuck, had learned enough of the trade to move to Mancelona and open his new business in March 1901. The Eclipse was a two-cycle engine; the scant information that survives about the company shows that six engines were ordered by a company in New Smyrna, Florida, in May 1903. A new 5-hp double was introduced in 1904, and one of these engines may have been installed in a Matthews Boat Company launch shipped to St. Petersburg, Russia, in March 1905. The new twin was apparently aimed at both boat and automotive applications, but none of the 10 or 11 cars named Eclipse used Starbuck's engine. Occasional Eclipse ads showed up in the boating press. The company ceased operation in early 1911.

Economic Engine Works
Utica, NY
Trade name: Economic

The 1910 Economic was a conventional two-cycle engine with a 4¼" bore and 5" stroke. The single-cylinder model developed 4 hp. One-, 2-, 3-, and 4-cylinder models were offered. All had jump-spark ignition and a removable, water-cooled heads.

Economy Gas Engine Works
Utica, NY
Trade name: Economy

Elbridge Engine Co.
Elbridge, NY (later Rochester, NY)
Trade name: Elbridge, Featherweight

Sometime around 1907, G. Edwin DeLong started the Elbridge Motor & Tool Co. It didn't take long—perhaps by 1908—for the company to become simply the Elbridge Motor Co. and, within a few years, the Elbridge Engine Co. By 1909 or 1910 the company had relocated west to Rochester. Its plant was on Culver Road just off the Erie Canal. Elbridge was an early champion of the advantages of "power for weight." While important in a racing boat, this approach was critical for an airplane engine, and, by 1910, the company was stressing that its 10- to 60-hp Featherweight models "made amateur aviation possible in America." Examples of the Featherweight may be seen at the Smithsonian and the New England Air Museum in Hartford, Connecticut.

Elbridge experimented exhaustively to come up with the right copper/aluminum alloy for its crankcases, which were much stronger than aluminum alloy only. The forged steel crankshaft was carefully balanced. Models included the A; B (heavier duty with nickel-plated heads and other parts painted black, dark blue, or maroon); C (speed boats); F "Featherweight" racing engines; and D, a 6-cylinder. Schebler carburetors were standard, and C, F, and D models were equipped with Atwater-Kent's "Unisparker" ignition. The A and B models had Bosch ignition. Dual battery/magneto systems were available. Depending on application and model, these were 3-port engines rated at 600 rpm or as high as 1,500. The company was outspoken in support of oil in fuel for lubrication, recommending one pint of Havoline "Crystal medium" [or similar] to five gallons of gasoline.

After the move to Rochester, Elbridge added two-cycle, air-cooled stationary engines to its product line. This was quite possibly an effort to diversify in the face of what had probably been recognized as a limited market for rather exotic two-cycle marine and aviation motors. Whatever the intent, the company was forced to alter its marketing approach; it appears likely that DeLong and Elbridge parted ways sometime in 1912. DeLong started another marine engine firm under his own name back in Elbridge. As for Elbridge, it bought time in 1917 by introducing a marine conversion of the Ford Model T engine, the "Wonder" (not to be confused with the two-cycle Wonder built in Syracuse). The chief claims for the conversion involved lubrication and "adjustability." The Wonder was said by Elbridge to be an addition to its line of two-cycle motors, but it is not likely that there was then any volume production of the latter. The motor appeared in *Motor Boating*'s 1915 list of engines suitable for racing, but Elbridge was a no-show at the big boat show down in New York the following year. Today, an Elbridge marine engine, especially one intact and original, is a rare find, an important and desirable example of high-technology, two-cycle engineering of its period.

Electric Launch Company (Elco), The Elco Works
Bayonne, NJ
Trade name: Elco
Best known to recreational boaters for its cruising motor boats, Elco offered a variety of engines over the years and seldom went out of its way to credit the engine manufacturer. In 1907, Elco installed a 60-hp air-starting Standard in a 40-foot cruiser. In early 1908, Elco showed a 35-foot motor boat powered by a 6-cylinder, 50-hp Peerless. More widely advertised that year were cruisers powered by A & B L-head, four-cycle engines. In 1920, Elco listed a series of 4-cylinder engines sourced from J.V.B. (as in Joe Van Blerck). The company was, by 1928, offering the F-series, a line of 6-cylinder engines of 65 to 160 hp. On the eve of the Depression in 1929, Elco was using the 125-hp F-62 in its 38-, 43-, and 51-foot cruisers, the Van Blerck–designed 48-hp WSM in the 34-foot Cruisette, an 85-hp Buda in the 31-foot Veedette, and a 27-hp Gray in its 26-foot runabout. With the sales declines of the '30s, Elco offered only two models, the F-42 (65 to 90 hp) and F-62 (100 to 145 hp) that came, as always, from an outside engine company.
See also American and British Mfg. Co. *and* J.V.B. Engine Co.

Elk Motor Co.
Muscatine, IN
Trade name: Elk
A 2-cylinder, 1,000 rpm, 10-hp Elk was noted as powering an 18½-ft. skiff in races at Peoria, Illinois, in 1910.

Ellsworth Foundry & Machine Works
Ellsworth, ME
Trade name: Ellsworth

Emeny, G. J.
Fulton, NY
Trade name: New Parker
See Parker, J. J. Co.

Emerson Engine Co.
Alexandria, VA
Trade name: Emerson
Emerson was in production of 1- to 6-cylinder, 10- to 60-hp two-cycle engines by 1909. These were high-grade engines produced using the era's best machinery. The company noted that it used specially made bolts and nuts from the Hartford Screw Company to be stronger than the typical commercial fastener. What's more, the company touted a workforce staffed "by the pick of workmen from the [local] U.S. Navy Yard and Gun shops." The Emerson was finished in "French gray enamel" highlighted by polished bronze and aluminum hand-hole plates. To escape a potentially troublesome head gasket joint, the engines were dome-head designs. Ignition was jump-spark.

For a period of about a half-dozen years, Emerson enjoyed some popularity both for competition and general use. In 1913 a mighty 6-cylinder (5" × 5") powered the 14-foot, single-step hydroplane *Little Leading Lady* to a Class C victory in the Mississippi Valley championship. Engine companies were always looking for superlatives that would distinguish them from the competition. In 1911, Emerson advertised, "We are equipping the largest lot of boats above thirty miles per hour ever produced by one firm." Noting its success with airplanes, the company advertised "Over One Hundred Flights in a Heavier-than-Air Machine without a single accident." Emerson, however, seems to have fallen prey by 1915 to the inexorable advance of easy-to-operate four-cycle engines.

Emery Bros.
Kennebunkport, ME
Trade name: Emery
The Emery brothers were building engines as of April 1911.

Emery, V. J.
Wollaston, MA
Trade name: Ideal
In business in 1910, Emery was then building engines of at least 40 hp, complete with a water-jacketed exhaust manifold. "A very substantially constructed and well-finished four-stroke engine made in Wollaston, near Boston," is how *The Rudder* described the Ideal.

Emmons Specialty Co.
Detroit, MI
Trade name: Oarsman

Empire Machine Works
Pittsburgh, PA
Trade name: N/A

Engineering Laboratory
Benton Harbor, MI
Trade name: N/A
These were marinized Ford engines.

Enterprise Engine and Foundry Co.
San Francisco, CA
Trade name: Enterprise
This company was founded in 1886 and grew to supply castings to most of the Bay Area engine manufacturers. In 1917, Enterprise built an engine that would run on either gasoline or distillate. This 6¾" × 8" 8-hp single-cylinder was a typical, heavy-duty California engine. It weighed some 1,750 lbs. and was rated at 400 rpm. A 20-hp, 3,200 double followed. Installed in a barge, the double was still serving reliably in 1955.

Like other Bay Area makers, Enterprise recognized the need for a diesel engine and introduced an 80 hp at 320 rpm 3-cylinder diesel in November 1920. It had Hvid fuel injection. In 1921, a license was obtained to use the Vickers common rail injection system. In 1924 Enterprise merged with the stationary engine maker Western Machinery Co. of Los Angeles. A new line of marine diesels introduced in 1933 used Bosch or Bendix solid injection.

Enterprise Machine Co.
Minneapolis, MN
Trade name: Westman
Westman sold its first engine circa 1901. Two-cycle, make-and-break and, later, four-cycle engines were offered. The latter were side-valve engines with one-piece cylinder castings and jump-spark ignition. Before 1910 the range included 1- to 3-cylinder models of up to 70 hp. By 1909, the company name

1913 Emerson

1913 Erd

1913 Erd

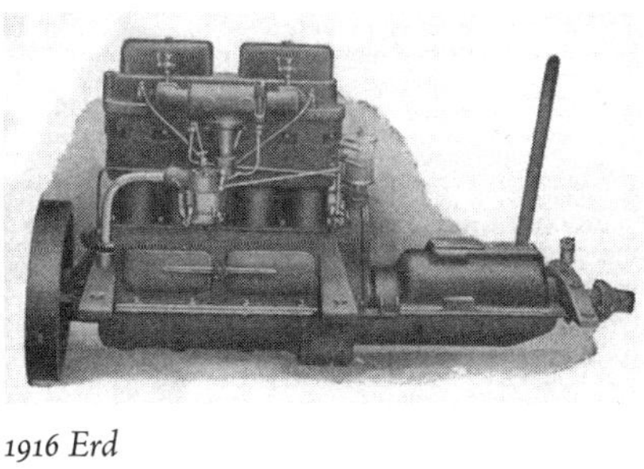

1916 Erd

1915 Essex

1906 Essex

Circa 1901 Fahrney (Courtesy of Joe Joyce)

1913 Fairbanks-Morse Type “G” with removable head

1913 Fairbanks-Morse Type “I” with dome head

1902 Fay & Bowen

may have evolved into the Enterprise Manufacturing Co.

Erd Motor Co.
Saginaw, MI
Trade name: Erd
Machinist John G. Erd and his son Harry founded their business in 1890 in a rented building at the Bliss and Van Auken lumberyard in Saginaw. There, they developed a 3″ × 3″, two-port, two-cycle engine with fuel intake by vaporizer. The sale of that engine in the spring of 1891 made Erd one of America's pioneer engine companies. Business increased enough that Erd moved its equipment to larger rented space at the corner of Hancock and Water streets, and they added new machinery. With the move came an expanded line of two-cycle engines with 1, 2, 3, 4, and 6 cylinders. The largest bore/stroke dimension was 5″ × 5″.

In 1897 or 1898, Erd again moved to larger quarters on North Water St. overlooking the Saginaw River. The "official" incorporation date of Erd was March 1909, backed by the considerable sum of $25,000. By 1911, Erd's two-cycle, jump-spark engines were available in six models of 2-, 3-, or 4 cylinders producing from 12 to 40 hp. These "second generation" two-strokes might be termed 2-/3-port motors with a port in the piston. That year, publisher Nelson Doubleday, the owner of a 22-foot W. H. Hand, Jr.–designed launch described his experience with his 3-cylinder, 22-hp Erd. "The engine I think is a real wonder. I have used it almost continually this summer with perfect results." He described running the boat for 60 miles in a day with "not a miss or a falter."

The Erd showed up on lists of motors suitable for restricted class racing with rpm ratings of 1,200 for the 286-cubic-inch, 40-hp 4-cylinder, and the 429 cubic-inch 6-cylinder. The lightweight "Erd Special" had polished water jackets atop the cylinder heads. The standard model had removable flat-top heads rather than the domed jackets. The Erds developed their first four-cycle engine, a 2-cylinder 5″ × 5″ T-head, shortly after the move to their new quarters. Although this engine was still in operation on the Saginaw River two decades later, the first series-built four-cycle engines, probably introduced in 1912 or 1913, were overhead valve, 4-cylinder, 4″ × 5″ models. The high-speed aluminum model weighed 450 lbs. while the standard iron-crankcase model weighed 650 lbs. and was intended for cruisers. Dual magneto ignition was standard.

In 1912, at about the time the Erds shifted to four-cycle engines, they built a new brick factory of their own at the corner of Mackinaw and Niagara streets. L-head "medium speed" motors (750 to 850 rpm) ranging from 1 to 6 cylinders and 3½ to 60 hp were added to the line. In 1917, the company sold more than 2,600 motors. Most of these were larger displacement four-cycle engines and they went into marine, industrial, and tractor applications. More than 3,200 motors were produced in 1918, and the company invested in $86,000 worth of new machinery. However, the postwar recession may have hit Erd very hard. By 1922, Erd listed only three four-cycle engines. The four-cycle S-4 was well known. Business, however, must have been poor because the company filed for dissolution on January 19, 1923.

Erds were painted dull black, the finish highlighted by polished-brass fittings. A number of Erd motors survive today, some in remarkably original condition. Such a motor would be a welcome addition to any collection.

Ericsson & Moon Mfg. Co.
Detroit, MI
Trade name: Ericsson
A line of two-cycle motors was built by this company beginning about 1906, and production continued through at least 1907.

Erie Pump & Engine Co.
Erie, PA
Trade name: Erie

Essex Engine Co.
Lynn, MA
Trade name: Essex
When Essex exhibited its two-cycle motors at the Boston Boat Show in 1910, it displayed both the Essex Standard and Essex Special models. The former were two-port models with make-and-break ignition rated at 450 or 500 rpm. The latter were said to be available in "either two-port or three-port type" with jump-spark ignition and were rated at 600 rpm. Single- and double-cylinder models were available in both series, producing 3 to 16 hp. Schebler carburetors were used. One could be forgiven for being confused by the model line, and it is hard to see how Essex's positioning would have helped it to fare well over the long term given such other New England brands as Palmer, Lathrop, or Bridgeport. The Essex seems to have been made from 1910 to 1913.

See Young, W. J. Machinery Co.

Evansville Gas Engine Works
Evansville IN
Trade name: Evansville

Everett Bros. Motor Co.
Brunswick, GA
Trade name: N/A

Evinrude Motor Co.
Milwaukee, WI
Trade name: Evinrude
Ole Evinrude had designed his first outboard in 1907. The company produced stationary engines, outboards (two- and four-cycle), and, later, some two-cycle inboards. Ole Evinrude, universally recognized as a mechanical genius, sold the company in 1913 to care for his ailing wife and make a transcontinental journey in his Packard. In 1916, he designed and built a V-8 engine to power his 42-foot cabin cruiser, and went on to form Evinrude Light Twin Outboard (ELTO). The 1916 Evinrude inboard Model CC was a 1-cylinder, 2-hp model; the DD a 2-cylinder, 5-hp model. Each had a magneto and clutch. Vertical, "canoe" versions that mounted

across the gunwales were also made. In 1917 the inboard models included three engines — singles of 2 hp at 800 rpm and 3½ hp at 650 rpm, 3½ hp, and an 1,100 rpm 5-hp twin. "The next year or two will see developments in internal combustion engines that up to this time people have only dared to imagine," predicted Evinrude's H. Piersach in *Motor Boating* in Dec. 1918 when 2-, 3½-, and 4-hp models were offered. In Jan. 1921, a new twin was introduced with en bloc cylinders. "The new type of unit construction has made possible the elimination of ninety-three parts, quite an item for a 4-5 hp powerplant," reported *MotorBoat* in Dec. 1921. Perhaps that was what permitted prices to be reduced. In 1924 the line consisted of two inboard models, a 2-hp single and the 5-hp twin.

See Marine Power Co.

F.A. Brownell Motor Co.
See Brownell, F. A. Motor Co.

F. A. Seitz & Co.
See Seitz, F. A. & Co.

F. A. Slitz
See Slitz, F. A.

F. J. Valentine
See Valentine, F. J.

F. M. Watkins Co.
See Watkins Motor Co.

Fadum, Frederick & Son
Baltimore, MD
Trade name: Fadum

German immigrant Fadum was well known in Baltimore for his skill as an engineer and machinist who had equipped many of the city's factories with everything from shafting to fans and steam heating systems. Local sources reported that Fadum was making gas and gasoline engines in 1898 although no details were provided. In 1911 Fadum introduced a four-cycle motor in 1-, 2-, 3-, and 4-cylinder models of from 4 to 30 hp. Jump-spark or make-and-break ignition was available on larger models, but the smaller motors appear to have been make-and-break. Schebler carburetors were used.

Fahrney
Chicago, IL
Trade name: Fahrney's Hydrocarbon Motor

Emmery Homer Fahrney, the son of an immensely wealthy Chicago doctor who'd made a fortune with patent medicines, received a patent in 1901 for an en bloc 2-cylinder engine design. How many of these engines were built and sold is currently unknown, but one has survived. The four-cycle, spark-plug ignition engine was rated at a comparatively high 800 rpm, and developed 8 hp. An unusual, boxlike float-feed "carburetor" at the engine's base fed vapor into the inlet pipe. Fahrney would become an important investor in Termaat & Monahan and later Universal. He received a number of patents related to engines and motor boats.

See also Termaat & Monahan *and* Universal Motor Co.

Fairbanks Company
Boston, MA
Trade name: Fairbanks Victor

In business by 1908 or 1909, this company built heavy-duty single- and double-cylinder motors of 1½ to 15 hp. An unusual feature for 1910 was the "substitution of a regular capped connecting rod with two bolts for the hinged-end connecting rod of previous models."

Fairbanks, Morse & Co.
Chicago, IL
Trade name: Fairbanks-Morse

Few engine companies grew to become major industrial corporations, but Fairbanks, Morse was one that did. In 1911 the company claimed that more than 100,000 of its engines were then in use. The presiding genius of the company was Charles Hosmer Morse, who had already made a fortune building scales, windmills, and electric motors when he focused on the potential of internal combustion. Morse acquired the rights to a horizontal gasoline stationary engine invented by a German immigrant named Franz Burger and produced in Sterling, Illinois, by the Charter Gas Engine Company. Those first engines were called Fairbanks-Charter after which time the name was changed to Fairbanks-Morse in 1895. Thus was launched the line of engines that would become known throughout the world.

By the early 1900s an extensive line of horizontal and vertical stationary engines was in production. The horizontal was adapted for driving the stern wheel of commercial riverboats. The first marine engine, the 20-horsepower, 4-cylinder, four-cycle Type "B," was introduced in 1900. The line was further developed to include a 100-hp, 4-cylinder version that, like other higher-horsepower Type "B" models, included a compressed-air starter. It's worth noting that the Fairbanks catalog for the Type "B" actually included installation drawings showing the drive system for a split-paddle-wheel stern-wheeler. What's more, the company noted in 1912 the availability of "horizontal engines of powers 2 to 50 with special devices for stern wheel boats."

In addition to the Type "B," Fairbanks developed a full lineup of 3½- to 15-hp, two-cycle motors before 1910. The "E" was a 3¾" × 3¾" single- or double-cylinder engine; the "G" was a 4½" × 4½" single, double, or triple; and the "J" was a 5½" × 6" single or double. All except the "E" had separate cylinders. The "E" and "G" had jump-spark while the "J" was make-and-break.

These two-cycle marine engines were offered together with an extensive line of four-cycle models, the medium-speed (800 rpm) "N" and "S" and the slow-speed "B." The latter were heavy-duty engines. The 20-hp, 2-cylinder (7½" × 8") double weighed 2,300 lbs. The 100 hp, 4-cylinder (11" × 12") weighed 18,000 lbs. Fairbanks, Morse finished its two-cycle models in glossy, baked black enamel highlighted by nickel-silver trim.

Company records showed that by 1915 close to 168,000 Fairbanks, Morse

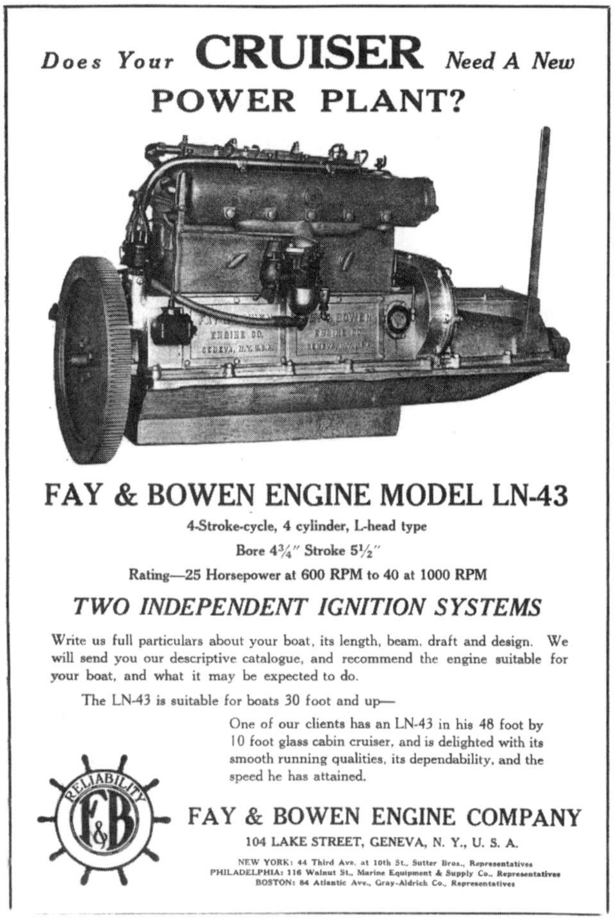

1922

1906

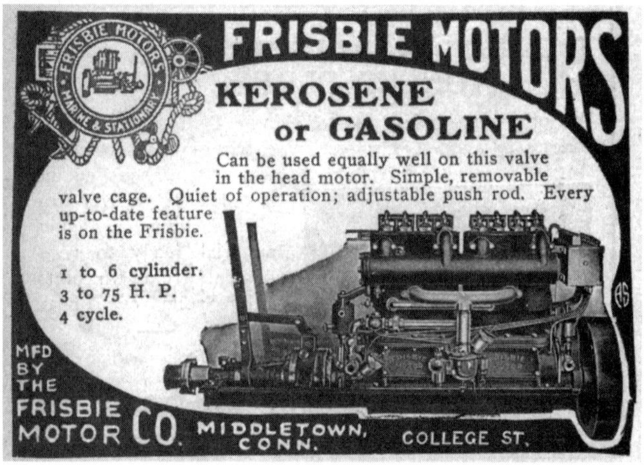

1913

1906

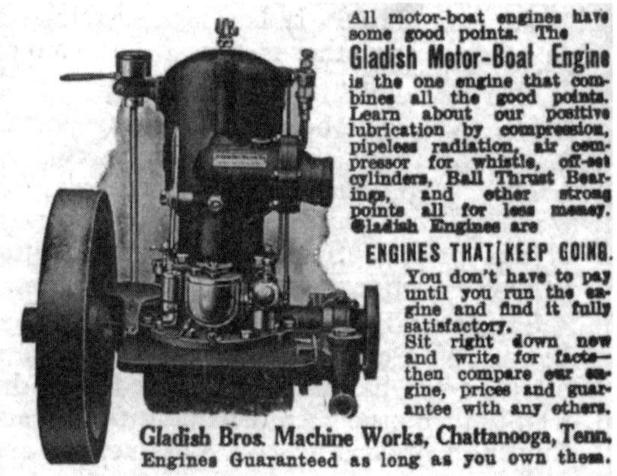

1913

engines had been sold. Many of the stationary engines were bought by farmers and railroads already familiar with the company's windmills. The level of production was such that engines were produced in Beloit and, with the acquisition of the Sheffield Power Works, makers of the two-cycle Simplex engine, in Three Rivers, Michigan.

By 1912, recognizing the commercial potential of such hot-bulb engines, company engineers began development of what would become the F-M Type Y oil engine. The introduction in 1914 of a marine version of this engine, the CO, gave a significant impetus to already strong sales. The CO was built in just the horsepower range then needed — 30 to 300 horsepower. It was the right engine at the right time, a time when gasoline prices were increasing and oil engines were becoming very popular. Models below 150 horsepower were equipped with reversing gears while more powerful engines were made directly reversible. "They filled such a need that they practically dominated their respective markets for years," said a Fairbanks engineer.

It was upon the CO that the first Fairbanks, Morse diesel was based. "One of the improvements consisted of the gradual raising of compression pressure with corresponding reduction of fuel consumption," recorded one Fairbanks, Morse retrospective. "This pressure by 1924 was so high that we obtained 'cold start' and we then called it a Diesel engine — our Model 32. It was followed in 1925 by a larger engine...our Model 33. Both models were developed in Beloit, Wisconsin, without the acquisition of foreign patents. They were designed specifically for the American market. With them, Fairbanks, Morse & Co. was firmly launched in the Diesel engine business."

Fairbanks, Morse was justifiably proud of its "home-grown" two-cycle diesel engine. The line of marine diesels introduced during the mid-'20s included such features as oil cooling of the pistons, pressurized crankcase lubrication, direct reversing, and improved performance, thanks to a separately mounted scavenging air pump. These would be features of subsequent models like the Type 37 offered in horsepower ratings up to 2,000. A smaller, four-cycle, high speed (1,200 rpm) model, the 36A was also introduced, fitted with a reduction gear. Then in the early '30s, development was begun on a thoroughly unique engine, a two-cycle with horizontally opposed pistons.

Fairbanks, Morse survived the Depression, built a variety of engines for the Navy during World War II, and still builds marine engines today as a division of EnPro Corporation. An old Fairbanks, Morse marine engine is a worthwhile addition to any collection, whether private or museum, as the brand represents the history of one of America's largest and most successful engine builders.

Fairbanks-Grant Mfg. Co.
Ithaca, NY
Trade name: Fairbanks-Grant
A boat builder, the company was in business as of 1905.

Fairbault Mfg. Co.
Fairbault, MN
Trade name: Fairbault

Fairfield Motor Co.
Bridgeport, CT
Trade name: Bull Pup
See Consolidated C.S.M.

Fairmont Engineering Works
Philadelphia, PA
Trade name: Chadwick

Farro
Baltimore, MD
Trade name: Farro
The company was begun circa 1902–1903.

Fay & Bowen Engine Co.
Geneva, NY
Trade name: Fay & Bowen
In the fall of 1895 Cornell-trained mechanical engineer Edward S. Bowen joined forces with an experienced salesman and finance specialist named Walter L. Fay and began a company named Fay & Bowen to manufacture bicycle spokes. Three years later, Bowen began to experiment with small marine engines. The initial model, a single-cylinder, two-cycle motor, proved itself in repeated endurance tests in a 25-foot boat on Owasco Lake near Auburn, NY. In July 1900 the two men sold their bicycle spoke business, and Fay & Bowen entered into the manufacture of marine engines and boats. The engines were built in a small brick factory in Auburn; the boats were built in a separate facility about two miles away. The inconvenience of this arrangement resulted in a move, in January 1904, to a waterfront site on Seneca Lake in Geneva, New York. "Fay & Bowen Engine Co.," read the sign, "Motors & Motor Boats."

From 1900 to 1911 the Fay & Bowen was strictly a two-cycle engine. These motors were medium-duty designs (500 rpm) with patented make-and-break ignition, "really the only perfect and satisfactory igniter." The 4-cylinder (6¾″ × 6¼″) developed 40 hp and weighed 2,480 lbs. What *The Rudder* called a "very comfortable runabout" was the highlight of Fay & Bowen's display at the 1910 New York Boat Show. This 25′ × 5′6″ boat was powered by one of the company's own 10-hp motors and was deemed a "sturdy family boat." By 1911 an engine able to run on kerosene or gasoline was available as were pumping sets and electric lighting sets. A T-head of 25 to 30 hp was the first Fay & Bowen four-cycle motor. This engine was introduced in 1911, two years before Bowen's death of typhoid fever at age 53. In 1912 the company positioned its four-cycle thusly: "In certain localities, there exists a prejudice against the 2-stroke engine, especially for heavy-duty, and it is to meet this prejudice that the company has developed this four-stroke line." Beginning about 1914, production was exclusively four-cycle. By 1916 a 6-cylinder was in production and the company installed one in a 30-foot raised-deck runabout to take to that year's National Motor Boat Show in New York. The "Big Six" was shown together with an en bloc, 1,400-rpm, 4-cylinder engine.

During World War I, Fay & Bowen was kept busy with a contract to build seaplane hulls for the Curtis HS-1. In 1919, Walter Fay retired, but the firm seemed well positioned with a line of two-cycle motors of from 5 to 20 hp and four-cycle engines of from 10 to 75 hp. By 1922, the line was four-cycle only. It included the LN-40 (7 to 12 hp), L-41 (14 to 23 hp), LN-42 (25 to 40 hp), LN-43 (25 to 40 hp), LNS-43 (33 to 50 hp), L-44 (30 to 45 hp), and the L-64, a 6-cylinder engine rated at 45 to 85 hp. The LN-40, LN-41, LN-43, and L-44 were of iron construction. The LN-42 and M-64 were available in iron or aluminum models. The LNS-43 was made in aluminum only. Schebler carburetors were used throughout. Lubrication was a combination of pump and splash.

Although it seems now to have been reasonably well conceived, the Fay & Bowen engine lineup was not competitive as the decade wore on. In Dec. 1928 the old standby L-series models were joined by new Fay & Bowens and a line of marine-adapted Continental engines — the Faybow Continental. These were 6-cylinder engines of 40, 65, 70, 80, and 88 hp, all with pressure lubrication. The new Fay & Bowens were 6-cylinder engines, the 2,800 rpm 60-hp 215-cubic-inch Challenger, the 2,200 rpm 75-hp Defender, and the 2,200 rpm 100-hp Conquerer. None of this was enough to permit the company to endure the Great Depression. When Fay & Bowen closed its doors, its buildings were demolished and the canal into which it had launched its boats was filled in to make a road.

Fears Motor Co.
St. Louis, MO
Trade name: Kastrup

In 1913 the Kastrup four-cycle engine was an open-base design offered in seven models of 1, 2, and 4 cylinders and of 4 to 24 hp. An unusual valve arrangement was key to the Kastrup. A three-way valve was located in a housing at the top of each cylinder. The goal was to keep the heat of combustion away from the valve head, "thereby avoiding pitting." The company also claimed that its unique inlet valve setup eliminated the need for periodic decarbonizing. Exhaust gases were expelled through a port uncovered on the piston's downstroke. A line of two-cycle engines was also built.

Ferro Machine & Foundry Co.
Cleveland, OH
Trade name: Ferro

Begun in a single wooden building before or during 1907, Ferro had grown by 1912 to employ some 1,300 people in a 275,000-sq.-ft. plant devoted to producing "the World's Standard 2-Cycle Motor." The company claimed to be the "largest marine gasoline engine works in the world" and it quite possibly was. In 1910, by comparison, Gray was employing 170 people. Ferro was also exporting widely by 1907. It also supplied steel-boat-builder Mullins circa 1907–1911. In 1911 alone, Ferro reported that Mullins had ordered 2,000 engines, which, if true, hints at the demand for small marine engines in that period. British entrepreneur Joseph Day was the British agent for the company. Charles H. Vincent, brother of the famous Packard engineer, Jesse Vincent, worked at Ferro as an experimental engineer. Charles B. Page, an 1899 graduate of MIT, was an important executive who would later join the Van Blerck Motor Co.

The Ferro was a three-port engine equipped with a float-feed carburetor of Ferro's own manufacture. Jump-spark or make-and-break ignition was available. RPM ratings ranged from 550 to 900. Unlike many engines of the time, the Ferro had a removable head secured by six studs and sealed — "a perfectly gas and water tight joint" — by a copper and asbestos gasket. The combustion chamber was hemispherical and the company pointed out the inherent efficiency of the design. Although the Ferro catalog was a work of art and highly detailed, it didn't discuss a basic design feature. That is, the cylinder was offset, presumably to reduce piston thrust and contribute to longevity. The engine was not intended to be reversible. Instead, a planetary-style reverse gear of Ferro's own manufacture was available as an option.

Having reaped the benefits of the two-stroke's heyday, Ferro ran into trouble when trying to convert to four-cycle engines and to develop a focus on the automobile industry. At the end of 1915, it unveiled a pair of overhead valve, four-cycle, 4-cylinder marine engines — 10 to 14 and 20 to 35 hp. These were en bloc engines with big bearings and an enclosed reverse gear. By 1915 the company was preparing to produce a four-cycle, overhead valve V-8 designed by Alanson Brush of Cadillac fame. The engine was selected to power the Scripps-Booth Model D chassis, making this little car the second production model, after Cadillac, to offer a V-8. The Scripps-Booth, however, was not a success. Page left to work at the Van Blerck Motor Company in 1916, while Scripps-Booth was purchased by General Motors at the end of 1917. Ferro's days as an engine maker were by then numbered. These engines, together with four two-cycle models and an outboard motor, were listed in Dec. 1917. After that — silence.

Fey, Lincoln H. Co.
Northfield, MN
Trade name: Lincoln

In 1908, Fey announced its engine was available either completed or as a kit.

Fifield Bros.
Augusta, ME
Trade name: Capital

The Fifields were in business as machinists by 1900 and offered a line of medium-speed, two-cycle engines during the years from about 1910 to 1915. The engine tag included the word "Builders," and one may assume that the brothers actually assembled the engines from castings and forgings supplied by vendors. A 3-hp Capital in the collection of the Owl's Head Transportation

1913 Kastrup by Fears

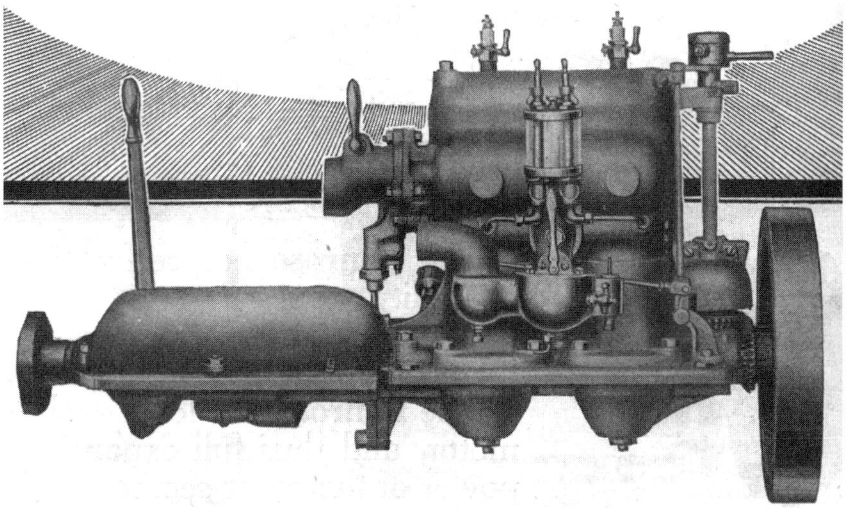

1915 Ferro two-cycle twin with Ferro reverse gear

1915 Ferro en bloc four-cycle

"1911 Ferro One Cylinder Jump Spark Engine in 4, 5½ & 7½ HP sizes"

1915 Fulton diesel

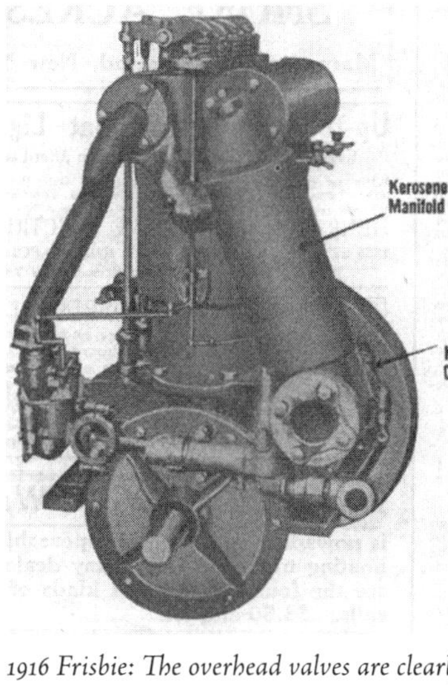

1916 Frisbie: The overhead valves are clearly evident.

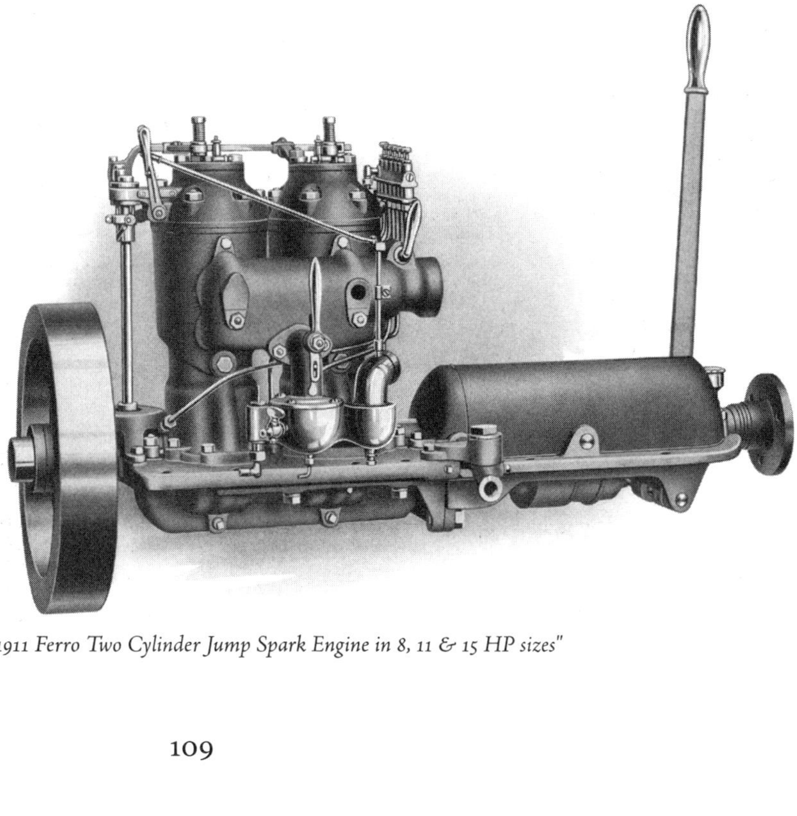

"1911 Ferro Two Cylinder Jump Spark Engine in 8, 11 & 15 HP sizes"

museum (and shown on the back cover) is a conventional-looking dome-head, jump-spark model. The 3½-hp Capital of 1910 was available with Orswell or Perfex ignition and a Holley float-feed carburetor. No incorporation papers for this company have been found until 1928 by which time engine building was probably long over. Subsequent to the engine business, the Fifields operated an automobile repair shop.

Fischbach Manufacturing Co.
Peoria, IL
Trade name: Fischbach
Two- and 4-cylinder, four-cycle engines were produced circa 1905 to 1908.

Fishback Motor Co.
Detroit, MI
Trade name: Fishback

Foster Motor Co.
New Haven, CT
Trade name: Foster

Fox Reversible Engine Works
Covington, KY
Trade name: Fox
See Dean Manufacturing Co.

Francis Mfg. Co.
New London, CT
Trade name: Francis

Fraser Machine & Manufacturing
Boston, MA
Trade name: Fraser

Frazer Bros. Co.
Adams, NY
Trade name: Frazer-Adams
By 1910, the company was building two-cycle, three-port engines in 1-, 2-, 3-, and 4-cylinder models of from 2 to 16 hp. The Frazer-Adams found a good market in small fishing boats in the Thousand Islands region. These engines had one-piece cylinders and flywheels with four large holes. Ignition was jump-spark. A Schebler carburetor was used.

Fred. J. Pease Co.
See Pease, Fred. J.

Frisbie Motor Co.
Middletown, CT
Trade name: Frisbie
The reliable and "friendly" Frisbie was the brainchild of a Cromwell, Connecticut, bicycle shop owner named Russell Abner Frisbie who built his first engine in a barn in 1900. That engine was an L-head. But when Frisbee began production, he used an overhead valve cylinder head. One engine soon won a race from New York up the Hudson to Albany. Increased demand led Frisbie to move from Crowell to Middletown in 1909 and the company became known as Frisbie-Heft. That early race win aside, Frisbie marketed primarily to the "average man" who needed a dependable engine for a cruiser or workboat. By the mid-teens, models ranged from 1 to 6 cylinders and 3 to 75 hp. During World War I, the factory built aircraft engine components. Afterwards, Frisbie sold his interest, but production continued under new ownership.

Frontier Engineering Co.
Detroit, MI
Trade name: N/A
Frontier went out of business in 1915.

Frontier Iron Works
Buffalo, NY
Trade name: Frontier
The company advertised a 350-lb., 60- to 70-hp V-8 in 1912. The target market was the lightweight hydroplane then making a big impact in races.

Fuller Foundry (*also* Geo. Fuller's Sons)
Hallowell, ME
Trade name: Fuller
The company made two-cycle, jump-spark engines.

Fulton Engine Co.
Erie, PA
Trade name: Fulton
The company began as the Art Machine Co. in 1904 in Baltimore. Ads claimed a New York location as well, although this may have been a factory sales branch. The Fulton was designed by engineer J. E. Baldwin. It was a two-cycle, three-port design available with standard make-and-break ignition or optional jump-spark. The business was purchased by Mr. Fulton in 1908 and moved to Erie. The engines sported a water-jacketed muffler by that time. A Schebler carburetor and Lunkenheimer fittings were used. Through 1912, 1- and 2-cylinder models were offered, producing from 5 to 15 hp. The company introduced both high-speed and heavy-duty four-cycle models of 4 and 6 cylinders circa 1913. These motors had individual T-head cylinders. By 1915 a big (8″ × 9″), 6,800-lb. diesel was introduced. It had air starting and separate fuel pumps for each cylinder. By 1918, the line included heavy-duty models up to 100 hp, the latter an oil-fueled engine.

Fulton Iron Works Co.
St. Louis, MO
Trade name: Fulton
Founded in 1854 to build marine steam engines, the company acquired the U.S. rights to a line of four-cycle stationary and marine diesels designed in Italy by Franco Tosi. The marine diesels, by the late '20s, ranged from 600 to 2,000 hp. These giants — they had bore/stroke dimensions of 23″ × 26″ and weighed from 114,000 to 318,000 lbs. — were started, as was usual, by compressed air. Fuel was injected mechanically. The company was still in business in the early post–World War II years.

G. E. DeLong
See DeLong Engine Co. *and* Elbridge.

G. Edward Shaw, Inc.
See Shaw, G. Edward Inc.

G. J. Emeny
See Emeny, G. J *and* Parker, J. J. Co.

Gaeth Motor Co.
Cleveland, OH
Trade name: Gaeth

Gaffga, J.S. & Co.
Greenport, NY
Trade name: Greenport

The Gaffga is an example of how a strictly local engine maker could prosper if he had an effective motor that appealed to a distinct market. John S. Gaffga of Greenport, New York, found that market in local fishermen. In 1902, Gaffga, who had worked for a period at Mianus, developed a two-cycle engine of his own. When Gaffga began producing the motor for sale in 1903, he was assisted by his brother, Walt. Castings were made in New London, Connecticut, and shipped across Long Island Sound on boats carrying potatoes or other cargo. The Gaffgas and a few workmen then machined the castings and finished the crankshafts that came from another supplier. Smaller parts, including the igniter, were made by the next-door blacksmith. Schebler carburetors, which were available in a variety of sizes, were used.

By 1910 Gaffgas were available in 1-cylinder models ranging from the $85 3-hp version to the $175 7½- hp model, and 2-cylinder versions of 6 to 15 hp, the latter selling for $350. The 4 hp was among the most popular models and was perfectly suited to the flat-bottom sharpies used in Great Peconic Bay and Gardiner's Bay. Said the little company's brochure in 1910: "Our engines are the two cycle type make and break ignition, which is considered by all boatmen to be best for marine purposes in open boats."

The Gaffga brothers were still making engines, four-cycle, T-head models, until the late '30s or even the early '40s. What's more, the engines did find a market outside Greenport. Gaffga-built motors powered boats in places as diverse as Alaska and Panama. John Gaffga once claimed, "We have paid particular attention to the fishing and working boats and have hundreds of our engines in boats in the fishing business in all parts of the country that are giving great satisfaction on account of their wide range of control, smooth and quiet running, that requires no attention from the operator. All of these features are very essential in the fishing trade."

Although John Gaffga named his product the "Greenport Motor," it was referred to by almost everyone else simply as a Gaffga.

Garrick, J. F.
Boston, MA
Trade name: Garrick

Gas Engine & Power Co.
Morris Heights, NY
Trade name: Speedway

It is believed that between 1885 and 1895 this company built some 3,000 naphtha engines designed by German-born inventor F. W. Ofeldt. The engines were installed in the company's launches. The firm also built steam yachts and, by 1905, gasoline engines. The company joined with Charles L. Seabury by 1907 to become The Gas Engine & Power Company and Charles L. Seabury & Co. The new firm managed to prosper despite its unwieldy name until, in 1919, the name was changed to the Consolidated Shipbuilding Company.

See also Consolidated Engine Co.

Gas Engine Supply Co.
Newport, KY
Trade name: Diamond

Gasoline Engine Equip. Co.
New York, NY
Trade name: N/A

Geer, Harry R. Co.
St. Louis, MO
Trade name: Geer

Circa 1905, Geer was touting its 2½-hp motor as the "Best light-weight 4-Cycle Engine made." The engine listed at $74 but castings and blueprints could be bought for $22.

General Machinery Co.
Bay City, MI
Trade name: Smalley

After General Machinery took over the bankrupt Smalley in 1907, it moved quickly to get the engines back into production and went to great lengths to drum up business and gain publicity. For the 1908 racing season, it supplied a 950-rpm, 100-hp 6-cylinder (5½″ × 5½″) motor to Oregon racer Johnny Wolff who installed it in a boat named *Wolff*. This unusual launch was 34 feet long but had a narrow 3′8″ beam and was said to weigh just 450 lbs. The fragile boat had some success but was soon replaced by *Wolff II*, 34′9″ × 4′5″ that, with the same engine, turned in average speeds of 32 mph. Subsequently, *Wolff V* had a 9-cylinder Smaller, each cylinder equipped with its own carburetor. The spark plugs were covered by what Smalley called a "hood," made of porcelain and designed to waterproof the plugs when the engine was installed in an open hull.

Smalley's 1909 catalog listed 11 other successful Smalley-powered race boats that competed from the West Coast to Detroit and Palm Beach where the 39′7″ *General* with her 6-cylinder 60- to 70-hp Smalley defeated competitors with much more powerful engines. That year, 24 men and one woman worked at the company, a reasonably robust number for a marine engine maker.

Such heroic motors were still based on Smalley's production models, three-port, two-cycle engines of 1 to 6 cylinders. Each model was available with different bore/stroke dimensions depending upon the power required. The six, for example could be had in 4½″ × 4½″ for 82 hp at a weight of 950 lbs. or 5½″ × 5½″ for 130 hp at a weight of 1,300 lbs. The engines were nicely finished. First, the cylinder castings were filled with powdered slate, which was hand-rubbed smooth, and then the castings were enameled black. All bare metal parts, whether steel, brass, or bronze, were highly polished.

In 1911, the company made news by going to what was then considered an extreme in the use of aluminum. Smalleys were offered with weight-reducing aluminum cylinders with iron liners. It was claimed such engines weighed 50 percent less than a model

1913

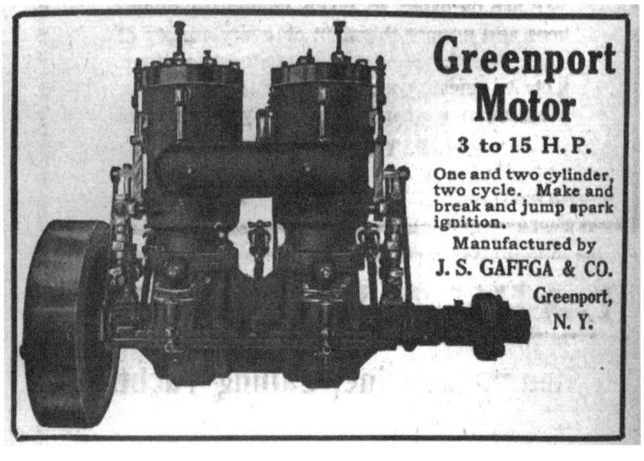

1915

This 1908 Gray ad depicted the company's line-up of two-cycle engines.

with cast cylinders. Committed as it was to two-cycle engines, and despite the Smalley's excellence, General Machinery had nothing to offer once four-cycle engines truly began to dominate the market. This domination was well under way by 1916 when the last Smalley ads seem to have appeared. The company filed for dissolution in Jan. 1924.

General Motors Corp.
Detroit, MI
Trade name: Detroit Diesel et al.
"One of the most interesting services rendered by the diesel engine," reported *Diesel Progress* in 1948, "was the ability of the same engine to power nearly all of the various types of landing craft. This engine was the General Motors Model 71 built by the Detroit Diesel Division. It was a 6-cylinder two-cycle engine rated at 225 hp at 2,100 rpm, with cylinders 4¼" × 5". It was fitted with marine reverse gear and reduction gear. One engine gave the necessary power for a 36-foot boat; and two, three and four engines were combined to provide for the larger craft; in some cases twin screws were used. Tens of thousands of these engines were built."

The Model 71 was the outgrowth of intense research that had begun in the late 1920s at GM's Research Laboratory where Charles "Boss Kett" Kettering presided. GM's own work into the development of unit fuel injectors was much bolstered when the corporation purchased Winton, which had itself been making great strides toward the development of a reasonably compact, lightweight diesel engine. Eventually, the "Jimmy" as it came to be known, was handed over to Gray Marine for conversion to marine applications. At GM, Detroit Diesel, Cleveland Diesel, and Electro-Motive would all produce diesel models that grew out of the company's acquisition of Winton and its own engineers' work. After World War II, the GM diesel revolutionized marine, rail, and road transportation.

See Winton *and* Gray Marine

Genesee Launch & Power Co.
Rochester, NY
Trade name: Genesee
In business as of 1903 building two-cycle engines.

George A. Lisk
See Lisk, George A.

George E. Belcher Machine Shop
See Belcher, George E. Machine Shop *and* Vanguard Engine Co.

Giddings, C.M.
Rockford, IL
Trade name: Sure Go
Giddings offered 1-, 2- and 3-cylinder make-and-break models in 1904.

Gierholtt Machinery Co.
Marine City, MI
Trade name: Gierholtt
James B. Giern and Anders P. Anholtt operated a number of businesses. The first, Giern & Anholtt, is believed to have been devoted to making dies and special tools in the 1918 period. During one of the company's evolutions, Gierholtt Gas Motor Company located on Belle River Ave. produced a 2-hp two-cycle engine in both inboard and outboard models. The specs for 1922 indicated that each of them was a 2¾" × 2½" engine fitted with a Lunkenheimer carburetor. They were rated at 900 rpm.

See also Hess Motors Corp.

Gifford-Wood Co.
Hudson, NY
Trade name: Midshipman II
This engine was an opposed, 4-cylinder L-head, according to 1934 ads.

Gile Boat & Engine Co.
Ludington, MI
Trade name: Gile
Incorporated in March 1909, this company was Willliam Gile's second foray into the business. The company advertised itself as builders of launches and marine and stationary engines. William Gile himself was the general manager; William L. Mercereau, perhaps a principal contributor of the $50,000 capitalization, was president. In 1910, the smallest Gile was a 2½-hp model. It seems unlikely that Gile would have had time after his 1909 departure from the C. T. Wright Co. to return to Ludington and design and tool a new line of engines for the boats he built. Thus, it is most likely that this "Gile" was a re-badged engine, perhaps a slightly re-spec'd T&M. The motors shared such features as a similar dome-head cylinder bolted to the base and a distinctive, rectangular wet exhaust manifold. The business is believed to have continued through 1915.

See also Gile Gas Engine Co. *and* Wright, C. T. Co.

Gile Gas Engine Co.
Ludington, MI
Trade name: Gile
Inventor William Gile was in production of two-cycle marine engines by as early as 1904 or 1905. The motors were offered in two model lines, one with dome-head cylinder castings and another with separately cast heads. It was the largest engines — 4" × 4" or 4" × 5" and 30 to 100 hp — that had the one-piece cylinders. Whether any, or many, of these big, multicylinder two-cycles were actually sold is now unknown. It's possible they didn't get beyond the realistic artist renderings of the brochure. Both models were equipped with a novel controller with positions for starting, stopping, and reversing. Gile also developed his own carburetor. In 1907 these fine motors caught the attention of Cass T. Wright, a wealthy businessman in Greenville, Michigan, and he purchased the company, retaining Gile to oversee the transfer and production. Gile returned to Ludington in 1909 as a key player in the Gile Boat & Engine Co.

See also Wright, C. T. Engine Co. *and* Gile Boat & Engine Co.

Gillespie-Boynton Co.
Paterson, NJ
Trade name: Watertown
This was the successor to the Watertown Motor Company. At the

New York Motorboat Show in 1915, Gillespie showed an attractive line of two-cycle motors including a single-cylinder (5″ × 5″) 7-hp "Fisherman," together with a 10-hp twin, 15-hp triple, and 20-hp four. These were en bloc engines with the intake and exhaust manifolds integral with the cylinder casting. That eliminated most piping and gaskets and impressed show-goers as being "strikingly neat and attractive." The motors were painted in French gray enamel (probably a glossy, medium-gray color) and had nickel-plated fittings. The pistons were ported so that gas passing from the base to the cylinders was carried by the piston until the appropriate cylinder port was uncovered. The result, said the company, was cooler operation. These 3¾″ × 4¼″ Watertowns were 1,000 rpm motors of reasonably light weight. The 10 hp weighed 200 lbs.; 15 hp, 270 lbs.; and the 20 hp, 340 lbs. Ignition was by jump-spark magneto or a special Watertown timer, driven by the crankshaft and mounted just behind the flywheel.

Gilmore Motor Mfg. Co.
Detroit, MI
Trade name: Gilmore
See Gilmore-Cragg

Gilmore-Cragg Motor Mfg. Co.
Detroit, MI
Trade name: Gilmore-Cragg (later Gilmore)
Machinist George A. Gilmore was producing a two-cycle marine engine by 1911, presumably at a shop behind his home, which was located at 352 Green Ave. It is conjectured that Gilmore needed financial and managerial assistance in his endeavor and that these needs were fulfilled by M. G. Cragg, the company president, and E. F. Alleman, the secretary-treasurer. They had an office at 602 Wayne County Bank Building. As of 1911, the company claimed to offer single- and double-cylinder two-cycle motors of 1½ to 30 hp. The motors had a solid flywheel and jump-spark ignition. Like many other early internal combustion engines, the first generation Gilmore had a circular base reminiscent of small "wine bottle" steam engines. (Portable, with a flat flange at the bottom of the base, such engines could be moved from one location in a factory to another.) An unusual but not entirely unique feature of this three-port motor was the location of a drip-feed oiler on the intake manifold. The original color may have been a light or medium blue.

Details of the business relationship between Gilmore and Cragg are now unknown but, since changes were made to both the company name and the engine name, it is likely that conflicts of some sort may have existed. Although the company's earliest known ad in 1911 referred to the Gilmore-Cragg Motor Mfg. Co., in 1913 ads were under the name of the Gilmore Motor Mfg. Co., the Cragg Motor Mfg. Co., and the Gilmore Motor Co., which then had an address at 404-406 24th St. By then, Gilmore had also developed a little "row boat & canoe motor" (an outboard), and claimed to offer four-cycle motors of 1 to 8 cylinders. By 1915, the Gilmore Motor Co. was producing a stationary two-stroke model with a spoked flywheel, the "Farmer's Friend." Changes of company names and changes of engine names were not a path to success in the crowded marine engine business where consistency of message and identity was important. It is believed that 1919 marked the final year of Gilmore production.

Glandish Bros. Mach. Works
Chattanooga, TN
Trade name: Glandish

Globe Gas Engine Co.
Philadelphia, PA
Trade name: Globe
See Pennsylvania Iron Works

Globe Iron Works
Minneapolis, MN
Trade name: White
This company had no connection to the Globe engine produced by the Pennsylvania Iron Works. Its engines were based on patents awarded to Clarence O. White beginning in 1899. By 1900, Globe was in production of marine and stationary engines. The marine engines were offered in two-cycle — below 6 hp — and four-cycle models. As was typical of most engines of that period, the Globes possessed a number of unique features. In January of 1901, *The Gas Engine* noted that the two-cycle White had both a water pump and a then-unusual fuel pump "for lifting the fuel to a constant-level reservoir from whence the supply is taken by the engine ... the gasoline is not taken into the crankcase, but it enters the by-pass between the crankcase and the cylinder.... The advantages claimed for this method are that an explosion in the crankcase is made impossible ..."

The four-cycle engines were offered as 8-, 12-, and 16-hp doubles or 4-cylinder models of 16-, 24-, and 32-hp. Rated speed was 380 rpm. All the engines had a patented make-and-break igniter.

Godshalk, E. H. & Co.
Philadelphia, PA
Trade name: Godshalk-Giant
A celebrated boat-building company famous for its launches, Godshalk collaborated with Packard in 1906 on a Gold Cup "autoboat," and the Navy was trying out a 10-hp Giant in a 20-foot whaleboat at the Norfolk Navy Yard. That said, the company's more usual product was its own boat powered by what the company intimated were its own three-port, two-cycle engines. Godshalk had introduced these motors by 1904, making it among the earliest builders to focus on the high-speed potential of the three-port two-stroke. Godshalk was one of at least ten companies to have acquired a license from Joseph Day to build a three-port engine. Two model lines were offered. The "A" had an iron base and cylinders. One- to 4-cylinder models were available in ratings of 5 to 40 hp. The "B" had spun-steel water jackets, lightweight steel cylinders, and a "bronze-nickel-aluminum" base — all aimed at reducing weight.

1901 Globe Iron Works White single-cylinder, 4-hp, two-cycle engine with make-and-break ignition. A fuel pump lifted gasoline to "a constant-level reservoir from whence the supply is taken by the engine." A reversing propeller was fitted.

1905 Godshalk, 5 hp

1901 White 2-cylinder, 16-hp, four-cycle engine. It was rated at 350 pm and weighed 1,450 lbs.

1905 Godshalk, 15 hp

1906 Geer

1911 Gilmore

1902 Grant-Ferris

In 1905 Godshalk moved across the Delaware to a new plant in Camden, New Jersey. The motor pictured in 1906 ads was a two-cycle with a spoked flywheel and removable cylinder head, entirely different from those previously offered. The company then advertised four-cycle 60- to 90-hp engines for "heavy work boats and large cruisers." Were all these engines really built by Godshalk? Just the two-strokes? These questions can't be answered with certainty at present. Godshalk appeared on no listings of marine engine manufacturers during the industry's expansion period of 1910 until about 1917.

Golden Gate Machine Works
San Francisco, CA
Trade name: Golden Gate
Established in 1887, this machine shop operated by Adam Schilling was located at 211 and 213 Main Street. The precise year when Schilling, assisted by another machinist whose last name was Hirsch, introduced the Golden Gate engine is presently unknown. However, engines of 1 hp and above were being marketed by 1892. The engine had electric ignition. Mixture entered the cylinder through an automatic intake valve. The Golden Gate was sold primarily for stationary applications, but marine models were offered. One rare survivor circa 1894 has an overhead camshaft. Research has yet to uncover what became of Golden Gate but it is quite possible that it was absorbed by another firm. A Golden Gate, or one of the other 1890s California marine engines, would be a treasure for any collector or museum.

Golden, Belknap & Swartz Co.
Detroit, MI
Trade name: G.B. & S.

Goldie, William Jr.
See Pierce-Budd

Goodwin
Trade name: Goodwin
In 1913, a single-cylinder, 4-hp motor was advertised.

Gooley & Edlund
Syracuse, NY
Trade name: Van Epps
The company traced its roots to 1903, but it became officially incorporated in 1914 with a capital stock value claimed to be $75,000. The principals were Malachi J. Gooley and John Arvid Edlund (1867–1931). The company was said to have been formed to manufacture gasoline engines and machinery. Neither Gooley nor Edlund is believed to have had any expertise in engine design, and their effort with the Van Epps engine was presumably that of manufacturer for a brief period.
See also American Gasoline Motor Company

Gorham Engineering Co.
Oakland, CA
Trade name: Gorham, Monarch
In business by 1912, Gorham built heavy-duty make-and-break and jump-spark engines up to 100 hp. "Hundreds of Gorhams earning big dividends for their owners," read a November 1913 ad. The Gorham had an open base with the cylinders supported on columns.

Goshen Motor Works
Goshen, IN
Trade name: Goshen
The 1908 brochure describes 1-, 2-, 3-cylinder, 3- to 50-hp jump-spark models with removable plates on the base to "adjust" connecting rods. The engines had a removable cylinder head and force-fed lubrication driven by a belt off a pulley attached to the gear-driven water pump. The 5″ × 6″ one-lunger displaced 117 cubic inches and produced 6 hp. The mighty 7″ × 8½″ 3-cylinder model displaced 981 cubic inches and produced 50 hp. Both "speed" and "heavy-duty" models were offered depending on application need. A line of V-bottom launches from 26 to 36 feet, was offered to customers who wanted to buy a complete Motor Boat. Goshen later became the White Manufacturing Co.

Graef, E. W.
New York, NY
Trade name: Graef
A mechanical engineer who became a regular contributor to *The Rudder* with articles about engine developments and technical matters, Ernest W. Graef produced his own engine starting in around 1905. According to boat designer William J. Deed, another contributor to *The Rudder*, Graef developed an engine that was initially started by an "electrical sparking device." Once the engine was warmed up, it relied for ignition on "a piece of metal [not a hot tube] kept hot by the exhaust." This seems to have been a somewhat dubious method because Deed noted that the instructions supplied with the engine advised that if the engine didn't get hot enough for the metal igniter to work, "then use electric igniter." The extent of the model line remains to be discovered. A moose of a one-lunger, an 8″ × 10″ 23-hp engine was advertised for sale used in 1912 for $200. There is currently nothing to indicate that the Graef achieved any long-term market success.
See also Reeves-Graef

Graham, Marcellus A.
San Francisco, CA
Trade name: Oriental
Machinist Marcellus Graham was born in Tennessee in 1838. His date of arrival in San Francisco is currently unknown but he was certainly there by 1889 or 1890 and appears as a registered voter in the city in the latter year. The Oriental — perhaps named to convey an image of hard work — was patented in 1891. The cylinder was mounted atop a crankcase whose shape would have seemed familiar to a steam engineer. Make-and-break ignition was used. A rare surviving example is a big 50-hp vertical twin.

Grand Rapids Engine Co.
Hillsdale, MI
Trade name: Monarch
In business by 1897, this firm introduced a line of two-cycle marine engines by 1899. These models were

replaced circa 1906 by splash-lubricated, jump-spark or make-and-break models. The most powerful of those in 1907 developed 80 hp. In 1910, at the Boston Boat Show, a 4-cylinder, 20-hp Monarch was displayed. By then, the company name had changed to the Grand Rapids Gas Engine & Yacht Co., and four-cycle marine engine models ranging from 5 to 160 hp were being produced. These were low-speed models (325 rpm) with individual cylinder castings. Paraffin and kerosene models were available to a market that likely was composed of a mix of workboat and pleasure cruiser buyers.

Grant Marine Motor Co.
Detroit, MI
Trade name: Grant Liberty, Grant Curtis
Grant had an ad in *Motor Boating* in July 1925.

Grant-Ferris Co.
Green Island, NY
Trade name: Howard and Brass Jacket
The company was in business by 1896 making agricultural equipment. Grant-Ferris listed its address as Green Island Street, Troy, although, at the time, Green Island was a small town in its own right across the Hudson from Troy. Doubtless, Troy was a name that would be familiar to readers of company ads. Gasoline engine production at some level seems to have begun in 1901, the engines designed by mechanical engineer William S. Howard. Grant-Ferris touted its products as "A LITTLE MORE than up to date" and claimed the headless Howard to be "as easily controlled as a steam engine." Two-cycle motors of 2 to 20 hp were offered with make-and-break ignition. Four-cycle models were also noted in ads in 1902 and said to be built in ratings of 2 to 24 hp.

The company advertised that it had a special arrangement with the Gas Engine & Power Co. and could deliver a complete boat, hull by Gas Engine and motor by Howard. Howards were also standard equipment in a variety of factory-built launches, and Grant-Ferris itself attempted boat/launch production when the Trojan Launch Company located to its address at the corners of George and Bleecker Street. Engineer Howard moved to New York City sometime in 1903 to work with the Gas Engine & Power Co. The last year in which the company appeared in the directory was 1912. It is possible that the Howard model built by Grant-Ferris and the Howard built in Philadelphia were both designed by William S. Howard.

See also Trojan Launch & Motor Works *and* Howard Engineering Works

Grasser Motor Co.
Toledo, OH
Trade name: Grasser
By 1910 Crasser was offering its Model A two-cycle, 2½-hp motor made at a shop at 102-4 Ottawa St. This was a jump-spark motor with a dome-head cylinder casting. It was sold complete with six dry cell batteries, coil, wire, on/off switch, shaft coupling, muffler, bronze propeller and stuffing box, and six feet of steel propeller shaft for $65.

Graves Motor Co.
St. Paul, MN
Trade name: N/A

Gray Motor Co. (later Gray Marine Motor Co.)
Detroit, MI
Trade name: Gray
Founded in 1905, Gray derived its name from vice president Paul Gray and secretary/treasurer David Gray. Their father, John Gray, had already been a wealthy banker before he loaned Henry Ford the money needed to start Ford Motor Company, of which Gray became the president. When John Gray died, he left his sons an estate valued at some $144 million. If the Grays were the money behind the new company, the visionary was Ora J. Mulford. He was an advertising genius who worked with, among others, Henry Ford, and he had an enormous enthusiasm for boats and marine engines. In 1905 Mulford sold his last interest in the pioneering Sintz Gas Engine Co. to focus on his new venture.

Initially, in 1906, Gray offered a Stanley Du Brie–designed two-cycle one-lunger. But it was engineer Oscar Kreis, hired by Gray soon after, who designed the two-cycle motors for which the company gained fame. These were the 1- and 2-cylinder Models "R" and "S," and the 1-, 2-, and 3-cylinder Model "T," the latter available in ratings of 7 to 36 hp. Early on, Gray established a policy of what it called "advertising prices in plain figures." That meant prices were given for a "complete outfit ready to install." In 1912, a 3-hp Gray could be bought for $60; the most expensive model sold for $528.

By 1912, with its factory turning out up to 50 motors a day, the Gray Motor Company faced the same decision that confronted all manufacturers of two-cycle engines — develop four-cycle engines or fail. In 1914, Gray introduced Continental-based 4- and 6-cylinder motors for marine applications. These were the 700-cubic-inch 4C (a single en bloc cylinder casting) and the 50-hp 840-cubic-inch 6C (cylinders cast in blocks of three). In 1915, Gray introduced the long-lived 281-cubic-inch, L-head Model D 4-cylinder (20 to 24 hp at 1,000 rpm) and the 2-cylinder Model D Jr., a 1,200 to 2,000-rpm, 6- to 7-hp motor that weighed 300 lbs. These motors offered the value with which Gray had been, and would remain, associated. The 4-cylinder model sold for $246 and was sufficiently robust to power a variety of 18- to 30-foot boats of various types. In 1916, a Gray ad showed a Hacker-designed, 21½-foot, V-bottom runabout built by the Mayea Boat Works in Detroit that could run at 20 mph.

By the mid-teens, Gray employed 170 people and claimed to be the "Largest Builders of Small Marine Motors in the World." Like all other engine makers, Gray was looking for superlatives and found them in the qualifier "small."

1911 Gray triple

The intake side of a Gray Marine diesel

1939 Gray 6-90 (6 cylinders, 90 hp)

1911 Gray

1923 Gray Model Z

1939 Gray Super 8

Gray's 4-cylinder, 281-cubic-inch Model D was introduced in 1915 and became a very important engine in the company's line.

1939 Gray Diesel – the GM engine handed over to Gray to for marine conversion and production during World War II

In 1917, it introduced the Model F, a 3″ × 4″ 12- to 15-hp en bloc engine that was doubtless a conversion. It remained in the lineup for only one year. During the war, Gray was kept busy building large quantities of its "Work" motors, which found application in pumping out Europe's trenches. Marine engines, too, were sold to the government. All this production occurred in a newly built plant in St. Clair Heights.

Gray faced the same downturn that confronted all makers after the war. In 1918, the company introduced the Gray Flyer, an en bloc, two-cycle twin (3½″ × 3½″) rated at 6 to 10 hp. By 1922, the model line included just the Model "U" two-cycle single and twin available in ratings of 3 to 8 hp (the twin was an en bloc engine), and a four-cycle, 4-cylinder, overhead valve engine. Dubbed the Victory, it was available as the 10- to 25-hp VM (3½″ × 5″) or the stroked 18- to 35-hp VE. The Victory was not exclusively a marine engine — it had been developed for sale to automobile manufacturers — but it was steadily improved upon and its designation was eventually changed to the Gray Model X.

With the marine engine business in its postwar malaise, Mulford began thinking seriously about building a car of his own. In fact, the Gray automobile was put into production, but things went no better for the Gray Motors Corporation than they did for thousands of other would-be car builders. In 1924, after the car effort had failed, Mulford bought back Gray's marine engine business and started the Gray Marine Motor Company. The new company's first plant was in an abandoned, two-story building that had once been home to the Northern Motor Car Company begun by automotive pioneer Charles Brady King.

Gray Marine's initial product was simply 3,000 Model Z engines, essentially the engine used for the Gray car. In marine form, it developed 14 to 20 hp at up to 2,200 rpm. It had a list price of $395 ($130 less than a complete Ford Model T Touring car). Fortunately, Mulford had demanded foresight in the engine's initial design phase. Its crankshaft bearings were made large enough (2″ or more in diameter) so it could be used for marine purposes. Even so, O. J. Mulford's son John, an executive with the company, recalled that the early days of Gray Marine were a time of struggle. He spent some time in New York in 1924, studying at Columbia and attempting to sell the Gray, but it was tough going. "Gray did not have much of a line to start with," he remembered, "and the old companies like Palmer and Lathrop with their slow-speed engines had the coast market at that time. It was decided they [Gray Marine] needed more than just a little, four-cylinder Gray car engine."

There now began a confusing but ultimately successful strategy of converting automotive blocks to be Gray Marine engines. Gray's chief engineer, Frank Fernstrum, was "a genius for converting engines and arranging for their production in the most economical way," recalled John Mulford. "Very little money was spent on these conversions and the results were wonderful. Gray began to get a reputation for producing a product that ran."

In 1926 the nature of Gray's struggle to become and remain a viable marine engine maker became evident in the model lineup. The headliner models that winter were two "new" sixes. The 50-hp Z6 (3⅜″ × 4½″) was a marine conversion of Pontiac's 241.6-cubic-inch six. Pontiac's larger 353.8-cubic-inch 75-hp 6-cylinder engine became Gray's A6. "This motor is not the product alone of automotive engineers," said Gray's ads that year. "It represents the best skill and efforts of engineers who are practical Motor Boatmen ..." Other models for 1926 included the familiar Model "Z"; the "ZB" (no reverse gear); the "V" ("Used by the United States Government"); the Model "O," a valve-in-head, one-lunger of 4 to 5 hp; the "U" 6- to 8-hp twin; a 5″x 7″; and the "H-50," a 50-hp four-cylinder "for cruisers and workboats," based on a Hercules industrial engine. The latter engines were sold in some quantity to Chris-Craft, and it was reported that after Gray ceased the Hercules conversions, Chris-Craft stepped in and did it themselves. (Although the Model Z may have had its limitations, all 3,000 were eventually sold at a good profit, many to the Elco Boat Company.) That spring, Gray announced the "new Imperial Six." This 3⅞″ × 5″, 75 hp at 2,100 rpm engine, painted in dark blue enamel that contrasted nicely with polished aluminum parts, was sourced from Studebaker.

In 1927 Studebaker continued to be a source for Gray. The Gray Marine Six-90 was "built in the great Studebaker plant in Detroit, and then turned over to the Gray Marine Motor Company, who supply a proper water pump, a marine type base, and water-cooled manifolds. Marine lubricating systems are built-in, but without changing the pressure lubricating system." John Mulford recalled that the converted Studebaker's light six, in particular, "turned out to be a very fine engine. It became one of the famous engines during the bootlegger-Guard wars."

The year 1927 marked the start of a new era for Gray. For one thing, production of two-cycle motors finally ceased, an event that had been rumored and reported, apparently incorrectly, three years earlier. This put Gray in the position of basing its *entire* model line on marine conversions of automobile engines. Another Pontiac engine was added to the line. The Pontiac 6-cylinder, 186.5-cubic-inch 40-hp engine became the Gray Six-40. John Mulford remembered that "the light Pontiac turned out to be a very fine export engine. Those who read the history of the evacuation of Dunkirk know that the Richardson cruisers played a tremendous part in this memorable sea operation. All the Richardson cruisers were powered with Gray-converted Pontiac sixes." In 1928, too, Gray introduced a version of Studebaker's new straight-eight. The Gray version had bore/stroke dimensions of 3⅜″ × 4½″ (322 cubic inches) and produced 100 hp at 2,400 rpm, 60 hp at lower speed, and a maximum of

1941 Gray marine single-cylinder diesel

1958 Gray Fireball V-8

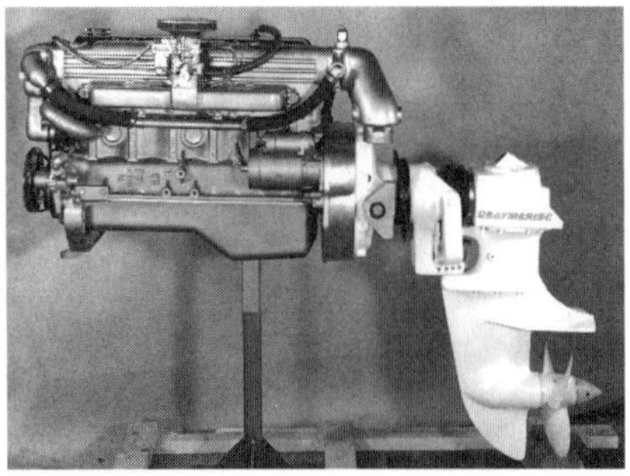

1969 AMC-based Gray six with inboard/ outboard drive and twin propellers

1947 Gray Lugger

1915 Greenport with reverse gear

1913 Grimm

114 hp at 3,300 rpm. The following year, in 1929, the company added a Pontiac-based 80-hp straight-eight as the Gray Eight.

While this method of building up a model line apparently worked reasonably well, it did have shortcomings. For example, Gray purchased its Pontiac engines in lots of 20 and had to pay for them with a certified check. The arrangement with Studebaker could have been similar. What's more, O. J. Mulford, who could well remember prewar years when the company actually designed and built marine engines in their entirety, was never quite comfortable at not having an exclusive supply of blocks. Mulford approached Ross Judson, president of Continental Motors Corporation and asked for an exclusive deal, which Judson refused, noting that he already had an arrangement with Joe Van Blerck. Van Blerck, the formidable pioneer of high-end, high-performance marine engines, was then converting and selling the Continental Van Blerck in a plant in New Jersey. Ultimately, however, Continental made an arrangement with Gray, and, according to John Mulford, "Thus began a fine and long exclusive arrangement with Continental Motors Corporation."

The development of an increasingly wide number of Continental-based engines during the early '30s saw Gray Marine's product line grow steadily to provide comprehensive coverage for cruisers, work boats, and auxiliaries. Model names familiar to a generation of boating enthusiasts now appeared, including the Sea Scout, the Six-51, Fireball, Phantom, and others. The Phantom models — 4- and 6-cylinder and a 150-hp 8-cylinder — announced in November 1931 were rated at between 2,800 and 3,800 rpm, and were intended, obviously, for performance-oriented boats. Weight, too, was kept to a minimum, with aluminum pistons optional in most models and standard in others. The Phantom Eight weighed 997 lbs at a time when a 150-horsepower Kermath Sea Farer, certainly not an overly heavy engine, was listed at 1,400 lbs. (Kermath would no doubt have pointed to its engine's heavier weight as proof that it was dedicated marine design, not a conversion.) All the motors were refined in their details with improvements in manifolding, carburetion, and lubrication.

At the time Gray announced its Phantom series in late 1931, it also announced "8 New but Proven 'High Torque' Grays." The most powerful, the 6-cylinder, 479-cubic-inch "480," produced 105 hp. These two model lines, High Torque and Phantom, would carry Gray through the '30s. All were available with or without reduction gearing. In 1930, too, Gray introduced the "Gray Vee-Drive Gear" capable of handling inputs of up to 200 hp. At the National Motor Boat Show in 1936, Gray emphasized that it was a modern, forward-looking company, displaying its engines under indirect lighting in front of a panoramic background that illustrated key design features.

All of Gray's experience at marine conversions led it into the diesel realm. According to John Mulford, the company engineered the marine conversion of the Hercules diesel for testing by the U.S. Navy. This project was canceled. However, an acquaintanceship between O. J. Mulford and Charles Kettering that dated back to 1911 would have far-reaching results. In 1936, Kettering decided to bring Gray into GM's project to build high-speed, lightweight diesel engines.

In 1938, the new diesel — it would become famous as the "Jimmy" — was introduced by Gray Marine. The line included 3-, 4-, and 6-cylinder models of 85, 110, and 165 hp, with a single-cylinder model was slated to follow. All were rated at 2,000 rpm and all shared bore and stroke dimensions of 4½" × 5" inches to permit complete standardization of parts. A unit fuel injector was used for each cylinder. Gray developed a freshwater cooling system for the engine, an oil cooler and filter, intake air silencers, a governor, clean-out plates for the base, a flywheel pulley, and an instrument panel. A transmission was added and 1.5:1 or 3:1 reduction gearing could be specified. The new diesel joined Gray's impressive lineup of gasoline engines available in medium- and heavy-duty models, the Phantom Series High Speed engines and the 6-cylinder Fireball offered in 244-cubic-inch models with a variety of carburetor configurations and a custom-built 225-cubic-inch racing version.

After Pearl Harbor, mass production of the Gray Marine diesel began at a new Gray facility, and the company was kept busy throughout the war. With the end of the war in sight, Gray Marine executives understood the engine market would soon change. In a remarkably poor bit of decision-making, they elected to be bought not by GM, which was ready and willing, but instead by Continental. The postwar lineup included models rated at different rpm that were suitable for work boats, cruisers, sailing auxiliaries, runabouts, and for racing. The Lugger was the workboat and cruiser-oriented engine. The Lugger Six-330 (330 cubic inches) found favor with oyster schooners in Delaware Bay and salmon fishermen in Bristol Bay. The 4-cylinder Lugger (Four-162) found its way into vessels as diverse as Hinckley sailboats and the Lyman Islander. The familiar old prewar names were retained: Express, Phantom, Fireball (now based on a variety of V-8 blocks). Gray engines became standard equipment in a variety of stock boats. The Century Boat Company in Manistee, Michigan, became Gray's best customer for a time. Richardson, which Gray had supplied prewar, again became a large purchaser postwar. Chris-Craft also relied on Gray as one of its vendors.

The '60s saw a decline at Gray, with engine production finally being curtailed altogether. In May 1968 Gray Marine was sold by what was then Continental-Teledyne to a trio of investors headed by Coldwater, Michigan, businessman Charles Stokes and, in Berkeley, California, by Bruce A. Wilson of Star Marine Engine Works.

Gray advertised its popular Fireball widely. This ad appeared in 1947.

1906 Hartford ad

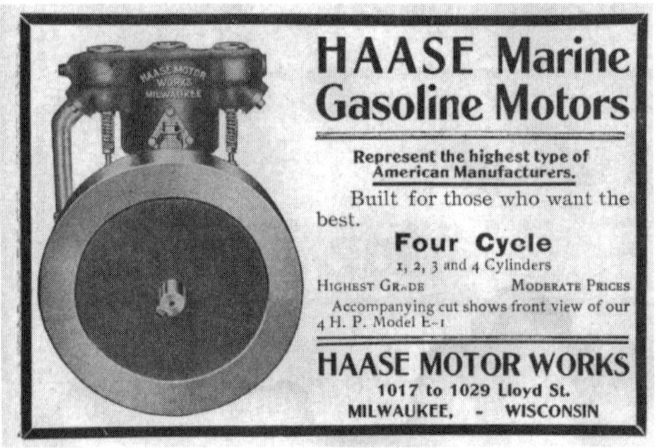

1905

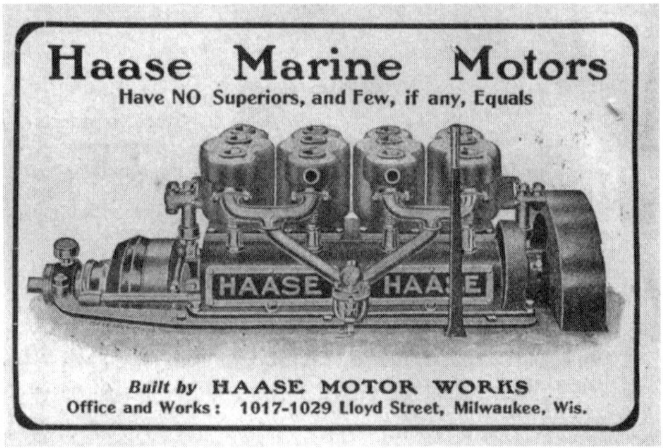

1906

Stokes had spent his career in the marine engine business and in 1944 was working in Coldwater at the old Regal company whose postwar plans did not include marine engines. So, Stokes left to found his Michigan Marine Motor Company, converting Hercules industrial engines into marine models. Soon, he added a line of bronze marine hardware to the company's offerings and then developed a reverse gear based on a Ford transmission. A popular mail-order catalog was begun, listing marine hardware and rebuilt engines including Grays, Chryslers, Kermaths, Scripps, Hall-Scotts, and Chris-Crafts.

The Gray Marine lineup for 1969 included four Continental-based 4-cylinder models: the 25-hp Sea Scout 91 (91 cubic inches), the 31-hp Four-112 (112 cubic inches), the 42-hp Four-162, and the 80-hp Four-80 162 (162 cubic inches). Production was begun on a 150-hp, 6-cylinder Gray based on an AMC block, and the famous Fireball V-8 (now rated at 280 hp at 4,400 rpm) continued based on a 401-cubic-inch Buick block. An inboard/outboard drive based on the AMC six was shown as were 4- and 6-cylinder diesels based on English Ford blocks. This ambitious lineup foundered when difficulties encountered in raising necessary capital ended the project. It was arranged for Teledyne to take back the inventory. Later, other companies bought components of the parts business, too, and today the sale of engine parts is all that remains of the company begun by O. J. Mulford and the Gray brothers in 1906.

See also Sintz Gas Engine Company

Gray & Prior Machine Co.
Hartford, CT
Trade name: Hartford (two-cycle), Gray-Prior (four-cycle)

In business from about 1904 until about 1930, the company produced the Hartford two-cycle engine and the Gray-Prior four-cycle. The commercial boatman, fisherman, oysterman, and lobsterman were the target buyers here as evidenced by an ad headline that read: "The Fisherman says: I prefer Hartford Marine Motors because they give real power all the time." The company always stressed the advantages of its make-and-break ignition on such engines as the Model X two-stroke for "use in open boats in salt water in all kinds of weather." In 1918 the firm added a heavy-duty four-cycle model to the line. This was a single-cylinder 8-hp engine with a 2½" diameter crankshaft. It came equipped with magneto and battery ignition, water pump, air pump for a whistle, and a bilge pump.

Gray-Hawley Mfg. Co.
Detroit, MI
Trade name: Skipper, Little Skipper

Gray-Hawley built a small line of low-priced — $39 to $94 — 2-, 4-, and 6-hp two-cycle engines from about 1908 to 1910 when the tooling was acquired by Monarch Tool Co.

See also Monarch Tool Co.

Green Bay Machine Co.
Green Bay, WI
Trade name: Sorge

Grimm Manufacturing Co.
Buffalo, NY
Trade name: Grimm

"The first real four-cycle marine engine within the reach of everyone," said Edward Grimm of his $250, 12- to 16-hp 4-cylinder engine in 1912. Thanks to the Grimm's aluminum base, the engines were comparatively lightweight. As the ads suggested, the engines were priced to appeal to the broadest market, $295 for a 25-hp 6-cylinder, $125 for an 8-hp twin. These side-valve engines had jump-spark ignition. Sometime in 1912, Grimm, who also manufactured woodworking machinery powered by an air-cooled one-lunger, decided the marine engine business might not be all he had hoped. He sold out to what became the Peerless Marine Motor Company.

See also Peerless Marine Motor Company

Gulowsen-Grei
Seattle, WA
Trade name: Gulowsen-Grei

Built in Christiana (Oslo), Norway, beginning circa 1903, these oil engines were also projected to be built in a factory on a six-acre site on Seattle's Salmon Bay, according to a Dec. 1918 press release. It does not appear that these heavy-duty engines were produced as planned.

Gunther-Wright Machine Co.
Owensboro, KY
Trade name: Rounds

H. A. Heinel
See Heinel, H. A.

H. C. Doman Co.
See also Doman, H. C. Co.; Universal Products Co.; *and* United States Motors Corporation

H. E. Dantzenbecker
See Dantzenbecker, H. E.

H. E. Hines Motor Co.
See Hines, H. E. Motor Co.

H. G. Trout Co.
See Trout, H. G. Co.

H. J. Leighton
See Leighton, H. J.

H.L.F. Trebert Engine Works
See Trebert, H.L.F. Engineering Works

Haase Motor Works
Milwaukee, WI
Trade name: Haase

Hall Brothers Gas Engine Works
Philadelphia, PA
Trade name: Hall Bros.

Hall Gas Engine Co.
Bridesburg, PA
Trade name: Hall

In 1909 Hall claimed victory in a number of important motorboat races including one from Bermuda to New York. The company's lineup included four-cycle models of 1 to 6 cylinders. It was a comprehensive line. In 1912 Hall's models were offered in both heavy- (7 to 85 hp) and light-duty (7 to 42 hp) models. The heavy-duty range had cylinders cast in pairs while the light-duty models had single cylinders. Ignition was make-and-break. Lubrication was force feed.

Hall Gas Engine Manufacturing Co.
Byesville, OH
Trade name: Hall

Born in 1860, Robert S. Hall became fascinated with internal combustion engines as a boy. As a young man, he studied every engine he could on Lake Michigan. Hall's own approach to engine design involved an effort to reduce or eliminate vibration. A brochure photo showed a pencil standing on the cylinder of a 25-hp model running at 600 rpm. Hall marine engines were horizontally opposed two- and four-cylinder machines that ranged from 3 to 50 hp. In addition to the low vibration resulting from the opposed configuration, Hall claimed — probably with some justification — that the engine offered additional advantages of a lower center of gravity, comparatively light weight, and space efficiency. The brochure also suggested that the horizontal configuration reduced the forces transmitted to the hull by comparison with a typical vertical cylinder.

These were L-head engines in which the cylinders were cast integrally with the valve chest. The company stressed the point that not a single gasket was used, and, while this was doubtless true, the long-term viability of the cylinder-to-crankcase joint, the end plates–to–crankcase joint, and the cylinder head–to–cylinder joint may well have been problematical. The engines were equipped with jump-spark ignition. Based on testimonials in the company's 1911 catalog, the first Halls were probably sold in 1909. It is unlikely that the company survived much past 1912.

Hall Gasoline Engine Co.
Portland, ME
Trade name: Hall

Hall was in business from 1902–1908. The nature of its engines remains unknown.

Hall-Scott Motor Car Co.
Berkeley, CA
Trade name: Hall-Scott

In an era when other companies were content to perfect the cast-iron, slow-turning marine engine and then manufacture it without further development, Elbert John Hall and Bert C. Scott created advanced engines of comparatively high performance and established one of the most progressive engine-building companies in the world. Born in Sacramento, California, in 1882, Elbert John Hall educated himself via correspondence and night-school courses and was part-owner of a San Francisco machine works by age 20. He gained experience with a variety of engines, including marine types and, in 1910, formed the Hall-Scott Motor Car Co. with Bert C. Scott. They produced aviation and gasoline locomotive engines.

In 1917 Hall collaborated with Packard's Jesse Vincent on the creation of the Liberty engine of World War I. With war's end and the decline in military production, Hall-Scott began adapting its experience to railcar and marine applications. In September 1919, the company unveiled its marine engine, "embodying the highest quality material and workmanship that made the Liberty Aircraft Engine unquestionably the present day premier aviation equipment."

To anyone familiar with the Liberty engine, the engine in Hall-Scott's photograph, the new LM-6, would have been immediately familiar. It had lightweight individual cylinders with welded-steel water jackets bolted to the aluminum alloy crankcase. An exposed overhead camshaft, driven by a gear on the crankshaft through a vertical shaft and bevel gears, operated the valves. As announced in 1919, the engine was rated at 200 horsepower at 1,750 rpm and weighed 988 lbs. The LM-6 was unveiled at the Motor Boat Show in New York in February 1920. The weight of the 6-cylinder had now increased to 1,290 lbs., as the steel cylinders had now been replaced with cast-iron units that could survive in salt water. The weight of the LM-6 would rise by several hundred more lbs. within a couple of years as a cover for the camshaft and other niceties were added. A 4-cylinder model, the 125-horsepower LM-4, was also offered. Still, the company did not give up on the steel cylinder version. It was designated the LM-6a. Intended for freshwater use only, the production "a" weighed 1,095 lbs and produced a mighty 250 horsepower.

By the early '20s, Hall-Scott had established itself as an important force in power boat racing. On the West Coast, an LM-6 powered *Miss Los Angeles*, a consistent race winner, but it was the John Hacker–built, 32-foot *Adieu*, owned by Chicagoan Webb Jay, that perhaps did most to publicize Hall-Scott. *Adieu* won the Fisher Trophy in 1921 and '22. This was a significant achievement, as Hall-Scott took every opportunity to point out.

Hall-Scott, meanwhile, introduced an en bloc engine, the HSM, in ratings up to 100 hp 1924. By 1926 Hall-Scott offered the LM with the choice of a sliding gear transmission with ball bearings and a multiple disc clutch, model "G," or a planetary gear, model "F." A "semi-supercharger" was added, too. It was based on vanes added to the enclosed flywheel that forced air into a plenum on the flywheel cover. From there, a pipe carried the air directly into the carburetor. The idea was to force air into the carburetor at greater than atmospheric pressure. More subtle changes that year to the LM included different gear ratios for water and oil pumps so they'd run at lower speed for improved durability. The semi-supercharger, together with revised intake manifolds and Zenith carburetors, improved performance as proven on the company dynamometer installed at the cost of $50,000 in 1921. The dyno made Hall-Scott the most sophisticated engine test lab on the Pacific coast.

Circa 1920 Hall-Scott LM-6a, 205 hp at 1,700 rpm and 1,600 lbs.

Circa 1930 Hall-Scott Invader

1902 Hercules

World War II era Hall-Scott 2268 V-12 as used in air-sea rescue boats

1906 Hercules

1913 Harbeck

1905 Hasbrouck

1915 Hermann

Hall-Scott engines did much to popularize and make more practical a wide variety of cabin cruisers, commuters, and speedboats. Said a company ad in *Motor Boating*: "Hall-Scott Marine Engines are especially suited for fast boats, because they are not too heavy and run dependably at any speed up to 1,800 R.P.M., giving perfect control and freedom from vibration throughout the entire range."

Engineer Charles Winslow, who had worked closely with E. J. Hall on the development of lubrication systems including the full-flow oil filter, called the LM-6 "the first really advanced, light-weight, powerful marine engine, which started fast pleasure boats being accepted by the general public."

Hall-Scott's collaboration with Winslow on oil filtration is worthy of note. According to Winslow, "The first oil filter that was ever built and applied directly to an engine was designed and built by [myself] in Vallejo, California, and applied to a Hall-Scott engine in Berkeley, California...We cooperatively designed full-pressure, full-flow filtering systems into a new engine, which was designed and built for Fageol Safety Coach and also another experimental engine for the Holt-built Caterpillar Tractors at Stockton." The Winslow Manufacturing Company produced a line of oil filters used widely and went on to produce oil and air filters, carburetors, and other devices well into the '50s.

Having established itself in the pleasure boat market, Hall-Scott moved aggressively on the commercial side, too. In the later '20s, Hall-Scott introduced a series of 100- and 175-horsepower marine engines with reduction gear drive for seine boats and tugs, which were intended to be viable alternatives to diesels in boats that operated fewer than 40,000 miles per year. "The fact that gasolene engines are easier to start and coupled with their durable qualities permitting them to operate over long periods without overhauling frequently causes them to be selected in preference to diesel equipment," claimed Hall-Scott. "The fact that the gasolene engines cost only about one-quarter as much as the diesel has been another factor to raise the demand and make an increased production necessary."

In 1936, however, the company introduced a diesel of its own, an overhead cam, 140-hp engine with mechanical fuel injection. It joined the Hall-Scott Fisher Jr., a 4-cylinder engine that could economically run on either gasoline or fuel oil. By the mid-'30s, thanks to Hall-Scott's reputation, it was reported that some fishermen actually were *forsaking the Hicks* for a Hall-Scott Fisher Jr. for use in crab boats.

All this experience with a variety of engines recognized for their high quality and performance, and Hall's widely publicized work on the Liberty in World War I, made Hall-Scott a logical choice to develop a new marine engine for naval use in fast air-sea rescue boats during World War II. The 6-cylinder Invader had been demonstrating its reliability and performance in Coast Guard picket boats and cutters since the early '30s. It was installed in a variety of landing craft and, after the war, was produced as the Hall-Scott 400. The company's response to the Navy's request was the new 2,200-cubic-inch V-12 Defender, considered by some to be the most outstanding marine engine of its class. More than 8,000 were built.

Both Elbert Hall and Bert Scott enjoyed successful careers, continuing with the company long after they sold it to the American Car and Foundry Co. in 1926. The Philadelphia maker of railcars, J. G. Brill, also owned a share of the company. Despite the advanced technology that Hall-Scott engines represented, the post–World War II era found the company unprepared to compete in a market that had forever changed. The big custom boats that had served as home to so many big Hall-Scott engines were now a rarity. They were replaced by smaller production vessels powered by marinized automotive blocks that included Hercules, Chris-Craft, and most of all, perhaps, Chrysler. Elbert John Hall, who had been granted the military rank of colonel and a Distinguished Service Medal for his role in the Liberty engine program, died in 1955. His old friend Bert Scott died in 1962.

Hallett Manufacturing Co.
Los Angeles, CA
Trade name: Hallett

Alfred Hallet, who died in 1944, claimed to have been involved with engines beginning in 1916. It was not until the late '20s, however, that Hallett advertised an engine that we know about. "Our motors are of the small bore high speed, four-cylinder type, which are justly superseding the heavier slow speed type heretofore used in marine work," stated the Hallett brochure. A photograph showed four small-bore pistons resting in a single hand. A reduction gear unit driven by silent chain was standard equipment. In 1928, Hallett offered its "Baby" in four 4-cylinder models. The L-4 and L-4-S (2 7/16" × 2 3/4") were rated at 8 to 10 hp at 2,000 rpm. The L-4 had Splitdorf ignition; the L-4-S, Auto-Lite. Both were splash lubricated. The M4 and M-4-S (3 3/16" × 4 3/8") developed 20 hp at 1,650 rpm. The M-4 had Splitdorf ignition; the M-4-S, Auto-Lite. These engines were pressure-lubricated. The potential market for such an engine was said to be yacht tenders, small boats, and as sailboat auxiliaries. Potentially, it might have had appeal to racers of small motor boats. However, Hallett did not reappear in *MotorBoat*'s 1929 listing of marine engine makers.

During World War II, Hallett produced a single-cylinder, four-cycle diesel engine (3 7/8" × 4 1/2") that developed 8 hp at 1,500 rpm. A 2-cylinder version was rated at 18 hp. These engines were offered in marine and generator versions. The company was then located in Irvine.

Hallin Engine Manuf. Co.
Tacoma, WA
Trade name: Hallin

Harriman Motor Works
South Glastonbury, CT
Trade name: Little Gem
Frank Harriman was an inventor and airplane designer whose efforts to build internal combustion engines for marine and stationary use date to at least 1909. Local history suggests he also attempted to build airplane engines and built an airplane that interested the Navy. However, Harriman lacked the business skills to turn any of these endeavors into a long-term success. The Little Gem was a four-cycle engine available in 1-, 2-, or 4-cylinder models producing 1½ to 35 hp.

Hardy, R. E. Co.
New York, NY
Trade name: Hardy

Harris Engine Co.
Wilmington, DE
Trade name: Harris
The company offered 10- to 225-hp models in 1918.

Harris Patent Co.
New York and Philadelphia
Trade name: Harris Valveless
The patent was for a two-stroke diesel in 1914.

Harry R. Geer Co.
See Geer, Harry R. Co.

Harvard Marine & Auto. Co.
Cambridge, MA
Trade name: The Harvard
By 1908 the company offered two model lines — the K (5″ × 5″) and L (4½″ × 4½″), 6- to 60-hp engines with jump-spark ignition. These were T-head engines with combined splash and pressure-feed lubrication. The one-piece base carried the reverse gear. The 4- and 6-cylinder models had a distributor and vibrating coil. The 2-cylinder had a commutator and double coil. Battery or magneto ignition was offered.

Harvey Marine Motor Co.
Rochester, NY
Trade name: Harvey Motors
In 1922, the company advertised a line of two-cycle engines including the 1-cylinder A (5 to 7 hp), the 2-cylinder B (10 to 16 hp), the 3-cylinder C (15 to 24 hp), the 4-cylinder D (20 to 22 hp), and the 6-cylinder E (30 to 48 hp). Each engine could be fitted to run on gasoline or kerosene, had Atwater-Kent ignition, and a Schebler carburetor. One must assume these engines had been developed by a previous company. Harvey's timing was poor for the two-cycle's day was largely over by 1922 and the company presumably achieved little or no success.

Hasbrouck Motor Works, Inc.
New York, NY
(later West Mystic, CT)
Trade name: Hasbrouck
In business by 1900 in Piermont-on-Hudson, Hasbrouck advertised its motor as "the most powerful and economical, and occupies less space than any gasoline engine on the market." These four-cycle engines had individual cylinders. The emphasis on fuel efficiency (less than ¾ of a pint per hour) was meant to distinguish the Hasbrouck from competing two-cycle engines. A 16-hp model for 1904 was rated at 1,000 rpm and weighed a competitive 500 lbs.

Hatch Oil Engine Co.
Brooklyn, NY
Trade name: N/A

Havana
Trade name: Havana
A listing appeared in *MotorBoat* in April 1924.

Hazard Motor Mfg. Co.
Rochester, NY
Trade name: Hazard
In 1911 Hazard advertised its "Celebrated Automobile Unit" for marine applications. While there seem to have been few takers, the engine itself was a modern-looking, fully enclosed four-cycle of 25 hp, probably a T-head.

Hebgen, A.G.
San Francisco, CA
Trade name: Liberty Kid
The company listed 1- and 2-cylinder models in the April 1921 issue of *MotorBoat*. The engine had a removable cylinder head with overhead valves.

Heer Engine Co.
Portsmouth, OH
Trade name: N/A
An ad for the company appeared in *Motor Boating* in Dec. 1911.

Hegeman Machine Works
Grand Rapids, MI
Trade name: Hegeman
In 1916 Hegeman listed nine models. They included three each of a 1-cylinder, 2-cylinder, and 3-cylinder. Ratings were from 3 to 18 hp at 700 rpm. Ignition was jump-spark. These Hegemans were lighter-duty motors priced at the low end of the market. A 3-hp Hegeman listed at $55 compared to $100 for a 3-hp, 190-lb. Hartford or $84 for a 210-lb. Hubbard.

Heinel, H. A.
Wilmington, DE
Trade name: Diamond
The Heinel side-valve engine was offered from 1905 through 1913. Models included 1, 2, and 4 cylinder.

Henry Keidel & Co.
Baltimore, MD
Trade name: Ardmore
See Keidel, Henry & Co.

Hercules Gas Engine Works
San Francisco (Alameda), CA
Trade name: Hercules
One of the very first San Francisco engine companies, Hercules began life having acquired rights to the engine first built by Palmer & Rey in the early 1890s. Hercules is believed to have been started in around 1895 with Jack Lorimer as factory superintendent, A. O. Warenskjold as design engineer, and Peter Barr as draftsman. By 1898, the company had its works at 405 Sansome St. Later, Hercules occupied a siz-

able building on Bay Street between Stockton and Powell Streets, a few blocks from the waterfront. In 1901, a Hercules ad claimed, "We have been building gasoline engines for nearly twenty years." That would suggest a starting date of 1881 or 1882 and was almost certainly an optimistic, dubious, and probably needless attempt to establish the company's credentials as an engine pioneer. The one man known to have been pioneering internal combustion in San Francisco at that early date was Daniel Regan. It's possible that Regan designed the first Palmer & Rey engine and Hercules wanted to capitalize on this work. Rhetoric aside, the company emerged for a time to gain considerable importance.

A 1904 ad revealed that Hercules offered heavy-duty 2½- to 500-hp engines rated at 400 rpm. The focus here was on big engines and commercial operators. Ernest W. Graef, an engineer who contributed regularly to *The Rudder*, wrote in May 1902, "The largest engines that are regularly on the market are the Hercules engines ... for business purposes. He noted that the engines were best adapted to heavily timbered sloops, schooners, and launches. "An Engine You Can Depend On" is the way Hercules described its products, and it boasted of the engine's low piston speeds. "All Hercules Engines are rated at moderate speed and can be safely worked up to considerably more power. Keep this in mind, because most engines are either overrated or will barely give the power claimed," said an ad.

After the earthquake in 1906, the company advertised that "Our plant was uninjured by the earthquake." If the earthquake didn't get Hercules, it is possible that labor problems or management issues did because key personnel left in 1904 and founded Atlas. Production continued through about 1909 when the company's "four-cycled engine" was available from 2½ to 500 hp. Some versions of the engines may have been built under license by Peerless Motor Company in Lansing, Michigan.

Hercules Motors Corp.
Canton, OH
Trade name: Hercules

This Hercules was founded in 1915 to produce engines for trucks and tractors. Eventually, Hercules blocks would be purchased for marine conversion by a variety of companies. Occasionally, Hercules itself advertised in the marine trade. In 1934, the company presented its DXI model, a 6-cylinder, 150-hp diesel equipped with a Joe's gear.

Hermann Engineering Co.
Detroit, MI
Trade name: Small Aristocrat

"Why buy a 2 cylinder 2 cycle motor when the SMALL ARISTOCRAT 9-12 H.P. 4 Cyl. 4 Cycle sells complete as shown for $150.00?" asked a Herrmann ad in 1915. The motor was equipped with a Kingston carburetor, a bronze water pump, and a reverse gear. A year later, the motor was advertised by a different company as the Aristocrat.

See also Carson Motor Co.

Herschell-Spillman Co.
N. Tonawanda, NY
Trade name: Herschell-Spillman

This company grew out of the Armitage-Herschell Company, established in 1876 to make amusement park carousels. After Allan Herschell married into the Spillman family, he and brother-in-law Edward Spillman — then involved in gasoline engine developments — formed Herschell-Spillman. Engines were adapted to power the carousels and a plan was made to build an automobile in addition to merry-go-rounds. The car-making idea resulted in production of only a few vehicles by 1904; the idea of building automobiles was then replaced by a plan to manufacture gasoline engines. These 4- and 6-cylinder engines, and even a V-8, were used by companies as various as the Fulton Motor Truck Company on Long Island, and car builders Daniels in Reading, Pennsylvania, and duPont in Wilmington, Delaware, among many others.

Located on the Niagara River just north of Buffalo, a hotbed of motor boating, Herschell-Spillman soon attempted to attract sales to the marine trade and advertised in *Motor Boat* in July 1908. They exhibited that year at the Buffalo Power Boat and Sportsmen's Show and had a company racing boat, the Herschell-Spillman. The company's marine engine efforts do not seem to have achieved anything like the success of that garnered in other markets. During World War I, the company contributed to Liberty engine manufacture, and engine production continued after the war under the auspices of Spillman Motors, which declared bankruptcy in 1924. Carousel manufacture continued under Spillman Engineering and later the Allan Herschell Co., and the rides were built into the 1970s.

Hess Motor Co.
Detroit, MI
Trade name: Hess Mono-Marine

Hess Motor Company was founded in Detroit in 1920 to produce a Model T–based single-cylinder marine engine. Why it sold out the following year is unknown.

See also Algonac Machine & Motor Works

Hess Motors Corp.
Algonac, MI
Trade Name: Hess Propel-Oro

Joseph DeBruyn incorporated the Hess Motors Corporation in Algonac in 1924. DeBruyn then acquired the motors developed by Gierholtt. Two-hp inboard and outboard versions were sold, and, as if to keep future collectors guessing, two names were applied: Hess Propel-Oro and Hess Baby Marine engine. Production is believed to have ceased after 1928.

See also Gierholtt Machinery Co. and Algonac Machine & Boat Works

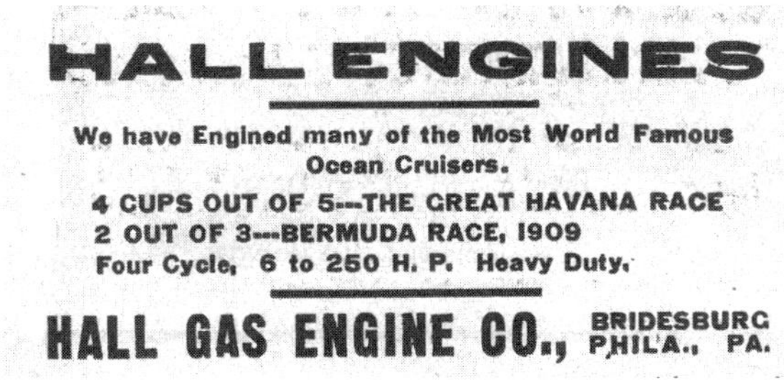

1913

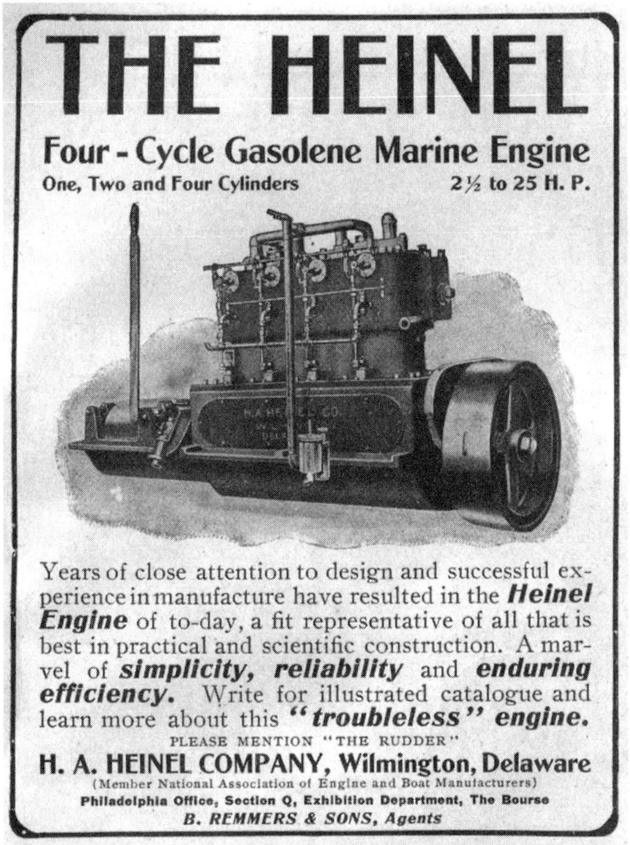

1906

1913

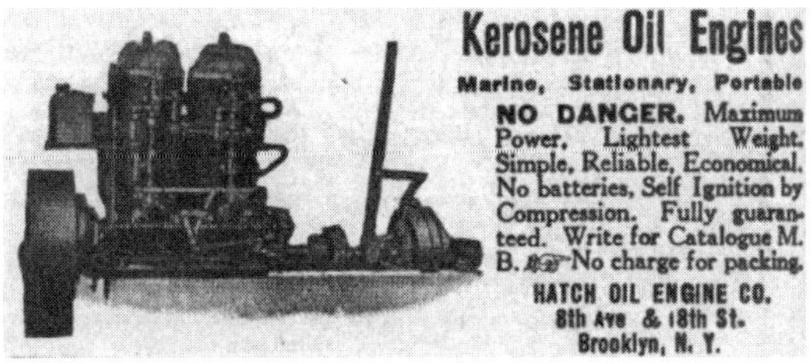

1913

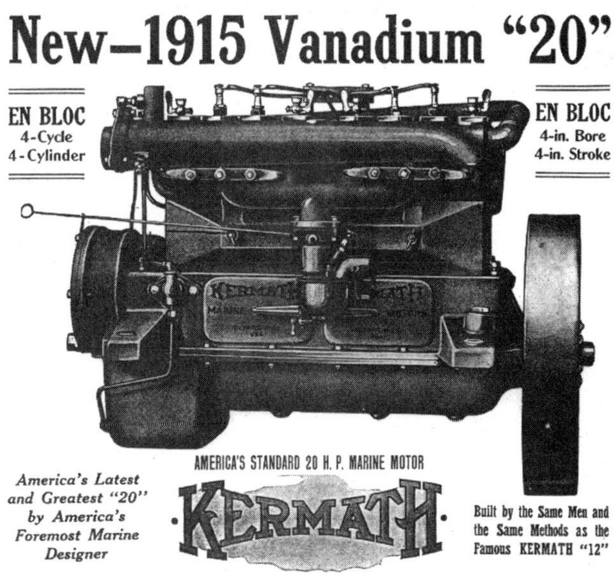

1915 ad for the new, more powerful Kermath "20"

Hettinger Engine Co.
Bridgeton, NJ
Trade name: Hettinger
Two-cycle motors were initially produced, but these were replaced by four-cycle engines. By 1916, eight Hettingers were listed, ranging from a 6-hp single to a 36-hp four, a hefty 7″ × 9″ engine that weighed 4,000 lbs. Atwater-Kent high-tension and Hettinger low-tension ignition systems were offered. Paragon and Hettinger reverse gears were used depending on model. The carburetor was a Schebler.

Hicks, J. L.
San Francisco, CA
Trade name: Hicks
Hicks was begun in San Francisco by James L. Hicks in 1889 to import and sell sewing machines. The focus changed to engines by 1892 when the company was listed as J. L. Hicks Machine Tools and Gas Engines. Single-cylinder Hicks engines were available in 6-, 8-, and 10-horsepower ratings. Of these, the 8 hp was the most popular. The six lacked power for work outside San Francisco Bay and was sometimes called the "Delta Hicks" while the 10 hp was found to produce excessive vibration.

In 1900 the company name was changed to J. L. H. Machine Tools and Gas Engines. It was sold around 1920 and renamed the Hicks Iron Works. By 1927, when the then-struggling Hicks firm was acquired by the Yuba Manufacturing Company, the sales brochure claimed, "Over 80% of the fishing boats using 6- to 35-hp engines in California waters are equipped with Hicks engines.... Hicks engines are designed for fishing boat use. They have the valve-in-head mechanism. Manufactured at Benicia, California, where river, bay and ocean fishing is a big business and the needs of fishermen are understood." Yuba expanded the model line to include 2-, 3-, and 4-cylinder models and abandoned the old factory on San Francisco's Howard Street for a plant in Benicia where Yuba also produced gold dredges and crawler tractors. Production ceased in 1932. A Hicks engine today is a sought-after collectible that evokes a long-gone era in West Coast fishing.

Hildreth Mfg. Co.
Lansing, MI
Trade name: Hildreth
"They make the round trip," claimed a Hildreth ad in 1906. What set this two-cycle motor apart from the then-typical two-port designs was its intake system. A float-feed carburetor fed mixture into the front of the crankcase where it was admitted to the cylinder through a disc on the crankshaft that served as the intake port. N. E. Hildreth was awarded a patent for the motor, which was available in 1- and 2-cylinder models of from 2 to 22 hp. Jump-spark ignition was used. The company seems to have been in business from 1905 or '06 to 1910, but sales volume was presumably low.

Hill Diesel Engine Co.
Lansing, MI
Trade name: Hill Diesel
Hill was a developer of high-speed marine diesels. Its PB engines of 1928 ranged from 20 to 75 hp at a speed of 800 to 1,000 rpm. A line of heavier-duty PA engines (600 to 800 rpm) was available. So was a 550 rpm "V." The 60- to 80-hp PA and the 90- to 120-hp PA were started by compressed air, the other models by air or electric. A single-cylinder (4″ × 4¾″) Model "V" was offered as well, with hand or electric starting. It was rated at 6 to 8 hp at 550 rpm and weighed 1,450 lbs. By 1936, the line had evolved. A 1,600 rpm, 200-hp eight had been added; the single had increased in power to 12 hp (5″ × 7″ at 800 rpm), and a compact, opposed piston engine "designed by R. E. Olds" was introduced. This 10 to 15 hp was likely the smallest (2½″ × 3½″) marine diesel built until well into the postwar era. Its acceptance as a marine engine is unknown.

Hill, Robert S. Co.
Detroit, MI
Trade name: Victor
Beginning in 1907, Hill specialized in little two-cycle, one-lunger engines of 1½ (3″ × 2½″) and 3 hp (4″ × 3½″). In this especially price-sensitive end of the market, Hill offered the motors "bare," with electrical, and with "boat fittings." As was typical, customers choosing the latter could get bronze "saltwater" fittings as a final option. Thus, prices for the 1½-hp Victor could range from a base $32.75 to $54.50. A 1909 ad claimed the Victor was the "only engine built with water-cooled bearings." Within the span of only a couple of years, Hill's published addresses included 78 E. Fort St., 23 East Woodbridge St., and 54 Woodbridge East.

Hines, H. E. Motor Co.
Traverse City, MI
Trade name: Hines
Owned by Harry E. Hines, the company was in business from approximately 1909 to 1913. Ads appeared in *Motor Boating* in Dec. 1912 and Dec. 1913.

Hinkley Motor Co.
Detroit, MI
Trade name: Hinkley
In Feb. 1924 Hinkley advertised six models, all with 4 cylinders. They were rated at 37½ to 81 hp. Hinkley claimed that every engine was test run for 12 hours, torn down for inspection, and then reassembled. It is difficult to imagine that such a costly process would have contributed to long-term viability in the marketplace. The company closed in 1926.

Hitchcock Gas Engine Co.
Bridgeport, CT
Trade name: Hitchcock
This company was established by Frank I. Hitchcock in 1905 at 130 Seaview Ave. Hitchcock claimed he entered the engine business in 1893 but didn't elaborate. Details of the first engines, presumably two-cycle, are currently unknown. On April 10, 1911, *MotorBoat* printed a nine-paragraph write-up on the "new" Hitchcock, an engine said to have an 18-year pedigree and designed "especially for cabin cruisers and run-

1927 Yuba-built Hicks make-and-break twin. The Hicks 2-cylinder could be had in ratings of from 10 to 22 hp. The engine weighed between 1,830 and 2,400 lbs.

1915 Hoosier

1915 Hubbard

1915 Holmes

This "first-generation" Hicks shows the typical round cylinder head with overhead valves. It may date to a period circa 1908.

1905 Hubbard

1932 Hill Diesel

abouts." That said, it is difficult to make out precisely what this engine really was. It had a 6″ bore and 7½″ stroke, and was available with either make-and-break or jump-spark ignition. To cite one example of the confusing copy relating to the engine: "There are no camshafts used ... each cylinder has its own two-to-one cam gear." Confusion aside, the Hitchcock valves seem to have been operated by gear-driven wheels rather than camshaft lobes, and by 1913 the engine was adapted to run on either gasoline, naphtha, or kerosene. One- to 5-cylinder models were offered, producing from 8 to 100 hp. By 1913, the engine's fuel efficiency prompted the company to tout its desirability for fish and workboats as well as pleasure craft. Hitchcock sold the company to John R. Brown, Sr., in 1920; it remained in the Brown family for four generations. One retrospective says diesel engines were produced, but details are lacking about these and about when engine-making ceased.

Hoch Bros.
Grand Rapids, MI
Trade name: All-Go
This small shop produced marine engines during, it is believed, the years 1910 to 1915.

Hoffman Launch & Motor Wks.
Chicago, IL
Trade name: Hoffman
Also appearing as simply the Hoffman Motor Works, this company was in production of a 2-hp, four-cycle launch engine by 1903. Like some other four-cycles of the period, the Hoffman had an automatic inlet valve. It was mounted in a valve chest on one side of the cylinder. The chest contained a mechanically operated exhaust valve driven by a camshaft that was itself driven by a gear wheel off the crankshaft.

Holderman Machine Co.
Toledo, OH
Trade name: Holderman

Holland Launch and Engine Co.
Holland, MI
Trade name: N/A
The company appeared in business listings from 1908 to 1914 when, in July, it was heavily damaged by fire. H. G. Garvelink was the company proprietor; he reported that seven launches were lost in the blaze. Details of the engine have thus far not surfaced.

Holliday Engineering Co.
Chicago, IL
Trade name: Holliday
A line of 1-, 2-, and 4-cylinder, four-cycle engines was offered in ratings up to 40 hp in 1911. Hollidays had the usual individual cylinders, jump-spark ignition, and a Schebler carburetor. They were no-nonsense "semi heavy-duty" engines intended for commercial boats and cruisers and possessed oversize bearings and pressure lubrication, features that meant something to Holliday customers. In 1912, it was reported that Holliday, which was located at 749 Bunker Street, had made a special effort to court overseas buyers and the export department had become a great success.

Holmes-Howard Motor Co.
Detroit, MI
Trade name: Holmes-Howard
Holmes offered nice-looking en bloc two-cycle models painted with gray enamel by 1914; later, four-cycle.

Holmes Motor Co.
West Mystic, CT
Trade name: Holmes
The "'Get-at-able' MARINE ENGINE," the open-base Holmes, apparently found some market with the United States Life Saving Service. That engine was sometimes called The Lifeboat Special. By 1910, however, Holmes was offering a handsome L-head engine with cover plates and cylinders cast in en bloc pairs. Both make-and-break and jump-spark ignition were offered. A Holmes was displayed at a 1911 boat shows alongside a 36′ "Government mahogany power lifeboat." In 1914 Holmes was issued a patent for the "Holmes-McClellan Life Boat Cruiser," a double-ender with the motor located in the stern. Company ads featured the 35- to 40-hp 6-cylinder K-6 as the "Life Boat Special."

Hoosier Motor Co.
Goshen, IN
Trade name: Hoosier

Houle, A. J. Motor Works
Holyoke, MA
Trade name: Houle
Located on the Connecticut River in central Massachusetts, Holyoke was home to machinist Arthur J. Houle, who claimed to have 15 years of experience in 1895. It is not clear when he produced his first engines. In 1907 Houle was advertising castings and drawings, but he also produced a line of two-cycle motors with several very service-oriented features. They were delivered with thin copper shims on both the upper and lower connecting rod bushings. These shims could be removed one by one as needed to take up wear. The main shaft journals were also easily adjusted by the owner for wear. It is now unclear what Houle's role was in the Motor Works, which emerged in 1910 under the proprietorship of A. H. Burgess. Presumably, Houle was the hands-on engine man. The 1910 model line featured enlarged intake and exhaust ports for improved performance. A company ad also claimed four-cycle engines were available although there is no evidence to support this.

Howard Engineering Works
New York, NY
Trade name: Howard
This company was presumably the successor to W. S. Howard. The 1911 Howard was a fine-looking 6-cylinder motor with the cylinders cast en bloc in pairs of three. The 4″ × 6″ T-head engine was said to be capable of 1,800 rpm. The plant was in Harrison, NJ.

Howard Motor Co.
Philadelphia, PA
Trade name: Howard

Howard, W. S. Co.
New York (Yonkers), NY
Trade name: Howard
After leaving Grant-Ferris, probably in 1903, William S. Howard got a job as chief of the gasoline engine department at the Gas Engine & Power Co./Charles L. Seabury. He left that position to start his own company. Howard's ads appeared in 1905. The engines were available either as kits or complete.
See also Grant-Ferris *and* Howard Engineering Works

Hub Motor Co.
Boston, MA
Trade name: Hub
A jump-spark engine was produced by Hub.

Hubbard Motor Co.
Middletown, CT
Trade name: Hubbard
The invention of H. W. Hubbard in 1902, this two-cycle engine was a heavy-duty (400 or 450 rpm) design available in either 1- or 2-cylinder models with removable heads. The 15-hp double, Hubbard's most powerful engine, had a 6″ bore and 6½″ stroke. Sight-feed oilers were still in evidence in 1914 on all Hubbards except the 1½-hp model. Hubbard offered customers a choice of three ignition methods. The one-lungers seem to have been make-and-break motors. The double was available with a high-tension magneto, make-and-break, or battery and coil. One version had the juice provided by a generator driven off the flywheel. A low-tension Sumter magneto system was also shown.

Hulse
Long Island, NY
Trade name: Hulse
Hulse produced only four or five engines, but one of them powered the Shelter Island Ferry for many years before it was replaced after World War II by a GMC diesel.

Humboldt Gas Engine Co.
San Francisco, CA
Trade name: Humboldt
These engines were made at the Orion Machine Works in Oakland.

Humphries, F. J.
Skaneateles, NY
Trade name: Gosper

Hunter, Everett Boat Co.
McHenry, IL
Trade name: Hunter
The Hunter was offering 10- and 12-hp models after World War I.

Hurd & Haggin
New York, NY
Trade name: H & H
This company was in production of an open-base, side-valve engine in 1906 at 316 Hudson St. "Motors designed to run every day in the year," read one ad. The engines had cast, one-piece cylinders and were available on 1- to 6-cylinder models.

Huron Motor Works
Detroit, MI
Trade name: Huron

Hydrocarbon Launch Co.
Although it didn't list where this company had been located, *The Gas Engine* reported in July 1911 that the firm had built a kerosene marine engine in ratings of ¾ to 12 hp and that the engines were available in the company's own launches that were built in models up to 50 feet in length.

Imperial Gas Engine Co.
San Francisco, CA
Trade name: Imperial
Here was a fine example of the heavy-duty gasoline engine one associates with the California makers. The factory was situated between Clementina and Tehama streets, a few blocks from San Francisco's waterfront. Production commenced in 1906 soon after the earthquake with a line of overhead camshaft 1- to 4-cylinder engines. "Well Designed and Well Made by Experienced People" read one of the ads. Of course, Imperials were equipped with a marvelously intricate make-and-break ignition system typical of California gasoline engines. An auxiliary inlet valve improved low-speed operation. By 1916, with gasoline prices having risen to 20 cents a gallon while fuel oil cost just a nickel, Imperial allied itself with its across-the-bay competitor in Oakland, Atlas, which was hard at work on a diesel engine.
See also Atlas-Imperial

Ingram-Hatch Motor Corp.
New York, NY
Trade name: Hatch
The company was producing kerosene marine and stationary engines in the 'teens.

International Gas Engine Co.
New York, NY
Trade name: International
The company's ad in *The Rudder* in July 1901 claimed the engines were available in 6- to 50-hp models. It was not uncommon at the time for engines to have a wine-bottle–shaped, open crankcase similar to that of a steam engine, and the International was among them.

International Mfg. Co.
Detroit, MI
Trade name: International-16
The 4-cylinder International was another marine engine based on the Ford Model T. In this case, a purpose-built block, which carried the reverse gear, housed the Model T vanadium-steel crankshaft and the usual Ford rods and pistons, giving the familiar 3¾″ bore and 4″ stroke. "Interchangeable Ford parts obtainable almost anywhere in the civilized world" is the way International touted the engine. The International had a custom, water-jacketed intake/exhaust manifold that warmed the mixture and was said to permit operation on kerosene. In 1923, this 4-cylinder, 10- to 18-hp motor sold for a competitive $365, a price point significantly higher than a marinized Model T engine like Southern's con-

version, which listed at $225 including reverse gear. The International was probably worth the added money. The Universal HM-2, a purpose-built, 2-cylinder marine engine rated at 12 to 15 hp, had a base price of $675 including a Paragon gear. International was located at 439 East Fort St. in Detroit and had as many as eight sales agents in strategically located parts of the country and in Toronto.

International Motor Co.
Detroit, MI
Trade name: Claus
International claimed in a 1908 ad that its two-cycle engine would better the fuel efficiency of the best four-cycle. Four- to 30-hp models were said to be offered. The plant was located at 868 Military Ave. Like so many engine companies, this one had a short life span. It was out of business as of May 1910.

International Power Vehicle Co.
Stamford, CT
Trade name: International
The company built marine, stationary, and portable kerosene-fueled engines. It was in business by 1905.

Iowa Marine Engine & Launch Works
Bellevue, IA
Trade name: Iowa Marine Engine
This company was begun in 1905 by 34-year-old Joseph Martin Brandt who first went to work at age 12 in a Dubuque buggy-top factory before moving on to gain experience in the manufacture of brass goods and pumps. In his late twenties, Brandt built a prototype two-cycle marine engine. In 1898 Joe moved to Bellevue and he began Iowa Marine Engine & Launch Works in 1899. The company's engines soon gained a reputation for durability and speed. A series of racing launches named *Red Top* took top honors in the competitive Mississippi Valley Power Boat Association races at Peoria. Brandt must have been a skilled engine builder: *Red Top II* and *III* were both powered by a successful 8-cylinder, 240-hp version of his original one-lunger. Brandt built his marine engines and boats until 1925 and then focused on automobile repair.

Isham Co.
Norwich, CT
Trade name: Thames
Isham noted that the Thames was "built without connecting rod" and claimed that single- and multicylinder models were available in 1905.

Ives Mfg. Co.
Baltimore, MD
Trade name: Farro, Ives

J. A. Holmes
Philadelphia, PA
Trade name: Facile
The Facile was a two-cycle oil engine, probably a hot-bulb-type, built to a British patent sometime before 1910.

J. F. Garrick
See Garrick, J. F.

J. J. Parker Co.
See Parker, J. J. Co.; Dutcher Machine Co.; *and* Emeny, G. J.

J.V.B. Engine Co.
Akron, OH.
Trade name: J.V.B.
The J.V.B. was a 425-cubic-inch engine developed by Joe Van Blerck and introduced in 1920 by the Wellman-Seaver-Morgan company, which did the actual manufacture in its Akron shop. It was an overhead-valve, mono-bloc design with removable cylinder liners, pressure lubrication, and detachable head. A positive-acting, foot-clutch-operated transmission was included. In 1923, an ad appeared under the J.V.B. Engine Co. logo, listing Joe Van Blerck as president. This, presumably, was a quickly passing phase in Van Blerck's confusing professional career.
See also Wellman-Seaver-Morgan Company and photo page 221

J. W. Lathrop & Company
See Lathrop, J. W. & Company

Jackson Co.
Trade name: Jackson

Jacobsen Gas Engine Co.
Sarasota Springs, NY
Trade name: Jacobsen

Jager (Jaeger) Chas. J. Co.
Boston, MA
Trade name: Mason, Mason-Jager
Charles J. Jager (the anglicized spelling) was awarded a patent in 1904 and was publishing testimonials shortly thereafter. Lightweight two-cycles ("yacht tender" engines) and L-head, four-cycle models were offered in 3- to 200-hp models. The heavy-duty 450- to 650-rpm Jager models were the core of Jager's business although a 25-hp, 1,200-rpm model had joined the line by 1915. The Jager had individual cylinders; the crankcase featured impressive access to bearings and the engines had elaborately conceived lubrication systems. The intake manifold was heated by coolant and the exhaust manifold had a cold-water jacket. Low-tension Bosch treble-ignition was standard. By 1906 a Jager was in use by the U.S. Navy, powering a 28-foot whaleboat. In 1908 Jager advertised that the Navy was using its engines in "Revenue Cutter Service" and that the Department of Commerce and Labor had selected Jagers for "special service in the Arctic." In 1910, Jager's display at the Boston Boat Show was notable for the showcase of trophies won by Jager-powered boats and the highly polished nickel hardware on the engines themselves. A year later, the company was stressing the popularity of the engines with fishermen in the Bay of Fundy who appreciated the ability to start on gasoline and switch to kerosene. In some cases, they also simply heated the "generator" with a "lamp" and both started and ran on kerosene. Like many companies of the time, Jager did not build its reverse gear to withstand the engine's full power, a requirement of the U.S. Navy. That changed in 1911 when the reverse gears and clutches were upgraded with "heavier and stronger working parts." Whatever

1907 Imperial (This is for Imperial Gas Engine Co., San Francisco)

1911 Jencick

Circa 1920s Kahlenberg 60-70 hp two-cycle oil engine

1905 Jager

1915 Jager

Circa 1910 Kahlenberg two-cycle

the virtues of its engines, Jager seems to have ceased production before or at the end of World War I.

James Craig Engine and Machine Works

See Craig, James Engine and Machine Works

Jencick Motor Co.

Port Chester, NY

Trade name: Jencick

Also known as the Jencick Motor Manufacturing Co. and Jencick Motor Corp., this firm introduced its wares at the New York Boat Show in 1908. The Jencick was the namesake of Stephen Jencick who had learned his trade as a youth in the shops of two European pioneers, Daimler and DeDion. Jencick's own motor was positioned as a "light weight, high speed [motor] of the highest grade." Initially, a line of 4- and 6-cylinder models of up to 80 hp was offered. They were based on individual cylinders mounted to a manganese bronze base. By 1911 a mighty 210 hp at 900 rpm, 7½" × 7½" 8-cylinder was available. A 12-cylinder appeared by 1914.

These four-cycle engines had a number of interesting features. Jencick was an early adopter of a one-piece camshaft; the shaft and lobes turned from a single forging, rather than the more typical cam of that era in which separate lobes were mechanically attached to the shaft. What's more, the camshaft rode not in plain bearings but in ball bearings for minimized wear and drag. To reduce weight, the crankshaft was "hollowed to the various bearings." Dual ignition was used, a low-tension make-and-break system and a secondary high-tension system with coil and distributor. The exhaust manifold was water jacketed. A carburetor of Jencick's own make was used. Air starting was standard, the compressor mounted right on the engine.

Impressive as its engines were, Jencick seems not to have made the required step of shifting from individual cylinders to an en bloc design. Van Blerck, to cite one other pioneering maker of high-grade engines, made the transition in 1914. Jencick appears to have ceased operation in 1915.

John V. Rice, Jr. & Co.

See Rice, John V., Jr.

John Stuart Co.

See Stuart, John, Co.

Johnson & Wells Mach. Co.

Neenah, WI

Trade name: Johnson

One-, 2-, 3-, and 4-cylinder models were offered by 1907.

Johnson Brothers Motor Co.

Terre Haute, IN

Trade name: Johnson

The sons of Danish immigrants, Lou, Harry, and Julius Johnson began their first company in a garage workshop and by 1908 had moved to a one-story brick building on the outskirts of Terre Haute. There, they built two-cycle, V-type marine and airplane engines in a variety of models before a tornado leveled the operation in 1913. By November 1915, the Johnsons were again listing their engines including a 50-hp V-4, a 75-hp V-6, a 100-hp V-8, and a 150-hp V-12. These motors were said to be the company's 1916 lineup. Production, if any, must have been limited.

See also Johnson Motor Co.

Johnson Motor Co.

South Bend, IN

Trade name: Johnson

In 1917 the Johnson brothers moved to South Bend and began the Johnson Motor Wheel Co., producing air-cooled, aluminum bicycle engines. A recession led to the company's demise, and the brothers, joined by brother-in-law Warren Conover, formed Johnson Motor Co. in 1921. The goal was to produce small marine engines that would appeal to the growing recreational boating market. Johnson built the first of 100 production outboards on Dec. 19, 1921. The Model E 2-hp inboard appeared in 1922, using the same 2-cylinder opposed powerhead as the 2-hp outboard. The Model E was a vertical-mount engine that powered a lower unit. The first of the outboards, restored under the direction of Warren Conover, is on display at the Antique Boat Museum in Clayton, NY.

See also Johnson Brothers Motor Co.

Johnson Motor Products

New York, NY

Trade name: Globe

This Globe was no relation to the famous Globes of the Pennsylvania Iron Works. The company was advertising in 1926.

Johnson, Carlisle, Machine Co.

Manchester, CT

Trade name: Bud-E

Jones, S. M. Co.

Toledo, OH

Trade name: Ralaco

S. M. Jones, under the direction of "Percy" Jones, became the successor to Rathbun-Lacy circa 1907. The Ralacos were heavy-duty engines — ratings of from 375 rpm to 600 rpm — known for clean lines and fine finish. One reason for the engine's clean looks was that the "inlet, exhaust, and water manifold piping has all been taken care of by passages within the housing." In 1912 the line included 2- and 4-cylinder models of 10 to 50 hp. The big (7" × 9") 2- and 4-cylinder engines had make-and-break ignition. The other models (4" × 6" or 5" × 7") were jump-spark. The engines had individually cast cylinders and overhead valves and relied on "perfected" splash lubrication. Holley carburetors were fitted. All this added up to a durable machine — the "silent, simplified engine" as the company was wont to say. In 1911 the company claimed the "world's record" for a 618-mile voyage from Ketchikan to Vancouver at 10.65 mpg. Ralacos were known for what one boating writer of the day called "their symmetry of line and beauty of exterior finish."

Joymotor Mfg. Co.
Chicago, IL
Trade name: Joymotor
This was a 2-hp engine.

Jule Motor Co.
Syracuse, NY
Trade name: Jule

Kahlenberg Bros. Co.
Two Rivers, WI
Trade name: Kahlenberg
In 1895 William and Otto Kahlenberg began repairing steam engines, working first in the backroom of their father's hardware store and then in a machine shop. They installed their first gasoline marine engine in a 35-foot, Mackinaw-type fishing boat in 1898. Its success became the basis for the company, and a line of two-cycle models followed. In a 1924 retrospective on the company, Will Kahlenberg pointed with pride to the fact that "from the time of building the first engine, parts were always made with jigs and templates and therefore standardized and interchangeable; an owner ordering from some remote corner of the earth or at home was always sure of a part that would fit."

Kahlenberg's engine was a husky, single-cylinder two-cycle with a make-and-break ignition system. It was a two-port engine. That is, the fuel/air mixture entered into the engine's base via a check valve, was compressed by the descending piston and forced through a transfer port to the cylinder. It probably developed peak power at around 400 rpm. With what proved to be great foresight, Kahlenberg equipped this first engine with a control lever that permitted the engine to be stopped and reversed, and it also adjusted the timing, which was important to speed control on a two-port engine.

Kahlenbergs soon became popular for tow boats and yachts as well as fishing vessels. On the Great Lakes, in those early days and for years afterward, people called the engines not a Kahlenberg but a "Two Rivers." Many boats had one Kahlenberg driving the boat and a smaller, single-cylinder rigged to operate the hoisting drum. By 1907, a fine new brick and concrete factory had been built.

Will Kahlenberg began experimenting with a single-cylinder, heavy-duty oil engine in 1914. This engine, which was started on gasoline but then switched over to run on fuel oil, powered factory machinery for a full year before a marine model was built. It was sold to a pair of Two Rivers fishermen in 1916. Although the Kahlenbergs always referred to the new engine as a "semi-diesel" or "heavy-duty crude oil engine," there was the typical reference in the lay press to the engine as being a *diesel*.

The 1916 catalog introduction to the engine made clear how the machine operated. "All engines of this type are started in the usual manner on either gasoline or distillate, and after the machine has been running two or three minutes, the fuel pumps are turned on and the engine then operates on the semi-Diesel principle — taking pure air only in the base, compressing it, and just before the piston reaches the top of its stroke, the fuel oil or kerosene is injected directly into the combustion chamber by the small injection pumps, one for each cylinder, timed exactly in rotation with the engine."

All these Kahlenbergs, whether gasoline or semi-diesel, were two-cycle engines. They remained so in 1922 when a new generation of engines was introduced. These were heavy-duty, crude-oil engines, "The New Kahlenberg Oil Engine." It was a hot-bulb engine with mechanically injected fuel. Gone was the previous model's ignition system. It was replaced with a torch that heated the hot tube — "starting tube" — within two or three minutes. Then, said Kahlenberg, turn on the air and fuel, "Pull the air throttle and the engine is agoing." The flames of the torches that were built into the cylinder heads to heat the starting tubes were ducted outside the engine compartment by a steel manifold. The new engine was available in 2-, 3-, and 4-cylinder models of 20 to 200 horsepower, and the catalog showed the line drawing of a Pacific Coast purse seiner with a 3-cylinder Kahlenberg in the engine room.

It was company practice to build each engine to a very high standard and then test it before shipment. Those who witnessed these tests reported that no engine left the factory until it was correct in every aspect. Sometimes, even after an engine had passed all its tests, Will Kahlenberg might give it his own personal tune-up. Said one owner: "This personal final inspection and regard for the individuality of the engine is refreshing in these days of quantity production, where parts are rushed through operations and inspections, and assembled and shot into the market with all possible speed."

The Kahlenberg brothers were early advertisers in *MotorBoat* magazine, and by the mid-'20s their ads had helped to generate an export business. There were some years, in fact, that company files revealed higher export than domestic sales. On the Great Lakes it was said, by the mid-'20s, that there were more fishing boats powered by Kahlenberg than with all other makes combined. Demand was such that the factory was doubled in size in 1924, its machinery driven, of course, by one of the new semi-diesels. The economy and easy operation of these Kahlenbergs made the earlier, gasoline models just about obsolete. The company, of course, still had the tooling for those engines. For a number of years after the new model semi-diesel was introduced, Kahlenberg catalogs retained an offer to build the gasoline engine for any customer requesting one. At least as of 1924, a few such special orders were received.

Like its predecessors, the new models were robust. Kahlenberg was known locally for being a "crank about crankshafts," and the dimensions of the crankshaft journals and of the main bearings were impressive. A 4-cylinder, 100-horsepower Kahlenberg had end bearings that were 10 inches long, the bearing shells cast of bronze with a layer of white metal lining that was bored to an exact fit. These 10-inch-diameter bearings weighed 17 lbs. apiece. The company liked to point out to customers that mass-produced engines — and one can pretty well assume they were referring to Fairbanks-Morse — had iron-shell main bearings with poured-in babbitt. Total main-bearing surface on the 4-cylinder was, the company calculated, more than 4½ *square feet*. That's probably why these bearings, although replaceable in less than a half-hour, so seldom needed replacement. Stories abounded of their long life. One owner ran his Kahlenberg for 16 years without a bearing replacement; another engine was run 21,000 hours on its original bearings.

Export business continued right up until World War II, the last shipment to the South Pacific involving engines destined for fishing boats in Dutch Indonesia. During World War II, the plant's entire output, produced by 100 workers, was devoted to the war effort. On September 14, 1944, Kahlenberg was awarded the Army-Navy Production Award and could then fly the special "E" award flag at the Two Rivers factory.

Like every other marine engine builder, Kahlenberg faced a new world in the aftermath of the war. Export sales continued, thanks in part to the Marshall plan, but the domestic business fell. A new 6-cylinder, turbo-charged engine developed by Will Kahlenberg's son, Roger, in 1954 was not enough to stem this tide. It was a 34,000-lb. engine that developed 600 to 900 horsepower at 300-400 rpm, and it sold for about $60,000 but cost $90,000 to build. The era of the light-weight, high-speed diesel had arrived, replete with war surplus Gray Marine diesels, and Kahlenberg ceased engine production to focus instead on the air horns and other accessories for which it remains well known.

Kahlenberg engines were painted with a pleasing "steel blue" enamel that contrasted nicely with the highly polished brass fittings.

Kalamazoo Gasoline Engine Works
Kalamazoo, MI
Trade name: Diamond
Kalamazoo built these medium-duty, medium-speed two-cycle motors from 1913 to 1916. The motors were rated at from 3 to 38 hp.

Keidel, Henry & Co.
Baltimore, MD
Trade name: Ardmore
In 1898, Keidel was famous in Baltimore as a wholesale hardware business said to be "the largest and most important concern of its kind in the entire South." Proprietor Henry Keidel was born in 1841 in Hanover, Germany, and emigrated at age 12. Dozens of companies' products were sold by Keidel including tools, hardware, cutlery, Eagle bicycles, and small arms. In the early 1900s, Keidel decided to add gasoline engines to the list and this meant finding a supplier to produce an engine to be sold under the house name "Ardmore." The company that supplied Keidel was C. L. Barker of Norwalk, Connecticut. Keidel advertised the Ardmore as a no-nonsense two-cycle engine intended for workboats. The 1904–1905 models were available in 1¼ to 10 hp. Like many engines of the time, the cylinder was headless with "no gasket or packing to blow out." A lever that moved in a slot cast in the flywheel could be set to ensure the motor would start in one direction or another. The company also claimed to offer a four-cycle model, but information about it and its supplier are currently unknown. By the time it began selling engines, Henry Keidel & Co. was located in six-story building at 405 and 407 West German Street.

Kemp Machine Works
Muncie, IN
Trade name: Airdrive

Kent Machine Works
Brooklyn, NY
Trade name: N/A
The company was in business by 1912.

Kent-Marvin Co.
Bellingham, WA
Trade name: KE-MA-CO
This four-cycle engine was made in 30-, 60-, and 90-hp models at a shop located at 1113-1115-1117 Railroad Ave. Kent-Marvin were general machinists, engineers, and pattern makers. The engine was available by 1912.

Kermath Manufacturing Co.
Detroit, MI
Trade name: Kermath
Founder of one of America's most long-lived and successful marine engine companies, James Kermath was born in London, Ontario, in 1875 and moved to Detroit in 1899. Having failed in an attempt to start an automobile company, Kermath entered the marine engine field. That project began and ended in 1906–1907 with the construction of a single 4-cylinder, 26-hp Kermath Speedway. With the Speedway's failure, it seems that Kermath joined the ranks of Detroit's many machine shops. Kermath, however, saw potential in volume production of an affordable, modest-horsepower, four-cycle marine engine, and he had such an engine ready for market in 1910.

Believing he possessed the right product, Kermath now turned to selling it, and his choice of a sales and marketing specialist proved to be a good one. The man Kermath hired was a methodical business executive with automobile experience named John "Jack" Farr. Prior to joining Kermath, Farr had taken increasing notice of the dynamic growth of small marine engine and automobile companies in Detroit. In 1910, Farr read a market survey suggesting that the majority of pleasure boats could be powered by a 12-hp engine, and the little Kermath, a sturdy, four-cycle, 12-hp model, seemed to be aimed squarely at the big market Farr had read about. The engine had

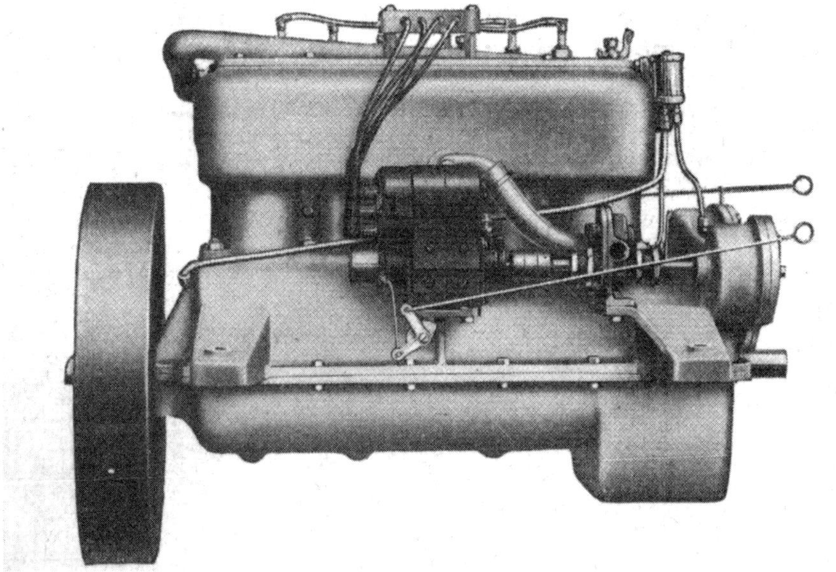

1915 Kermath Vanadium "20"

1922 3-hp Kermath

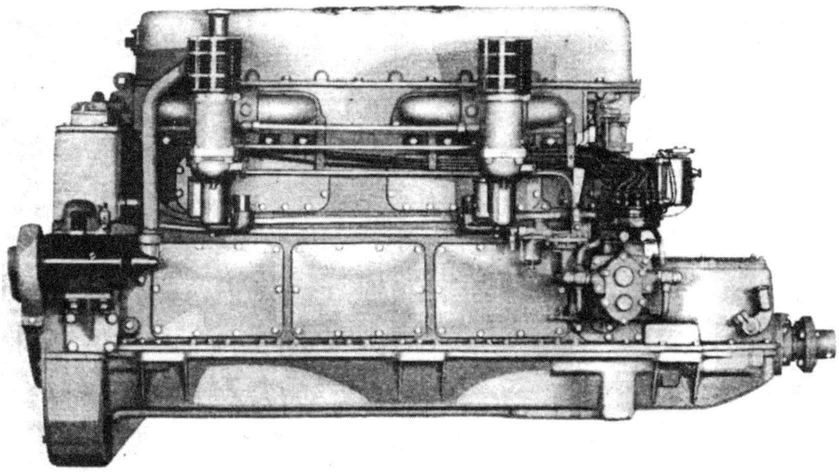

Kermath Sea Wolf

1905 K&V "Uncle Sam"

Kermath 8–10 hp Heavy-Duty twin

1905 K&V "Uncle Sam"

cylinders cast in pairs. Bore/stroke was 3½″ × 4″, and the 12 hp were produced at the then-high rate of 1,000 rpm. Ignition equipment was supplied by Briggs-Stratton (battery) or a Bosch magneto. The engine weighed 360 lbs. A reverse gear was included, mounted on a separate base.

Thirteen years after he joined Kermath, Farr gave an interview to journalist Norman Beasley, which was published in a *Motor Boating* article in March 1924, and recalled the early challenges. "It was tough work in the beginning," Farr remembered of his first days at Kermath. "Lack of sales — great cost of getting business — manufacturing problems — building one product and refining it — standardizing our motor — getting the product down to a point where we could truthfully admit to ourselves that it was dependable."

Initially, the strategy was to iron out manufacturing issues and simply build the engines efficiently. Sales and marketing were to be conducted by the Morton Sales Company, under an exclusive arrangement. The details of the Kermath/Morton relationship remain to be uncovered. Morton was incorporated in 1911 and located at 45-49 Fort Street East. Kermath's address was 47 East Fort, suggesting a literally close relationship. Harry D. Morton was president of Morton, and Ward D. Morton was secretary-treasurer. Farr was listed as vice president. (The Kermath plant was located in the Boydell Building on E. Congress, the same place where the Dodge brothers built engines for Henry Ford.)

Viewed from the distance of almost a century, the Kermath-Morton arrangement seems to have resulted in the possibility of immediate name-recognition issues. When *Power Boating* magazine published its list of marine engine companies for 1912, Kermath was listed as a brand, but the company in the listing was Morton. The arrangement lasted 18 months before Farr decided that Kermath would be better served without another firm's name in the mix, and responsibility for sales and marketing was brought entirely in house.

Looking back on that time in a 1929 magazine article, Farr recalled frequent discouragement: "Few people would be even courteous enough to listen to the arguments put forth in favor of a four-cycle design." Farr persevered, and Kermath sales increased steadily. In terms of weight, price — the 1912 price was $190 — and compactness, there really was nothing quite like this Kermath on the market at that time. To cite but two comparisons, Termaat & Monahan offered a 2-cylinder, 12-hp two-cycle engine that weighed 480 lbs. and cost $278. A four-cycle, 10-hp, 4-cylinder Niagara weighed 525 lbs. and cost $485. Helping to build the engine at such a competitive price was the focus on a single model and the volume ordering of key parts such as block castings and crankshafts — never less than 500 at a time.

Farr methodically built the business with personal sales calls, ads in the boating press, and by establishing a dealer network whose shops were required to stock low-cost replacement parts for the convenience of customers. All this soon paid off, and in 1915 the company added a 20-hp 4-cylinder (4″ × 4″) aimed at the runabout and small cruiser and "built by the same men and in the same methods as the famous Kermath 12." The "Kermath Vanadium Twenty" was a winner that predicted marine engine architecture that would prevail for the next two decades. Instead of the Twelve's two en bloc cylinder castings, the Twenty had a one-piece block. It was an L-head engine and the valve train was fully enclosed for neatness and quietness.

By the time the new Kermath was introduced, and because of the initial 12-hp model's success, James Kermath was being hailed in the press as "Twelve-Horsepower-Jim." Internally, however, some now unknown event resulted in James Kermath's leaving his own company at a date currently not known. Possibly, it was 1915 or early 1916. In 1916, Kermath became a partner in the Kermath-Whitcomb Co. — Arthur C. Whitcomb was secretary/treasurer — a machine shop located at 5626 McGraw. In 1924, Kermath-Whitcomb was contracted by Stanley R. DuBrie to be the sole production facility for the 5-hp, single-cylinder, four-cycle marine engine that DuBrie unveiled in 1923. It was reported the engine would be built "under the personal supervision of James Kermath." Apparently, "Twelve-Horsepower Jim's" success — or the success of his former company — had been noticed because a press note on the new DuBrie said, "This company devotes its entire attention to building one size motor in large quantities, the only way to build a good motor at low cost." As for Kermath-Whitcomb, it disappeared from business listings in 1928. In 1929, James Kermath went to work for the Fisher Body Division of General Motors.

During World War I, Kermath supplied components for the Liberty engine. After the war the company found itself with exactly the sorts of marine motors that the postwar market wanted. What's more, the rise of production boat-building companies looking for affordable marine engines also benefited Kermath. Farr remembered that his first such customer was the International Shipbuilding and Marine Engine Company of Nyack, New York. It purchased the Kermath 20 for its 32-foot cruiser.

In the postwar period, Kermath introduced 1- and 2-cylinder (3½″ × 4″) engines of 3 and 6 hp that sold for $135 to $315 and became very popular. In 1919, 73 people worked at Kermath, up from 44 in 1914. By 1922, the line had grown to include 4-cylinder models of 35, 40, and 50 hp. These Kermaths were all fitted with Kingston carburetors and Bosch ignition and starting equipment. Ever the marketing man, Jack Farr conceived of the "Kermath Rainbow" for New York's 1924 Motor Boat Show. The "rainbow" was an eye-catching display of six Kermaths: a purple 4 hp, a blue 8 hp, a green 12 hp, a yellow 16 hp, an orange 20 hp, and a pink 50 hp.

One of the little Kermaths was also shown in a launch built by Canada's Peterborough Canoe Co.

With the company solidly entrenched and accepted, more powerful engines were added including the 150-hp "O" (for overhead camshaft) in 1926. Remarkably for a production engine, the 648-cubic-inch "O" featured 4 valves per cylinder and hemispherical combustion chambers. Late in 1927, a second version of the "O" was produced. The 175-hp Sea Hawk featured revised exhaust manifolds and dual, heated intake manifolds. A 1½" Schebler carburetor was used; this was later changed to a 1¾" version. The block and crankcase side covers were also revised by comparison to the "O." In 1928, an increase in stroke to 5¾" resulted in a 678-cubic-inch displacement and a horsepower boost to 200 hp. Improvements were also made to the oil pressure by-pass and camshaft oiling.

The final year for the original 150-hp "O" was 1929, but its basic structure was continued on models produced into the 1950s. Late in 1929, Kermath unveiled its long-lived Sea Wolf. This 225-hp engine was an upgraded version of the 200-hp model. It featured an oil cooler, auto-advance Delco or Autolite distributor, and minor upgrades. Both the 200 and the Sea Wolf could be ordered with an optional Holley DDV-5 downdraft carburetor instead of dual 1¾" Scheblers.

During 1930–1931, the 175-hp Sea Hawk was discontinued, and a 190-hp version of the "200" was made available. The standard carburetion was now dual Stromberg 1¾" updrafts. Ongoing cosmetic changes involved the valve covers, oil reservoirs, generator mounts, and flywheel cover. The new 1,414-cubic-inch Sea Raider V-12 made big news in 1931. The Sea Raider was based on the Sea Wolf but it had a much stronger crankshaft, crankcase, and transmission. It was available in 300- or 450-hp ratings.

The ongoing depression did not stop Kermath from steadily improving its engines. Counter-weighted crankshafts were introduced during the 1933–1937 period for all the overhead-valve engines, which resulted in a mild rise in compression ratios. New water and oil pumps were introduced and an increased number of models could be had with closed cooling systems. The L-head models also were upgraded, and ratings attained as much as 225 hp. In 1936, the old 495-cubic-inch standby, the medium-heavy-duty, 100-hp Sea Captain, was completely revised so that all that remained of the previous version was the same bore/stroke dimensions. Five new models were introduced. These included the 4-cylinder, 2⅝" × 3", 10-hp Sea Cub; the 6-cylinder 3¼" × 4⅛", 85-hp Sea Flyer; the 4" × 4¼" 95-hp Sea Rover, a medium-duty, cruiser-oriented engine; and a pair of lightweight, 6-cylinder diesels bringing the total number of diesels to three. The new gasoline models and the diesels were Hercules-based, and Kermath had also used Pontiac blocks as a basis for marine conversion. In the mid-'30s, "after a long series of tests and development work under various boating conditions," Kermath introduced the 110-hp Sea Zephyr, a converted Lincoln Zephyr V-12. A 90-hp version of Ford's flat-head V-8 was also offered. In 1938, the 779-cubic-inch Sea Raider 6 (5¼" × 6") was introduced. The 250-hp engine had the same lower end as the Heavy Duty V-12.

Kermath found customers for its engines among a variety of production boat building companies such as Sea Lyon, Matthews, and Chris-Craft. John Hacker, whose experience with boats and marine engines dated back to the early 1900s, remained an advocate of "true" marine engines, rather than conversions, and made Kermaths standard equipment for more than 20 years. Jack Farr, a longtime advocate of production motor boat building, remained careful to note that Kermath engines were also well adapted as sailboat auxiliaries. The 1937 catalog singled out the famous circumnavigator William Albert Robinson, noting that Robinson's 32-foot *Svaap* had been equipped with a Kermath 8- to 10-hp Heavy Duty engine.

During World War II, Kermath produced its 225-hp Sea Wolf for use in Coast Guard picket boats. Also produced were a variety of V-12 engines including a 1,560-cubic-inch version of the Sea Raider whose initial 550-hp rating eventually climbed to 580. The original Sea Raider 6 was upgraded from 250 to 275 hp and later to 280 hp. Lack of production capacity required Kermath to outsource some V-12 production to the Red Wing Engine Company.

Kermath emerged from World War II to reintroduce its prewar models. In 1946 company ads touted the "World's Greatest Array of Diesel & Gasoline Marine Engines." These ranged from the 25-hp Sea Cub to the 550-hp Sea Raider Special. A 134-cubic-inch, 61-hp Jeep-based engine, the Sea Jeep, was introduced in several models with direct- or reduction-gear drive. So was the innovative Screwball, a unit-drive system in which Kermath 1-, 2-, 4-, and 8-cylinder engines were vertically mounted atop a gearbox. Lubrication was full pressure from a dry sump tank. In 1947 two new small engines were introduced. These were the "16-1" (later called the Sea Pup) and the Sea Twin (later called the Sea Kitten). These 5- and 10-hp engines were intended for use as sailboat auxiliaries and for small power boats. Production of the single and twin was turned over to Ballantine Industries, but the engines continued to appear in Kermath sales literature.

Despite its best efforts, Kermath could not escape the fast-changing postwar boating world with its production fiberglass boats and trailerable, outboard-powered runabouts. In 1953, Kermath became "one of 16 integrated companies" of the Barium Steel Corporation, which had been heavily fined for its business practices in 1952. Why Barium wanted a marine engine builder along with a variety of steel mills is unknown. The condition of Kermath's financial health must have been nearing critical for some time before Barium took over the company, and the ownership change did nothing positive. Kermath ads disappeared by 1956. Kermath's last known activity occurred in the mid-'50s when an ad appeared for the Hydrojet Division of the Kermath Manufacturing Company. But no mention is made of an engine's being part of the drive system. As for James Kermath, he retired from GM in 1952 at age 78. The inventive Kermath had been awarded 100 patents during his long career. Kermath died at age 92 at Detroit's Henry Ford Hospital in May 1967.

Always sturdy and practical, Kermaths remain a good choice for use in antique powerboats today although spare parts for many are, at the time of writing, becoming scarce. Those restoring a Kermath are in luck when it comes to information regarding correct paint colors. From 1920 until World War II, the engines were painted medium gray. The color is currently available from Bill Hirsch Auto. During World War II, Kermath engines produced for the Navy were painted dark gray. Postwar, the medium shade reemerged and, in fact, leftover surplus engines were repainted with it. The Ballantine-built 1- and 2-cylinder engines were painted a light aqua shade. Kermaths supplied to Gar Wood were repainted red by that company, but the original gray remains underneath. At Hacker, the engines were painted a pea-green color. Kermaths used by Chris-Craft were said by that company to be a special gray but were, in fact, the standard factory color. The correct color of Kermaths built prior to 1920 is not currently known, but light gray is presumed to be a likely choice.

See also Morton Motor Co.

Kermath-Whitcomb Co.
Detroit, MI
Trade name: N/A
See Kermath Manufacturing Co.

Kerosene Power Co.
Minneapolis, MN
Trade name: Viking

Kershaw-Williams Co.
Boston, MA
Trade name: K-W Special
In business by 1908 or 1909, the company offered 1-, 2-, and 3-cylinder two-cycle engines with make-and-break ignition. These ranged from 1½ to 18 hp. In 1910 a 5″ × 5″ "heavy-duty" single-cylinder model intended for the local fishing fleet was announced. It emerged in 1911 as a 4½″ × 5″ that was rated at a presumably conservative 6 hp at 500 rpm. A three-cylinder version produced 15 to 18 hp. Orswell jump-spark ignition and Schebler carburetors were used. The company also built dories and launches equipped with its own engines. No more was heard from the company after 1912.

Kimble Bruns & Co.
Jersey City, NJ
Trade name: American Boy
The company produced a 2-hp engine with make-and-break ignition.

King Brothers
Syracuse, NY
Trade name: Barber
See Barber Brothers *and* Wonder Mfg. Co.

King, Charles B. Co.
Detroit, MI
Trade name: King
On March 6, 1896, Charles Brady King is said to have become the first person to drive a gasoline-powered horseless carriage through the streets of Detroit. Following him on a bicycle was his friend Henry Ford. Immensely gifted, a trained artist with great natural engineering talents, King experimented with two-cycle marine engines before settling on a four-cycle with "electric igniters." King Patent Gas Engines for Vehicles and Launches may have been in business by some point in 1896. An 1898 catalog illustrated by King himself includes photos of a very clean-looking 2-cylinder engine painted a dark color but with a polished brass cylinder head cover and intake pipes. The engine was splash lubricated, the oil supply maintained in a "supply chamber" on the crankcase. Starting was said to be accomplished easily with two slow revolutions of the spoked flywheel. After service aboard the *U.S.S. Yosemite* in the Spanish-American war, King returned to Detroit and, in 1899, sold his company to Olds Motor Works. Subsequently, he focused his efforts on the automobile industry with his King Motor Car Company.

Kirkham Motor Mfg. Co.
Bath, NY
Trade name: N/A
The company was in business in 1912.

Knockdown Motor Company of America
New York, NY
Trade name: Unknown
In 1908, this firm offered an engine kit that "You assemble, paint and test."

Knox Motors Co.
Springfield, MA
Trade name: Knox-Springfield
Knox built 20- and 40-hp 4-cylinder, four-cycle engines with overhead valves and make-and-break or jump-spark ignition. All oil passages were cast or drilled into the crankcase. The engines appear to have been in production from 1919 to 1922 by which time the company was called Knox Motors Associates.

Koban Mfg. Co.
Milwaukee, WI
Trade name: Koban
A 3-hp model was listed in 1918.

Kowalski Motor Works
See Kowalsky Engine Company

Kowalsky Engine Company (*also* J. Kowalsky Engine Co. and Kowalski Motor Works)
Verona, PA
Trade name: Kowalsky Simplex
The marine engine industry attracted its share of brilliant men with the ability to excel in various fields of endeavor, men who preferred the pursuit of new challenges to founding and running a

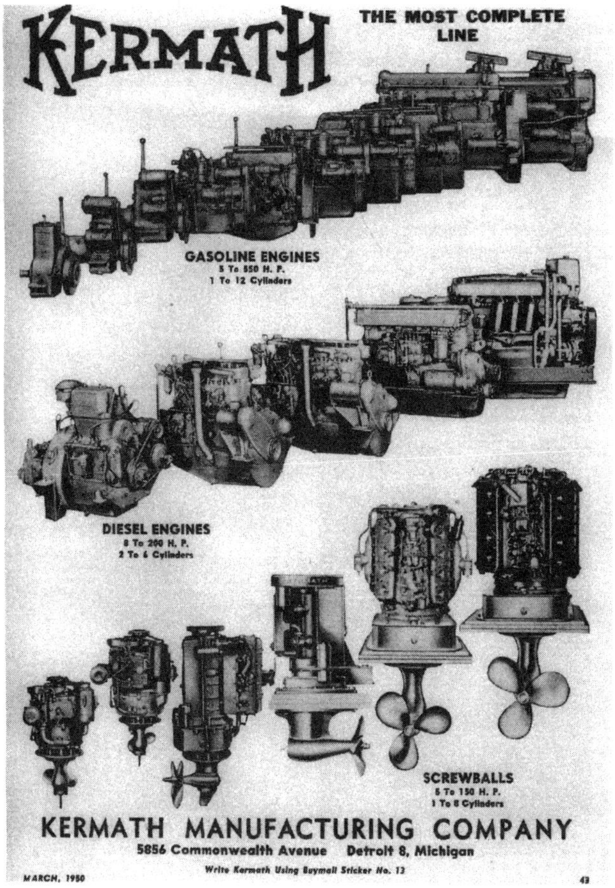

1950 Kermath ad showing the complete model line

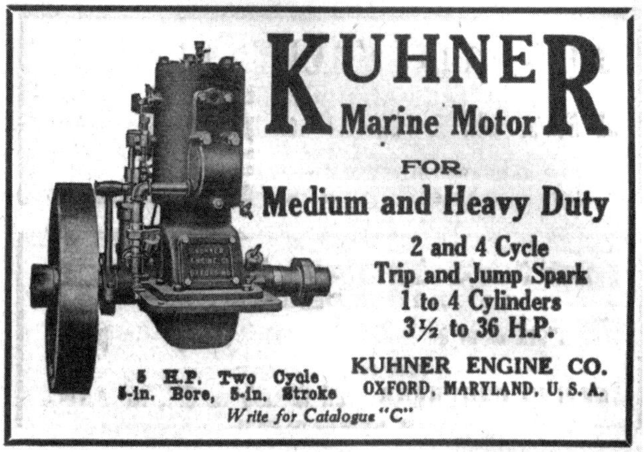

1915

1905

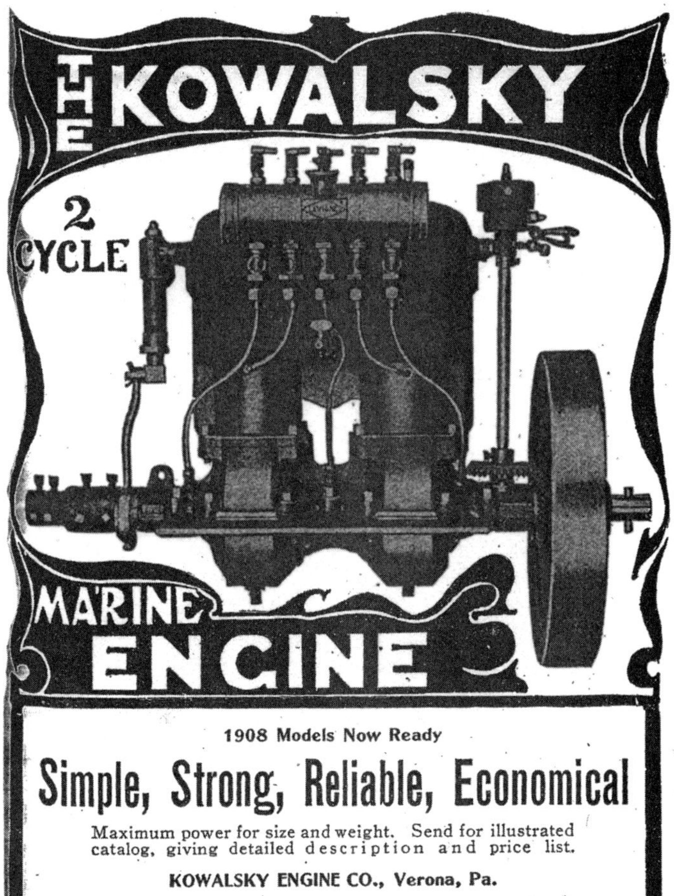

1913

focused, long-term business operation. John Kowalsky (1867–1935) was one such man. A gifted inventor, Kowalsky produced a line of two-cycle marine and stationary engines. In 1905–1906, $75 could buy a two-cycle, 2½-hp Kowalsky Simplex. This was a two-port motor with a make-and-break igniter. A jump-spark 3-hp model was also built. The engines were manufactured in Pittsburgh by Force & Briggs.

The two Kowalsky engines listed for 1906–1907 were 40-lb. 1-hp. and 95-lb. 3-hp models. Kowalsky described his engine like this: "The Cylinder is water-jacketed. The head is cast integral. The cylinder is bored and scraped with line-bladed reamer, making a true and uniform surface ... The piston is carefully fitted with three snap rings ... one at the lower end. Many builders omit the lower ring resulting in loss of compression in crank case, which means loss of power." Like many engine builders, Kowalsky offered boats ready to run starting at $125 and, for his own amusement, built some exceptionally fast boats for use on Pittsburgh's three rivers. In 1908, a modified version of the Kowalsky engine went into production at Waterman. By then Kowalsky had succumbed to his fascination with aviation. He is best remembered for his three airplanes, the first of which he created in 1911. A more successful plane followed in 1913. These flying machines were a not-unfamiliar sight to residents of Verona. He also designed an electric motor so small it was best viewed under a microscope. It attracted the interest of Henry Ford as something of potential interest for his museum. None of Kowalsky's efforts resulted in financial success, and the gifted inventor died penniless at age 73.

See also Waterman

Krupp Marine Iron Works
Detroit, MI
Trade name: Krupp
Two- and four-cycle engines — the latter at ratings up to 100 hp — were produced as of 1912.

Kuhner Engine & Machine Co.
Rock Island, IL, later Oxford, MD
Trade name: Kuhner
Founded in 1899 at East Davenport, Iowa, as the East Davenport Machine and Novelty Company, Kuhner moved across the Mississippi River to Rock Island in 1906. Kuhner offered a 1-cylinder, two-cycle engine and 1- and 2-cylinder, four-cycle engines. The company was reestablished in Oxford in 1911, ceased engine manufacture in World War I, and closed after the war.

Kuhnert, L. C. Jr. Co.
Chicago, IL
Trade name: Kt. Duplex Opposed
The company advertised its engine in *Motor Boating* in 1912. Apparently, few noticed.

Kumberger & Vreeland Inc. (K & V **Engine Manufacturers**)
New York, NY
Trade name: Uncle Sam
Located at 33 Sullivan Street, K & V had, by the end of 1905, manufactured and sold more than 1,800 Vreeland stationary engines to customers in New York City alone. When the company decided to enter the marine engine business, probably in 1904, they contracted with engineer E. W. Graef to design for them a two-cycle motor. This was probably a three-port engine since K & V was among the companies licensed by Joseph Day, the British inventor who then held a patent on the design.

The patriotically named Uncle Sam probably entered production late in 1904 or at some point in early to mid-1905. This line of 1- and 2-cylinder, make-and-break ignition engines ranged from 4 to 12 hp. The ads touted that the engine was for "Any man desiring to get an honest engine and not looking for the impossible."

Commercial fishermen were reported to be among the first customers, and the engine became popular with working watermen. By 1908–1909, the company literature also referred to four-stroke engines in "larger" sizes as well as the Vreeland stationary engines.

L & D Motor Works
Hartford, CT
Trade name: L & D
This two-cycle motor had a 3½" bore and 4" stroke and produced 3 hp. Both the crankcase and water jacket were aluminum, doubtless useful features in a part of the market where light weight was important. The company was in business as of 1919.

L. C. Kuhnert, Jr. Co.
See Kuhnert, L. C. Jr. Co.

L. J. Wing & Co.
See Wing, L. J. & Co.

Lachine Motor Works, Ltd.
Lachine, WI
Trade name: Armstrong

Lackawanna Mfg. Co.
Newburgh, NY
Trade name: Lackawanna
Lackawanna claimed 1898 as the year it started business, but the company's origin and general history is sufficiently convoluted to suggest caution in attempting to interpret events. The company's roots in engines may derive from its attempt to revitalize the Conrad Motor Carriage Co. in Buffalo under the auspices of the Lackawanna Motor Co. In 1904, Lackawanna displayed a two-stroke-powered car at that year's New York Automobile Show. It currently seems likely that after the failure of this car-building effort, the company became the Lackawanna Mfg. Co. and the focus turned to marine engines. The company appeared for the first time in the Newburgh directory in 1906 with a location on Lander St. at the corner of South William.

Lackawanna was a licensee for the Day 3-port patent and touted its twin-cylinder jump-spark, two-cycle marine engines as being up-to-date, durable, and easy to start. The Lackawanna twin with its en bloc upper cylinder casting was, and remains, a very distinctive-looking engine. In 1909, apparently in an attempt to expand its market, Lackawanna introduced an optional make-and-break ignition system. The shaft driving the igniter mechanism was driven by bevel gears off a shaft extending up from the crank. A petcock fitted to base allowed handy elimination of excess gasoline if the engine became flooded. Water-cooled manifolds were used. A rear-starting device was an option as was a reverse gear. Customers could also choose from the standard gear-driven water pump or an optional plunger pump.

The Lackawanna model line was extensive. By 1912, it included the following: 1-cylinder models K (2 hp) and L (4 hp); 2-cylinder models M (4 to 5 hp), O (9 hp), P (12 hp), and R (15 hp); the 3-cylinder A (22 to 25 hp); the 4-cylinder F (27 to 35 hp) and G (30 to 35 hp); and the 6-cylinder H (45 to 55 hp). A Herz timer, Perfex spark plugs, and Columbia batteries accompanied the engines. Lackawanna disappeared from Newburgh at some point in 1912 and in 1913 was advertising a "new factory and new management" and listed its address as Ballston Spa, NY. The new factory apparently was that of the Davison-Namack Foundry Co. at the corner of Chester and Lewis streets. By now, the two-stroke engine's popularity had seriously waned. The company advertised marine and stationary motors but appears to have ceased operating in 1915.

Lacy Brothers
Toledo, OH
Trade name: Lacy
This company was in business by as early as 1903. It offered removable-head, two-cycle models of 2 to 30 hp and four-cycle models of 8 to 100 hp.

Lacy Marine Motor Co.
Rochester, NY
Trade name: Lacy
At the New York Boat Show in 1916, Lacy introduced an advanced, high-speed V-8 based on cylinders cast in fours en bloc. These, like most of the other key components, were aluminum castings. The engine had a 4½" bore × 6" stroke and removable heads. It produced 100 hp. Lacy also showed its engine at the New York Automobile Show. A 200-hp version was also said to be available. Production quantities remain unknown.

Lakeland Motor Co.
Lakeland, MI
Trade name: N/A
The company was in business in 1927.

Lake Shore Engine Works
Marquette, MI
Trade name: Superior
The company was an outgrowth of the foundry begun in 1858 by Charles T. Harvey. The first gasoline engine experiments were conducted in 1895 and these were followed by more successful work conducted by engineer C. H. Blomstrom, who developed a 4-hp "Superior" engine intended for automotive use in 1900. When the company, by then known as the Lake Shore Iron Works, experienced financial difficulties, it was reorganized as the Lake Shore Engine Works on Jan. 1, 1900. Engine patents in the name of a former manager, S. H. Holley, were purchased and orders for stationary and marine engines initially were received at a promising volume.

The Superior was offered in stationary and marine models. These were two-cycle, 400-rpm engines with make-and-break ignition that the company called "touch spark." Single- and double-cylinder (6" × 6") engines were offered, each with forged, one-piece crankshafts. The cylinders and bases were said to be made of an alloy the company referred to as "Lake Superior charcoal iron." The twin's cylinder casting was a single piece. In 1900, a 12-hp Superior was installed in a boat of the Life Saving Service and arranged to drive two 18-inch propellers. Those who tested the boat reported it cruised at 7½ mph using two gallons per hour and performed well under sail or power in winds to 40 mph. Five Superiors were displayed at the Sportsman's Exposition in Chicago in 1901, and the engine gained enough interest that export sales were made to England and Denmark. In 1902, the Life Saving Service ordered the Superior for installation in a number of boats. Despite this promising effort, engine production was halted in 1905, reportedly because the plant was "so far from the fertile sales field in this phase of the business." Thereafter, the company built a variety of mining equipment and machinery for Henry Ford's factories, among other endeavors.

Lamb Boat and Engine Co. (later Lamb Engine Co.)
Clinton, IA
Trade name: Lamb
The Lambs were a prominent Clinton family when, in the late 1890s, they acquired a downtown factory at 18th Avenue South and 4th Street and began manufacture of cream separators. In 1900 Dwight Lamb converted the plant to the manufacture of "High Grade Marine Gasoline Engines." Although a single-cylinder two cycle was said to be available in 1½- 3-, and 5-hp models, the focus was on — in the parlance of the time — four-cycle engines designated as Medium, Medium Heavy Duty, Heavy Duty, and High Speed. A single bore/stroke dimension of 5¼" × 6" was used for 2-, 3-, 4-, and 6-cylinder models. Early models had polished brass water jackets and were often known as "Brass Jackets." Most were rated at 450 or 500 rpm although a 4-cylinder (35 to 40 hp) and a 6-cylinder (65 to 70 hp) were 1,000-rpm engines. One of these latter may have powered the *Lamb IV*, which was at one time claimed to be "the fastest racing boat on the Mississippi River." (Such claims were, it must be noted, not altogether unusual.) Through at least 1910, the Lamb engines were L-heads, but the others were said to be

19??

1911

1911

1908

"combined overhead valve and plunger valve type[s]." The show engines displayed in 1910 appear to have been painted a medium gray color. The next year, the paint was a dark color. The Lamb flywheel had four small holes.

Growth was strong, and the company found sales agents on the East Coast where Lambs were regularly displayed at boat shows. Agents were also located in the Northwest and San Diego. The big Lambs found favor in quite a few large motor yachts. A pair of 18-hp Lambs was installed in a 40′ × 9′ cruiser and drove it along at 10 mph. A 67′ × 12′ twin-screw clipper-bow cruiser built in 1907 by the New York Yacht, Launch & Engine Company was equipped with two 4-cylinder, 40-hp Lambs that would drive it easily at 13½ to 15 mph. Besides such domestic sales, some Lamb engines were exported. In 1916, at the Boat Show in New York, Lamb's principal exhibit was a new model intended for fast runabouts. This was the Lambkin Model G, a T-head, 4-cylinder (3¾″ × 5¾″) that developed 30 hp at 1,200 rpm. The Model G never got a chance to show what it could do. When Lamb died in 1916, production immediately ceased. The factory was purchased by the Climax Engine Company, also in Clinton, and Climax stationary engines were produced there into the 1940s.

Lathrop, J. W. & Company
Mystic, CT
Trade name: Lathrop

One of New England's quiet stalwarts, Lathrop began before 1900 and kept going as steadily as one its slow-turning engines for decade after decade. Along the way, the company earned a fortune for an entrepreneurial one-time needle peddler from Worcester, Massachusetts, named James W. Lathrop. Lathrop's first exposure to engines came aboard a Palmer-powered launch at Mystic, Connecticut. When the Palmer's igniter developed a problem, Lathrop watched as the boat owner sketched out an improved version. As things developed, the two men then designed an engine to go with the igniter. They sold that engine, and, as J. W. himself remembered the event, "they got enough to buy a lathe and built two more engines. From then on, they were in the engine business." The year was 1897.

The first Lathrops probably owed quite a debt to Palmer, but they soon emerged with their own unique character. That is to say, a Lathrop was recognized as a conservatively designed, heavy-duty motor of proven merit that was carefully built to last a long time. The Lathrop's reputation as a sturdy engine spread quickly and, by 1901, the company was shipping off engines to Seattle, Jacksonville, and Bath, Maine. By 1905, Lathrop claimed to have sold more than 2,600 motors. J. W. Lathrop exercised hands-on control over all aspects of his shop, which he insisted be kept "neat and orderly at all times."

Ever laconic, James Lathrop was given the opportunity by *MotorBoat* magazine in 1924 to describe his company's emergence, and he provided only four sentences. The gist was that a fisherman who had bought a motor in 1900 had called again in March 1924 to order a new 30-hp Lathrop. What else needed to be said? The company's advertising was similarly low key, seldom more than a few words unadorned by a picture. If ever there was a triumph of product over marketing, it was the Lathrop engine.

Lathrops were positioned to appeal primarily to fishermen and the commercial boatman. The two-port, two-cycle's moderate 500 rpm, make-and-break ignition, and "solid, substantial" castings were exactly what such users needed. The market was such that, long after the introduction of four-cycle models, Lathrop was still offering a line of two-cycle motors. In 1928, for example, the two-cycle models outnumbered the four-cycle 13 to nine.

In its early years, Lathrop claimed to have pioneered two features in particular. One was a drip feed oiler located on the starboard side of the base to supply oil direct to the crankcase. The value of the device would be eclipsed once it was determined that an oil/gas mixture was the best way to lubricate a two-cycle engine. Typically, Lathrop continued to build its two-cycle models with the oiler for years after mixing oil and gasoline had become accepted. The other feature was that "the cylinder is so made that the motor can be operated on either the two- or three-port principle at will of operator, the change only requiring a few minutes." This feature is unlikely to have been viewed positively by Lathrop owners. It is almost certain that most of these motors were operated as two ports, with the intake mixture flowing from the carburetor, through the check valve, and into the base.

By 1905, the single-cylinder Lathrops were available in models ranging from 2½ to 10 hp; the doubles from 7 to 20 hp. By the '20s, the two-cycle motors ranged from 3 to 48 hp. By then, Lathrop had buckled to the pressures of technology and offered jump-spark as well as make-and-break ignition. Heavy-duty, four-cycle T-head Lathrops were introduced in 1915. The 3-cylinder was typical of the line. It was a 400-rpm, 2,000-lb. machine with individual cylinders (5½″ × 8″) and removable cylinder heads. Lubrication was force-feed. These Lathrops had large water jackets and big hand-hold plates in the crankcase so sediment could be readily cleaned out.

War-oriented production demands required Lathrop to increase his shop's size and production capacity. Postwar, that made it easier for the company to fulfill demand, and distribution in the Northwest increased in 1922 when the Fairbanks-Morse Co. of Seattle became the Lathrop distributor there. The 2-cylinder, 475-rpm, 16-hp T-head was expected to generate a wide following among Pacific Coast trollers. At the plant, meanwhile, Lathrop had by now installed the latest machine tools and could count on a stable group of workmen, many of whom had been with the company for from 12 to 15 years.

This conservative line of engines continued pretty much as it had been conceived well into the 1930s. James Lathrop died in 1935; production was then overseen by his son and, later, a grandson. The LH series was introduced, the "H" standing for Hercules, as the engine was based on a Hercules block. Lathrop equipped the engine with a reverse gear and offered a choice of 12″ or 18″ flywheel options. Still, the old flavor was gone, and so were the old days. After a succession of owners in the mid-'50s, the company ceased operation. For years thereafter, Lathrop parts and advice were provided by Edward Welles, Sr., who had worked at the company beginning in 1916. Welles had a little, white-painted wooden building in Mystic. A big, dark red Lathrop greeted visitors from its post inside the door.

Lauson, A. H. Co.
Milwaukee, WI
Trade name: Lauson
Designed by Andrew H. Lauson, the Lauson was a four-cycle engine available in 1-, 2, and 4-cylinder models. The company was housed at several addresses between 1906 and 1913, the last one being 917-919 3rd. The company was renamed Badger Motor Company in 1912 and the motor became the Badger.
See also Badger Motor Company

Lauth Auto & Engine Co.
Port Huron, MI
Trade name: Lauth
The company had an ad in *Motor Boating* in Dec. 1911 and filed for dissolution in Feb. 1913.

Le Jeal Motor Works
Erie, PA
Trade name: Le Jeal

Le Roi Co.
Milwaukee, WI
Trade name: Le Roi
In 1922, Le Roi brought out a thoroughly contemporary, enclosed 4-cylinder (3½″ × 4″) 15-hp motor for use in runabouts and other small boats. It sported a Zenith carburetor and Atwater-Kent battery or a magneto ignition system.

Leary Gasolene Engine Co.
Rochester, NY
Trade name: Leary
The company was founded circa 1910 to build a two-cycle, three-port motor designed by John J. Leary who claimed to have 17 years' experience in 1913. A fourth port was fed by a separate carburetor. "One carburetor and manifold is adjusted to give a perfect mixture at extremely slow and medium speeds. The other carburetor is set so that when both are in use, the correct mixture is made that gives full speed and the Acknowledged High Power." By 1916, the line included 1- through 6-cylinder engines of from 5 to 30 hp, all rated at 800 rpm and equipped with Atwater Kent ignition.

Leighton, H. J.
Syracuse, NY
Trade name: Leighton
Herbert Jackson Leighton was among the first men to focus on development of a lightweight, high-performance, two-cycle marine engine. Born in Vestal, New York, in 1858, Leighton received his degree in mechanical engineering from Cornell in 1880. Upon graduation, he went to work at the Syracuse steam engine shop of John Sweet — the Straight Line Engine Company. (In addition to being president of the company, Sweet had also been a professor of engineering at Cornell.) In 1885 Leighton and a colleague, J. Coffin, were granted a patent for a scale. Presumably it was this patent that led to the establishment of Coffin and Leighton, which remained in the Syracuse city directory until 1900.

In 1898, Leighton, who was referred to in some local newspapers on occasion as "Captain Leighton," joined the first yacht club in Syracuse. Having become a founding member of the Syracuse Steam Yacht Club, Leighton promptly began working not on a steam launch but on a gasoline engine. This 2-cylinder, presumably two-cycle engine, was installed in a launch named *Echo*, a 15-mph boat that signaled the future at the SSYC.

By 1900 Leighton was immersed in the engine building that would make him well known locally and for a time regionally. By 1902 he was linked with gasoline engines, not with Coffin and Leighton. Details of early Leighton motors are scanty. In large part, this is because — the rare press story to the contrary — they appear never to have been placed in series production. Neither did Leighton submit material to the key boating magazines of the period, which published annual lists of marine engine companies and their products. In fact, had not Leighton-powered launches won three successive Gold Cups (*Chip* — short for Chippewa, as in Chippewa Yacht Club — in 1905, *Chip II* in 1906 and 1907), we might know even less about Leighton than we do about dozens of other marine engine builders who marketed their wares on a strictly local basis.

It should be pointed out that the 1906 and 1907 Gold Cup victories were based on corrected time calculations. Also, the 1907 engine, at least, was a 251-cubic-inch, 2-cylinder "rule beater." It had a 4″ bore and 10″ stroke to circumvent the rules, which were based on bore diameter. The cylinders were charged not by a carburetor but, instead, by a third cylinder that transferred the intake mixture to the power cylinders under slight (about 5 psi) pressure in order to ensure that burnt exhaust gases were thoroughly expelled. This was an idea originated in England by Dugald Clerk, a pioneer of two-stroke engines, who conceived of the "displacer" (displacing the burnt gases) cylinder idea. Clerk had patented his engine in 1881, and Leighton might well have studied the principle in college.

With all that as background, it must be said that Leighton was an important innovator. By 1902, he had built what may have been the first 8-cylinder engine ever installed in a boat. *The Rudder* sent engine designer E. W. Graef to Syracuse in the late fall of

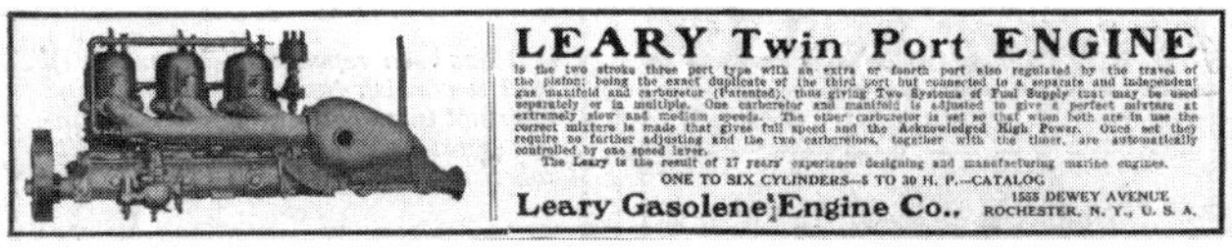

1913

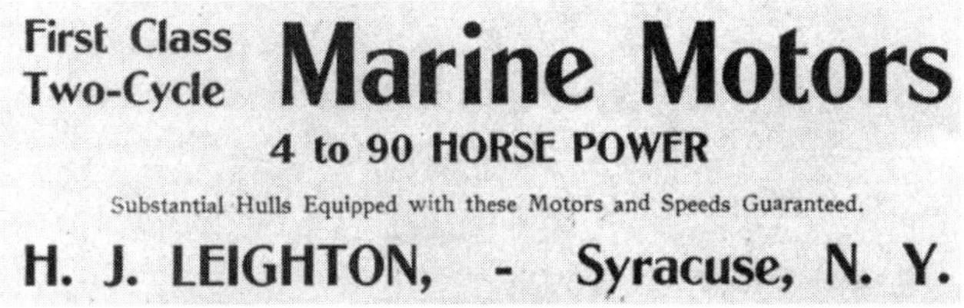

1906 Leighton ad

1915

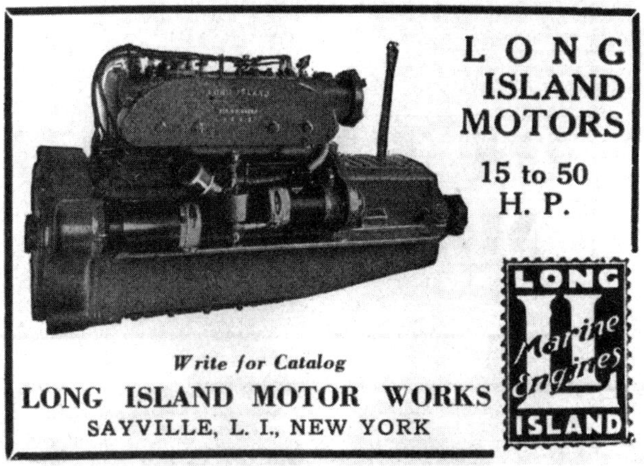

1929

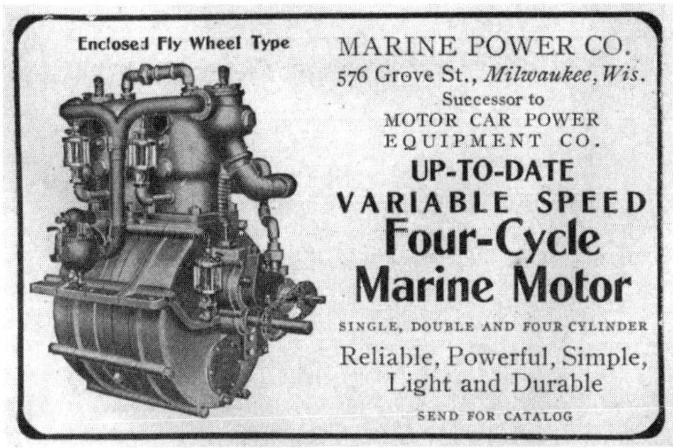

1906

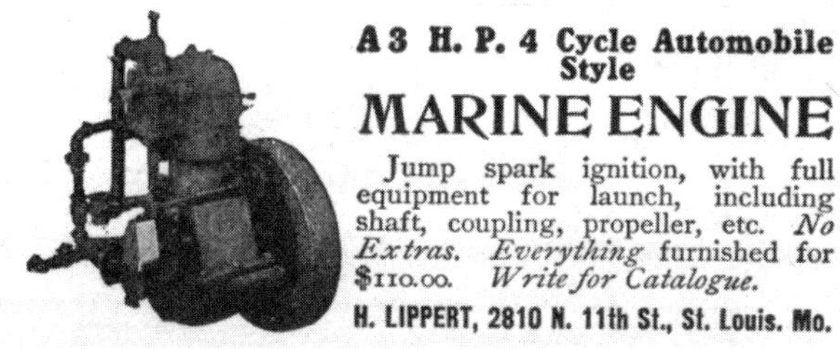

Circa 1907

1913

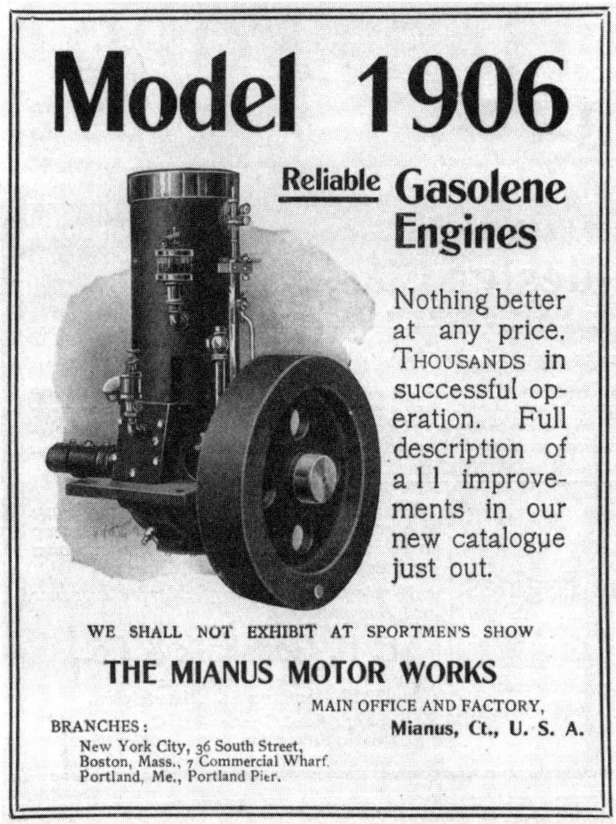

1906

1902 to see this engine and the 55-foot launch Leighton had built. Leighton took Graef to the Syracuse Yacht Club and, after some initial battery trouble, got the 120-hp, 8-cylinder engine started and switched over to magneto. Leighton proved the boat could run at 23 mph, and Graef came away from his visit believing Leighton had "the fastest gasoline launch afloat."

Leighton attracted wider attention in Sept. 1903 when his 55-footer, now named *Adios* and powered by a 6-cylinder motor, defeated the 60-foot *Standard* powered by a Standard 6-cylinder engine that developed 110 hp. The following year, *Standard* won the first Gold Cup event. By way of comparison, a few years later, Leighton claimed that his "light weight, high-speed" 110-hp six, weighed 870 lbs. to the Standard six's 3,200 lbs. It should be noted that the race between *Adios* and *Standard* was marred when the latter broke down and several minutes were required to get her going again. That said, the 23-mph top speed of *Adios* was very impressive in 1903. *Standard*, of course, went on to a few decades of prominence while Leighton's engines became interesting curiosities for a few years and then disappeared.

Among the best surviving photographs of a Leighton is one published in *The Rudder* in December 1908. This engine was one of a pair of 6-cylinder Leightons from *Chip III*, the boat that placed second in the 1908 Gold Cup. Each cylinder (bore/stroke was 6″ × 5″) was fed by a Schebler carburetor. Although the engine's exhaust side showed a single manifold, as installed in *Chip III*, the motor previously had been equipped with an exhaust stack for each cylinder. These were some 6 feet tall to get the exhaust well above the heads of the driver and crew. The cylinders were cast with large rectangular openings covered by brass plates to reduce weight. The engine was described as a "combination of the two and three port systems." Given the 1,100 rpm at which the engine could turn, it seems unlikely that a check valve was present, but this can't be ruled out. Perhaps an auxiliary air port was present.

In 1911 the "regular line" was said to consist of 2-cylinder 10- and 20-hp motors, 3-cylinder 15- and 30-hp motors, and 4-cylinder 20- and 40-hp motors. The lightweight, high-speed models included an 80-hp four, a 110-hp six, and the 160-hp eight. In 1908 Leighton built a pair of 200-hp, 6-cylinder motors (6″ × 5″). Installed in the 4,000-lb., Leighton-designed-and-built *Chip III*, these motors were expected to make the boat a serious threat in the 1908 Gold Cup. In the race, however, overheating halted *Chip III*, and the Clinton Crane V-8-powered *Dixie II* won the cup.

A Leighton ad appearing in *The Rudder* in 1913 showed an engine completely different from the earlier models. It was a 4 cylinder (5″ × 4½″) with what appears to be shrouds over the cylinders. Was it simply a shrouded version of the 1907 engine in the photograph? Whatever it was, it produced an impressive 50 hp at a weight of 350 lbs. and was advertised at $750. Doubtless, this Leighton was among the lightest and most powerful marine engines available at the time. One of the few comparable makers, Pierce-Budd, offered a 40-60 hp 6-cylinder 4″ × 4″ that weighed 390 lbs. and sold, in 1910, for about $1,100. Pierce-Budd, however, achieved a market position and longevity that eluded Leighton.

In May 1915 *Motor Boating* published a list of motors suitable for Restricted Class Racing. A pair of Leightons made the list: a 4-cylinder, 336-cubic-inch model, rated at 50 hp at 1,000 rpm, and a 6-cylinder, 530-cubic-inch six rated at 75 hp. Both shared the over-square bore/stroke dimensions (5″ × 4½″) of the motor advertised in 1907. Doubtless, Leighton was aware of the advantage of reduced piston speed, and apparently his porting arrangements permitted this approach to be workable. Were these engines three-port designs? Was a charging cylinder involved? We don't currently have answers to those questions.

Leighton-designed "torpedo-boat hull" (curved-stern) launches became a reasonably common sight on Oneida Lake, Onondaga Lake, and, doubtless, others in the region. Some of these were big boats. A used-boat ad in the local paper in 1911 was for an 80′ × 5′ Leighton launch planked with ¾″ cedar and powered by a 4-cylinder, 12-hp engine. Presumably, that horsepower rating was on the conservative side.

By 1920 Syracuse directories listed Leighton as a mechanical engineer but did not link him to gasoline engines. For a while, he built the oil pumps used by the Franklin Automobile Company in its wonderful cars, which were also built in Syracuse. By the time Herbert Leighton died in October 1937, he had become a well-respected and civic-minded citizen of Syracuse. Although at least one Leighton-designed launch is known to survive, not a single Leighton engine has thus far been discovered.

Lillis, Thomas
Grassy Point, NY
Trade name: Uncle Sam
A *Motor Boating* ad in 1911 and the engine's name suggest this shop may have acquired the patterns for Uncle Sam engines from Kumberger & Vreeland.

Lincoln H. Fey Co.
See Fey, Lincoln H. Co.

Lindley Boat & Marine Engines
Canastota, NY
Trade name: N/A

Lion Engine Works
Camden, NJ
Trade name: N/A

Lippert, H. Co.
St, Louis, MO
Trade name: Missouri
Machinist Hugo Lippert was proprietor of a "gasoline engine shop" located at 2810 N. 11th St. from 1908 to 1913 although his first ad appeared in 1907. The company was reorganized as the Missouri Engine Company in 1914.

See also Missouri Engine Co.

Lisk, George A.
Detroit, MI
Trade name: Lisk Marine Engine
Introduced circa 1904, the Lisk was a four-cycle engine. These Lisks were comparatively high-rpm designs, all rated at 700 rpm according to 1911 specifications. Bore/stroke dimensions of 4⅝″ × 5″ were used for 1-, 2-, 3-, and 4-cylinder and 30- to 40-hp 6-cylinder models. One model, a 350-lb., 12- to 15-hp 4 cylinder, had dimensions of 3¾″ × 4″. That was 175 lbs. lighter than the 15- to 18-hp Lisk 3 cylinder. In 1916 the Lisk models ranged from 5 to 60 hp. The plant was at 1190 West Jefferson Ave.

Livingston & Van Epps
Syracuse, NY
Trade name: Van Epps
The company was listed in the 1909 Syracuse City Directory. Presumably, it was a short-lived effort to continue production of the Van Epps engine.
See also Van Epps Motors

Loane-Hiltz Engineering Co.
Baltimore, MD
Trade name: Fisherman, Dredger, Crabber
In business by 1912, the company then offered four models, all based on 6″ × 7″ cylinder dimensions. Two-, 3-, 4-, and 6-cylinder 450-rpm models were available. Horsepower ranged from 12½ to 38. Ignition was jump-spark. The engines were labeled the Coaker in Newfoundland. A later model based on Ford components was named the Fisherman Ford. The company was still in business in the '20s with the heavy-duty 6- and 12-hp Fisherman, "Not Pretty Enough for Yachts." By 1915, the company was known as Loane-Trask and later became Loane Engineering.

Lockwood-Ash Motor Co.
Jackson, MI
Trade name: L-A
Fred and Arthur Lockwood, who specialized in repairing bicycles and electrical equipment on East Courtland Street, started the company in 1901 as Lockwood Brothers. By 1903 the brothers had become dealers for Pope-Waverly and Oldsmobile, the latter an especially lucrative franchise thanks to the series production of the single-cylinder model that became famous as the "curved dash Olds." In 1906, the Olds magazine, *Chug*, reported an anticipated $2 million in orders. The increase in business resulted, in 1906, in the Lockwood brothers moving to larger quarters. They also took in a partner, W. L. Ash, who had a business making spark plugs. Now, as the Lockwood-Ash Motor Company, the firm introduced two small marine engines. These motors were an outgrowth of experiments conducted by the Lockwood brothers beginning around 1904 or 1905.

The lineup contained a single-cylinder 2-hp, and a 2-cylinder, 5-hp jump-spark motor. These were soon joined by the 4-cylinder, 10-hp Model D. These little motors, produced in a wood-framed factory building on the corner of East Main and Horton Streets in Jackson, found widespread acceptance. In 1909, a new line was introduced. It included 1-, 2-, and 4-cylinder models of 6 to 25 hp. Cast manifolds mounted on studs and secured by nuts were used instead of threaded pipe connections. In 1912 the company reported production of 8,000 motors. Sears-Roebuck sold them as the Motor-Go, noting in its ad copy that the motors were "Shipped from factory in SOUTHERN MICHIGAN." Collectors have unearthed motors on which the tag reads "Lockwood-Ash Motorgo." How long this affiliation lasted is not recorded. Sears was still advertising the Motor-Go in 1927, but it is entirely possible, even likely, that the motors were then being manufactured by another company that had acquired the tooling from Lockwood-Ash.

The Model 44, a 4-cylinder, four-cycle engine of 16 to 20 hp, was introduced around 1918. By then, however, the company was focusing equal attention on outboard motors, the first of which it had offered in 1914. When W. L. Ash died during the winter of 1921, his son had no interest in continuing in the company. The Ash family sold its stock to the Lockwoods, and the company reverted to the original Lockwood Brothers name by the mid- to late '20s. In 1923, the company introduced a new single-cylinder, 5-hp, four-cycle engine based on Ford Model T parts, a "Ford-part motor," as it was called. This L-A Model 41 was also sold by Sears under the name "Wonder." Ultimately, the focus on outboards grew — a new L-A "rowboat motor was introduced in 1923 alongside the Model 41 inboard — and Lockwood became a familiar name on the racing circuit. This led in 1929 to the merger of Lockwood Brothers with the Outboard Marine Corporation (OMC) and the closure of the Jackson factory. The Lockwood Division of OMC, however, was short-lived. It fell victim to the Depression.

All this did not mean the end of the L-A marine engine. The Nadler Foundry and Machine Company, a Lockwood dealer in Plaquemine, Louisiana, purchased the rights and patterns and began manufacturing the engines. They were now known as the Nadler. "Why Old-Timers Choose the NADLER," was the headline on the company brochure. It was a business that would continue for some 20 years. In 1974, the company's then–vice president, J. Allan Nadler, Jr., recalled that the engines "were produced in somewhat limited quantities for approximately 20 years and were used primarily by the loggers and fishermen in South Louisiana."

Loew Manufacturing Company
Cleveland, OH
Trade name: Loew-Victor
Founded circa 1900 to manufacture tools and machinery, Loew entered the marine engine business in time to display its first model at the 1910 Boston Boat Show. This 25-hp Loew-Victor was a T-head, dual-ignition motor installed in a 30′ cruiser built by the

1905 Lamb

The Lamb four cylinder 24 h.p.

1911 Lamb

1908 Leighton intake side, engine of Chip III *Gold Cup boat*

1916 Duesenberg manufactured by Loew-Victor

Racine Boat Manufacturing Company. The show engines, at least, appear to have been painted an off-white color. The lineup included a 6-hp single-cylinder, a 12-hp double, an 18-hp triple, two 4-cylinder models, and a six. An aluminum crankcase was optional on the fours and standard on the six. The single- and double-cylinder engines had gravity lubrication; the others, pressure.

The company claimed it had spent more than $20,000 on tooling for the engines and that there was "absolute similarity" in each part, permitting easy repair or replacement. It was deemed desirable, on the first lineup, at least, to fit every motor with an air compressor so that a boat with a whistle would give "the same satisfaction as a whistle connected to a steam boiler."

The 20- and 30-hp Loew-Victor engines were popular in launches of 20′ to 26′ feet, and more powerful models went into cruisers. In addition to building motors for sale through its own network, in 1911 Loew also supplied motors to the W. H. Mullins Co. That year, too, the Loew name was associated with the National Boat & Engine Co. although evidence of a connection to that questionable effort to form a conglomerate of troubled companies is lacking. The 1913 models were equipped with standard electric starters and a generator, features that would be increasingly demanded by boaters no matter what their budgets. That fall, Loew became part of a new organization and would shortly thereafter leave its plant at 1908 Madison Ave. in Cleveland for Chicago. The Loew Manufacturing Co. appeared briefly again in 1926. At that time, the company took a full page ad in the June issue of *MotorBoat* where it showed the 6-cylinder Loew-Knight sleeve-valve engine. The sleeve-valve presumably worked even less well afloat than in an automobile.

See also Loew-Victor Engine Company

Loew-Victor Engine Company
Chicago, IL
Trade name: Loew-Victor, Duesenberg

In Oct. 1913 it was announced that the Loew Manufacturing Company of Cleveland would be consolidated with Harbeck Motors of Chicago and that the wealthy Jervis R. Harbeck would become president of what was now called the Loew-Victor Engine Company. Harbeck was among those Chicagoans who had made a fortune developing adding machines and other such equipment. He was also a speedboat enthusiast, who in 1914 commissioned the Harmsworth Trophy racer *Disturber IV*, a 60-mph boat powered by two Duesenberg engines that together produced at least 600 hp.

Capitalization was a very hefty $250,000, but the confusion caused by the "consolidation" of Loew and Harbeck was immediate and lasting. *Power Boating* reported that "all the present Loew-Victor models with improvements for 1914 will be put on the market, as well as the four- and six-cylinder, which Harbeck introduced last season." Exactly what those Harbeck engines were remains unclear, other than that they were referred to as being "four-cycle, heavy-duty engines."

Jervis R. Harbeck, meanwhile, promptly engaged Fred Duesenberg to develop two engines. "The new Duesenberg engines of the Loew-Victor Engine Co," is how *Motor Boating* referred to the engines when they were shown in Sept. 1916. "... [I]t remained for the Loew-Victor Company this year to develop the Duesenberg engine in a commercial way and place machines of this type on the open market where yachtsmen in general might obtain them for their express cruisers." That December, a West Coast magazine referred to the "Duesenberg express cruiser engine displayed by the Loew-Victor Engine Company." *The Rudder* called the engine "the new horizontal-valve Harbeck, a creation of F. S. Dusenberg [sic]."

The Loew-Victor–built Duesenberg engine had unusual horizontally mounted overhead valves. This was, according to the report in *The Rudder*, doubtless based on discussions with Harbeck or Duesenberg, said to "give exceptionally free gas passages, and small area of combustion chamber walls [that] minimizes the dissipation of the heat units, and while thus directly increasing the horsepower, permits the use of high compressive pressures that cannot be employed with advantage in the ordinary 'T' or 'L' head type." The article noted that the valve design did away with push rods and tappets and that all the valve-gear was hidden beneath aluminum housings. The engine's cylinders were cast in pairs of two; pistons were aluminum. The $5,500 260- to 300-hp six emerged in fall 1916, followed by a $7,000 360- to 400-hp eight. Thanks to its popularity for boats intended as submarine chasers, the eight was sometimes dubbed "The Patrol." These magnificent 6¾" × 7¾" engines were Art Deco in style and doubtless impressed enthusiasts of their era as remarkably exciting creations, if beyond the reach of all but a select number of boat owners.

Although the original Loew-Victor line of marine engines supposedly remained available, the ultimate result of the Loew-Harbeck merger was that the Loew-Victor name had pretty much disappeared by the end of 1916. Some called the motors "Harbeck engines." By the time of the 1917 Power Boat Show in New York, the trade name Loew-Victor was heard no more. As for the Duesenberg marine engine, after gaining much publicity during late 1916 and into 1918, its production appears to have ceased by 1919. None of those magnificent engines are known to survive.

Long Island Motor Works
Sayville, NY
Trade name: L.I.M.
Founded by a mechanical genius named Karl F. Pausawang, the company began building marine engines in 1925 and continued until about 1934. The line included 1- and 2-cylinder, four-cycle engines with overhead valves. The 10-hp single, with a Joe's gear, sold for $325 in 1929. The company also marinized Model T engines and converted some with its own overhead-valve heads. When it ceased engine building, the company continued to do a variety of marine and automotive work. During World War II, it built the prop shafts used in PT boats.

Lord, Fred E.
New York, NY
Trade name: Lord-de-Luxe
Naval architect Frederick K. Lord also turned his hand to engine design and, in 1916, showed a thoroughly contemporary, enclosed four-cycle engine. It seems unlikely that more than a few examples were ever built.

Lorimer
Oakland, CA
Trade name: Lorimer
In the early '30s one of the original trio that had founded Atlas, Jack Lorimer, retired. This retirement, however, was short-lived. In 1936, Lorimer, together with his old associate Peter Bahr and Lorimer's son, Ralph, started a new company, the Lorimer Diesel Engine Company. Although these Lorimers were conceived as compact, lightweight engines, they retained the objective of simplicity so characteristic of the Atlas. But they employed a welded-steel crankcase and were fully enclosed. Some smaller models were designed to start on gasoline and then be switched over to diesel operation. This, however, caused problems with insurance, and the company was dissolved shortly after World War II.

Louvet, E. & Son
Woodhaven, NY
Trade name: Louvet
The company showed 1½- to 8-hp, three-port, two-cycle motors at the 1905 National Motor Boat and Sportsman's Show in New York. Production, if any, was probably very limited.

Loveland Manuf. Co.
Bridgeton, NJ
Trade name: Loveland
One of a number of makers of heavy-duty engines in New Jersey, Loveland in 1910 offered a 60- to 65-hp at 350 rpm, 4 cylinder (8¼" × 10") with a 32-inch flywheel and jump-spark ignition. The smallest company offering was a two-cycle, 3 hp with a 4" bore and 3½" stroke.

Lozier Motor Co.
Plattsburgh, NY
Trade name: Lozier
Born near Wrightsville, Indiana, in 1842, Henry A. Lozier, Sr., became a successful sewing machine salesman. In 1887 Lozier invested a portion of his savings in the emerging bicycle industry. That initial investment became a four-factory enterprise in Toledo, Ohio, whose output was the Cleveland bicycle. When he decided to get out of the business in 1897, Lozier had already made a fortune, and he decided to invest some of it in what he believed would be the next big craze, the automobile and the motor boat. After experiments with a motorized tricycle and several steam car prototypes, Lozier and engineer George R. Burwell, who had designed the company's bicycles, focused on marine engines.

Burwell designed a two-cycle, single-cylinder engine and a four-cycle twin. Believing a big boom in motors and boats was imminent, Lozier now determined to move from Toledo to New York state and a location that would give him greater access to eastern markets. Plattsburgh was the city chosen, in part because local businessmen invested in the company and because a dam at nearby Indian Pass would be a ready source of electrical power for the big new factory. By spring of 1901, Lozier was in its new home.

The two-cycle Lozier engine, available in single- and 2-cylinder models, was built entirely within the company's own shops. Lozier made its own drop-forged steel crankshaft, the gray iron connecting rod, and the vaporizer. The latter, noted the company, was easily serviced in case a waxy deposit left by lower grade gasoline should impair its seating. Many parts of the engines carried the company logo. Each cylinder was tested to 80 lbs. hydraulic pressure. Completed engines were run in using a factory power source. Then they were run under full load for six to 12 hours, driving a propeller submerged in a 6-foot-deep steel tank. These were two-port engines rated at between 310 and 375 rpm. Speed control was largely managed using a lever-controlled butterfly valve installed in the transfer port. The engines had a muffler designed around a hot-air drum that fed air to the vaporizer to warm it for improved performance. While batteries were used to fire the make-and-break igniter during starting, a magneto system then took over. "Always Ready" is how Lozier headlined its engine ads. The Lozier engine was finished in oven-baked, glossy black enamel, the nickel fittings highly polished.

When the Lozier was introduced in 1901, the gas/oil mixture that would eventually become the accepted method of lubricating a two-cycle engine was unknown. The Lozier had an oiling system that forced oil into the cylinder under slight pressure generated by the outflowing exhaust gases. Oil was directed through cast-in passages to the piston rings and combustion chamber. The rod bearing was splash lubricated, the oil poured into the base through

The Loew Victor four cylinder engine.

1911 Loew-Victor

1902 Lozier

1913 Loew-Victor

1905 Lozier

Mianus 15 h. p. double cylinder.

1911 Mianus

Circa 1944 Lorimer diesel (starting was accomplished on gasoline)

In 1916, recognizing the era of the two-cycle engine was at an end, Mianus introduced the second of its four-cycle models. This 450 rpm, 2-cylinder, overhead valve engine produced 16 hp. Its inlet manifold was cast inside the exhaust manifold so that the fuel would be warmed before entering the combustion chamber.

a special fitting. Grease cups were installed to lubricate the main bearings. Ball-type thrust bearings, rather than discs, were fitted to increase the bearings' durability. The water pump was driven by chain off a sprocket on the crankshaft.

Lozier offered its single-cylinder engines in ratings of 1½ to 6 hp, the twins at 8 to 20 hp. The engines were available in the company's launches that ranged in size from 16 to 40 feet. The boats were built in the company shop in Bascom, Ohio, and soon proved very popular. Not long after the move to Plattsburgh, the boats underwent a redesign by Frederick Milo Miller who had gone to Glasgow as a young man to learn boatbuilding. Upon his return, Miller worked for Charles Seabury in New York before opening his own shop in Malone. He joined Lozier when the Plattsburgh plant was opened and ultimately oversaw the construction of some 4,000 boats.

The Lozier enterprise was initially prosperous and the boats and engines sold in every greater numbers during the 1901–1905 period. A second factory was opened on the East River in New York City. Branch offices were established in Europe and the Far East, and H. A. Lozier, Jr., garnered favorable publicity for the company when he drove his launch *Shooting Star* to a Gold Challenge Cup win in 1904. Through it all, Henry, Sr., never forgot his interest in the automobile business. Before he died suddenly in 1903, the senior Lozier had authorized further research into entering automobile production. By 1907 Lozier was building 50 to 60 cars a year, while boat production diminished. In 1908, Lozier boatbuilding was transferred to Traverse City, Michigan, to make room for more auto production in Plattsburgh. Marine operations ceased about 1910. By 1915, Lozier's automobile business was in deep trouble and the refinanced company failed three years later. The well-equipped plant in Plattsburgh was still for sale as of 1921.

Lycoming Manufacturing Co.
Williamsport, PA
Trade name: Lycoming
Lycoming got its start making a variety of machinery in a factory it had bought from the Demarest Manufacturing Company, a maker of sewing machines and bicycles, in 1908. Lycoming's entry into the engine business came in 1910 as a supplier to the Velie Motor Vehicle Company of Moline, Illinois. A near-disastrous association with Charles Frederick Herreshoff's failed car-making effort followed, succeeded by a more fruitful relationship with Dort. Perhaps Lycoming's ultimate product was the engine it produced for the Duesenberg Model J. By then, the company was known as the Lycoming Manufacturing Company and was owned by E. L. Cord.

Lycoming's marine engine division was formed in 1928. Ads touted that "using modern production machinery produces a uniformly superior product with complete interchangeability of parts." A lineup of 4-, 6-, 8-, and 12-cylinder motors was soon available. These engines were an immediate success, the 299-cubic-inch straight-eight being selected for use in the 1929 Dodge 26-foot runabout. Thus began what was among the most sophisticated efforts to finally combine a mass-produced marine engine with the vision for mass-produced boats. A year later, Dodge announced that all its boats "will be equipped with new Lycoming marine motors, designed and built after the most exhaustive research by the Lycoming branch of the great Auburn automobile organization, makers of Auburn, Cord front-wheel-drive and the 265 horsepower Duesenberg automobiles under the supervision of two of the world's greatest motor car designers, Fred S. Duesenberg and Harry A. Miller, whose leadership in the automobile racing world is undisputed."

In 1934, with the Depression still strong, Lycoming ads stressed its marine engines were "The Greatest Value on the Market." By then, this was a comprehensive line of 35- to 165-hp engines available in 4-, 6-, and 8-cylinder versions, not to mention a 1,010-cubic-inch 325 hp V-12. Performance models could also be ordered. In 1936, for example, the UHE 223-cubic-inch six was a racing version of the UH. The UHE had a somewhat larger bore, twin carburetors, a hotter camshaft, and weight-reducing aluminum parts. Reduction gears were optional.

Ultimately, Lycoming focused on airplane engines. It remains in business today as a division of Textron.

Lynch, Ball & Co.
Grand Rapids, MI
Trade name: Termaat & Monahan, Capital
The company was begun in 1906 to sell gasoline engines and mill machinery at 31 Ionia St. The engines sold included the Termaat & Monahan and those built by the Capital Gas Engine Co.

M & S Engine Co.
Seattle, WA
Trade name: M & S
The company was listed in 1918.

M. O. Cross Engine Co.
Detroit, MI
Trade name: Cross
See Cross, M. O. Engine Co.

Macheck, A.J. & Co.
Milwaukee, WI
Trade name: Amphion

Magaan Air Brake Co.
Detroit, MI
Trade name: Harrison

Major Marine Gas Engine Co.
Detroit, MI
Trade name: Major Jones
The company was located at 972 4th Ave. At least one ad has been found from 1908. The company was still listed in 1909–1910.

Mallinson, J. H.
Rochester, NY
Trade name: Economic

One of at least a dozen engine makers in Rochester, Mallison was building 3-port, two-cycle motors as of 1913. The model line included single-, double-, and 6-cylinder engines.

Manhattan Manufacturing Co.
Manhattan, NY
Trade name: Manhattan
Introduced in 1897, the Manhattan was available in 2- to 15-hp models. These were four-cycle engines with an automatic inlet valve, an exposed camshaft and gears, and a ball-type thrust bearing. The engine had a governor intended to prevent over-speed if the propeller left the water in a seaway. The engine also was equipped with a fuel pump, a potential selling point because boats with gravity-fed fuel systems were not necessarily insurable. Apparently, this feature did little to make the Manhattan a success.

Mantha, Wm. L. Co.
Bayport, NY
Trade name: Baymotor

Manzel & Co.
Buffalo, NY
Trade name: M & B

Marathon
Trade name: Marathon

Marblehead Machine Co.
Marblehead, MA
Trade name: Colchester
Born in Windsor, Nova Scotia, in 1874, Arthur E. Colchester moved to the United States as a young man and learned about gasoline engines while employed at a New Bedford shipyard. In 1905 or '06, Colchester moved to Marblehead and opened a machine shop — the Marblehead Machine Company — on Lee Street, just above the boatyards that crowded the town's harbor. "Gasoline Engines Repaired & Adjusted," read Colchester's sign. At some point, Colchester developed his own two-cycle, one-lunger marine engine with make-and-break ignition. Colchester sold at least a few engines to local fishermen. In 1919, Colchester returned to Nova Scotia, settling in Bridgewater. There he designed a four-cycle, 4-cylinder engine for Acadia Gas Engines.
See also Acadia Gas Engines, Limited

Marine Iron Works
Chicago, IL
Trade name: N/A
The company was in business by 1906, located on Dominick Street, and advertising engines of 23 to 144 hp.

Marine Power Co.
Milwaukee, WI
Trade name: N/A
Located at 576 Grove St., this company was the successor to Ole Evinrude's failed Motor Car Power Equipment Co. That firm had been founded with the idea of supplying automobile makers with a standardized, four-cycle engine. The new company was up and running in 1906, offering four-cycle motors with a big, circular base that enclosed the flywheel. Single-, double-, and 4-cylinder motors were offered. The company failed, presumably because of inadequate capital. It was followed by an enterprise that enjoyed more success — Evinrude.

Marion
Trade name: Marion

Mark Motor Co.
Cleveland, OH
Trade name: Mark
A 4-cylinder model was announced in 1908, but production is uncertain.

Marks, Saml.
Oxford, MD
Trade name: Pilot
This Pilot was an open-base, four-cycle motor. The cylinder was supported on a casting with four legs giving the Pilot a steam engine–like appearance.

Mason Machine Works
Taunton, MA
Trade name: Mason

Massnick Manufacturing Co.
Detroit, MI
Trade name: Massnick
The company was begun in 1907 and was located at 1567-1569 River. Massnick filed for dissolution in 1909, and that may have marked the end of its marine engine efforts. It reemerged as the Michigan Auto Parts Company, which remained in business until March 1923.

Matheson
Trade name: Matheson
The 4-cylinder, four-cycle Matheson was a 90-hp engine, possibly a marine conversion. An ad appeared in *Motor Boating* in April 1925.

Mathewson
Trade name: Mathewson
No information has emerged about this 6-cylinder, 50-hp engine that was advertised in 1912.

Maus Marine Motor Co.
Buffalo, NY
Trade name: Maus
The Maus was a two-cycle motor equipped with jump-spark ignition.

Maximotor Makers
Detroit, MI
Trade name: Maximotor
This was a lightweight design with an aluminum crankcase. The 1913 range included the Model A (40 to 50 hp), Model B (60 to 70 hp), Model C (70 to 80 hp), and Model D (90 to 100 hp), with E and F models as special orders. The Maximotor was the brainchild of Max Dingfelder who located the business in the plant occupied a few years earlier by automaker Pungs-Finch at 1524-1528 Jefferson Avenue East. The company appears to have been in business for only one year.
See also Dingfelder Motor Co.

Maxwell & Fitch
Rome, NY
Trade name: Maxwell & Fitch

Maynard-Adams Engine Co.
Berkeley, CA
Trade name: Maynard-Adams
This 3½-hp engine used Ford parts.

McDuff, W. J. Machine Co.
Lakeport, NH
Trade name: Bumble Bee, McDuff
Born in 1870, William McDuff became a machinist who produced a line of two-cycle marine and stationary engines. McDuff developed his first single-cylinder engine into multicylinder models that included twins, possibly triples, and, according to at least one man who knew McDuff, a 4-cylinder version. Low-tension, make-and-break ignition was followed by high-tension, jump-spark systems.

Bumble Bee was said to derive from the black-painted cylinder casting highlighted by the brass fittings. Details of these motors remain to be unearthed, but it is thought that production began circa 1902. The McDuff Building in Lakeport became a well-known address in the town. William McDuff died at age 75 in Fernandina, Florida.

McFadden Mfg. Co.
Havana, IL
Trade name: Havana
McFadden was building a line of 1-, 2-, and 3-cylinder, two-cycle motors of 2½ to 15 hp and advertising them in 1908.

McIntosh & Seymour Corp.
Auburn, NY
Trade name: McIntosh
Generally thought of in regard to big ships' diesels (The company offered models of up to 1,350 hp in 1918), McIntosh also had some success in selling into the big motor yacht market, typically dominated by Winton. The company maintained an office in New York's Chrysler Building where its "Exclusive Yacht Representative" was available for sales consultations. By the early '30s, the engine builder was a division of the American Locomotive Company and was offering 60 models ranging from 250 to 2,000 hp.

Mead Gas Engine Co.
Providence, RI
Trade name: Mead
In 1901, Mead advertised engines of from 2 to 100 hp.

Mecco Engine Co.
Philadelphia, PA
Trade name: Mecco

Mechanical Engineers Investing Co.
Los Angeles, CA
Trade name: FW

Mechanics Foundry & Machine (*also* Mechanics Foundry)
Fall River, MA
Trade name: Mechanics
The shop was located on Davol Street. Ads in 1906–1907 mentioned the company's being the maker of the "DAVIS SINK," but a 1915 ad identified it as maker of a two-cycle marine engine. That was quite possibly the engine formerly built by Chester S. Belcher who, it is believed, then worked at the company.

Mercury Motor Co.
Long Island City, NY
Trade name: Mercury
The "Mighty Mercury" was a T-head engine with jump-spark ignition. The 60- to 65-hp model of 1910 was a 4-cylinder 6″ × 6½″ engine with the reverse gear mounted on the engine frame. By 1912, the Mercury was winning races in the region of New York City, Long Island, and New Jersey. In 1913, the line included a 5½″ × 5″ 52-hp 4-cylinder; a 6½″ × 7″ 125-hp 4-cylinder; and a 6½″ × 7″ 250-hp 8-cylinder. By 1916, the line included 1-, 2-, 4-, and 8-cylinder engines. The one-lunger developed 6 hp while the larger of the two eights (6½″ × 7″) developed 220 hp. All were jump-spark ignition and came equipped with a Paragon reverse gear. Mercury stressed that its engines had die-cast bearings with greater surface area than many competitors.

Metalweld Service Co.
Philadelphia, PA
Trade name: Fordson Marine Engine

Metz Engines
Trade name: N/A
The company advertised in *MotorBoat* in Feb. 1912.

Mianus Diesel Engine Co.
Stamford, CT
Trade name: Mianus
See Mianus Motor Works

Mianus Motor Works
Mianus, CT
Trade name: Mianus
This pioneer maker of high-quality two-cycle engines was begun in 1897 and soon became an accepted staple of the market. The company offered very successful two-cycle, two-port designs with make-and-break or jump-spark ignition. After a redesign in 1908 — the motor remained a two-port, two-cycle — the engines underwent modest refinements as deemed necessary. For example, in 1911 it was found that the make-and-break igniter could be improved somewhat in terms of adjustability and ease of access. At the same time, Mianus offered the option of snazzy nickel trim rather than brass. Force-feed oilers were also made optional since the practicality of mixing oil and gasoline had been found entirely sufficient. The 1911 line ranged from a single-cylinder (3¼″ × 3″), 700-rpm, 2-hp model to a 4-cylinder (5¾″ × 5½″) 600-rpm, 40-hp model that weighed 750 lbs. That year, Mianus claimed 15,000 engines in use on the East Coast alone, with many sold to the government. Demand then kept busy two factories, the original Mianus facility and another in Stamford. Ten distributors were located on the East Coast and in Seattle and St. John, New Brunswick. Four-cycle models were added in 1915. These were T-head engines with low-tension, make-and-break igniters. They were offered in 2-, 3-, and 4-cylinder versions, the latter rated at 32 hp. In 1916, the company introduced

Circa 1906 Model

1915 Nelseco (New London Ship & Engine Co.)

1902 Motor Vehicle Power Co. ad

1911 Mohawk (see S-R Manufacturing Co.)

1911 Murray & Tregurtha

1923 Missouri hot-bulb engine

1906 20th Century by New York Yacht, Launch & Engine Co.

a 450-rpm, 2-cylinder, overhead-valve model. Its inlet manifold was cast inside the exhaust manifold so that the fuel would be warmed before entering the combustion chamber. Also in 1916, Mianus introduced a flat-bottom skiff and a yacht tender powered by a little 2-hp one-lunger. The boats were built for Mianus by Luders. Two-cycle production continued into the late '20s, but by 1928 the company had been renamed the Mianus Diesel Engine Co., and a line of 1- to 6-cylinder diesels was the centerpiece. The largest (7¼″ × 9⅞″), was a direct reversing (sliding the camshaft reversed direction of rotation) model that was rated at 400 rpm and weighed 9,500 lbs.

Today, a pre-1918 Mianus in good original or authentically restored condition represents a fine example of the classic Connecticut-built marine engine.

Michigan City Motor Works
Michigan City, MI
Trade name: Junior

Michigan Marine Motor Corp.
Detroit, MI
Trade name: Michigan Marine Motor

Introduced around 1920, the Michigan was a single-cylinder, four-cycle engine with overhead valves. Because the engine was based on Ford parts, the company noted "repairs at any Ford garage." The 4-hp Michigan was aimed at 16- to 24-foot boats, but it is likely that anything much over 18 feet would have been under-powered. The firm went out of business in April 1924. A Canadian Michigan Marine Corp. was located in Ford City, Ontario.

Michigan Motor Co.
Grand Rapids, MI
Trade name: Michigan

In business by 1904, Michigan advertised its four-cycle, 1,000-rpm motors as "modern motors at modern prices."

Michigan Motor Works
Detroit, MI
Trade name: Junior

This 4-hp, single-cylinder engine was based on Ford parts. It was also made in Canada.

Michigan Yacht & Power Co.
Detroit, MI
Trade name: Sintz
See Sintz Gas Engine Co.

Mickelson Motor Works
Milwaukee, WI
Trade name:

The company was in business in 1911.

Middleditch, Benjamin
Detroit, MI
Trade name: Middleditch

Benjamin Middleditch was born in Amherstburg, Ontario, in 1859. He moved to Detroit in 1880 and established a machine shop. In 1885, he assisted his older brother Phillip to establish the Petoskey Iron Works in Saginaw and then returned to Detroit to continue his business. Although Middleditch earned a reputation as an engine expert, there is no evidence that he ever called his shop an engine company although the name Middleditch Engine Company did appear on name plates or castings of the engines he built. Instead, his City Directory ads read "Machinist and Engines" or "Engineer and Manufacturer of Machinery." Middleditch built a 1½-hp marine engine and a twin-flywheel stationary version. Because those engines were similar to the one built by Phillip Middleditch at the Petoskey Iron Works, it is likely that Ben Middleditch designed both. Among the items Middleditch made as a vendor was a "fuel injector" used by the Detroit Engine Works on its kerosene models. Possibly, Middleditch also had a hand in the design of several other Detroit-made 2-cycle marine engines, although this remains, at present, informed conjecture rather than documented fact. While renowned in Detroit for his ability to build machine tools and engines, Middleditch also earned an apparently substantial sum through real estate investments. Ben Middleditch died in 1916. His British-born wife, Lucy, lived until 1960. One of their sons, Leigh, carried on the family specialty and became president of an iron foundry.

See also Petoskey Iron Works

Mietz, August Corp.
New York, NY
Trade name: Mietz & Weiss (later Mietz)

In addition to its highly regarded and extensive line of stationary engines, which were sold on an international basis, this company produced a respected line of heavy-duty, hot-bulb marine engines. August Mietz was born in Weige, Germany, on Dec. 1, 1834, and became a foundryman. After emigrating to America in 1859, Mietz started the August Mietz Iron Foundry. In 1894, he became associated with another German immigrant, inventor Carl W. Weiss, who was then perfecting a three-port, two-cycle, hot-bulb oil engine. That meeting led to the establishment of a company to build Weiss's engine. Soon, a line of horizontal- and vertical-cylinder engines came on the market, ranging from 2 to 80 hp. It was followed by a line of gasoline stationary engines.

By 1908, the hot-bulb line included 4-cylinder engines of 30, 60, and 120 hp. The 30-hp version weighed 3,000 lbs. and cost $1,800. When Mietz died in February 1915, Weiss left the firm to found the Carl W. Weiss Engineering Company, and the Mietz family continued engine production. That year, a 200-hp, 4-cylinder Mietz & Weiss oil engine was installed in Lightship No. 98; in 1916 Lightships No. 101 and 102 (the *Brenton Reef*) were equipped with a similar engine. The 1922 model line included 2-, 3-, and 4-cylinder models of from 15 to 350 hp. While the 15 model weighed 1,905 lbs., the biggest four (16″ × 21″) weighed 65,000 lbs. These engines were built primarily for commercial applications.

In 1922, the engines' trade name was listed simply as Mietz, and it is possible that this change had occurred some years earlier. By the early '20s, too, the trend of engine design was always toward more compact, lighter machines and this fact, coupled with the gradual appearance of modern diesels and the increasing competition among makers of smaller, kerosene-fuel engines, must have hit the company hard. Present research gives no indication that Mietz attempted to innovate to meet changing times. By the mid-'20s, Mietz no longer appeared in annual listings of marine engine manufacturers.

Miller Gas and Vacuum Engine Co.
Chicago, IL
Trade name: Miller
See Miller Engine Company

Miller Engine Co.
Chicago, IL
Trade name: Miller
This North Talman Avenue firm traced its origins to either 1901 or 1903 (depending on sources), when Joe and Bernard Miller became interested in the possibilities of the marine engine business and began Miller Brothers. By 1911, the brothers offered a line of nine 2-cylinder engines of from 6 to 25 hp, and five 4-cylinder engines of 18 to 50 hp. These Millers were medium- (500 rpm) or heavy-duty (300 to 400 rpm) machines. The 40-hp four had dimensions of 7″ × 7″ and weighed 2,400 lbs. The 50-hp four had dimensions of 7″ × 10″ and weighed 3,500 lbs. Cylinders were cast separately. Ignition was jump-spark. Schebler carburetors were used. In 1914, the company added a 2½-hp outboard, the "Miller Portable Motor" to the line. Its designation had become the P-1 by 1922.

At some point in the early-to-late teens, the company became known as the Miller Gas and Vacuum Engine Co. In 1917, Miller displayed two 4-cylinder engines at the 1917 Power Boat Show under the auspices of its agent in the city, Consolidated. The smaller model (3½″ × 5″) was rated at 12 to 20 hp at 600 to 1,000 rpm. The larger engine (5¼″ × 7½″) produced 28 to 35 hp at 400 to 550 rpm. Both had removable heads. The smaller had a five-bearing crankshaft, the larger had six main bearings. That was unusually ample support. Both engines were en bloc designs. These fours were clean-looking engines with Bosch magneto ignition, ideal for a runabout or small cruiser. By 1922, the company name had been shortened to Miller Engine Company. In 1922, models were available in 1-, 2-, 3-, and 4-cylinder models of from 4 to 50 hp "For Every Size and Type Boat." These remained the medium- to heavy-duty L-head motors for which the company was known. Joe Miller pointed to lower initial cost, lower cost of operation, and greater demand as the reasons his company focused on such engines. The Miller was marketed as "The Motor of Renown" and was positioned as a common-sense design that offered high value per dollar. By the late '20s, the company name evolved yet again when it became the Millers Motor Corporation, but the location remained on North Talman Avenue. It is likely that business ceased during the Depression.

Milwaukee Motor Co.
Milwaukee, WI
Trade name: Imperial
The word "imperial," like "peerless," was one that must have sounded good to engine makers. This particular Imperial was described in *Motor Boating* in December 1910 as a 4-cylinder that developed 80 hp at 800 rpm. The crankcase was fully enclosed, "insuring an oil bath at all times and keeping out all dust."

Milwaukee Motor Mfg. Co.
Milwaukee, WI
Trade name: Wisconsin, Consistent

Mingst, C.P.
Evansville, IN
Trade name: N/A

Missouri Engine Co.
St. Louis, MO
Trade name: Missouri
A builder of two-cycle, "hot bulb" (semi-diesel) models, Missouri was one of a number of firms that hoped to capitalize on the rising cost of gasoline that had begun by the early teens when a gallon of gasoline cost 20 cents but a gallon of fuel oil only five cents. "Oil Engines Pay Big Dividends," said company ads. Ignition required to start the engine was provided by heating a ball-shaped projection in the combustion chamber with a blow torch attached to the cylinder head. (Later, the bulbs might be heated electrically.) Once the engine was running, the heat of combustion kept the ball at the needed temperature. Missouri was begun in 1914 by machinist Hugo E. Lippert, who became the vice president, and Adolph Winkelmeyer, president. In 1918, Missouri offered 1-, 2-, 3-, and 4-cylinder engines to run on kerosene, fuel oil, or crude oil. The firm was located at 2810 N. 11th Street. The market here was presumably commercial boats. The company was listed in the City Directory through 1929.
See also Lippert, H. Co.

Model Gas Engine Works
Peru, IN
Trade name: Model
This company was acquired in 1914 by the Pittsburgh Model Engine Company.

Modern Machinery Co.
Wilmington, DE
Trade name: N/A
An ad for the company appeared in *Motor Boating* in Dec. 1911.

Modern Marine Equip. Co.
Cincinnati, OH
Trade name: Modern
Modern was in business in 1914.

Mohler & DeGress
Astoria, NY
Trade name: Mohler & DeGress

Monarch Tool Co.
Cincinnati, OH
Trade name: Little Skipper, Skipper
Monarch opened for business in 1905 as a tool and die maker, metal pattern maker, and model maker. Probably in

1909, Monarch acquired the tooling for the Gray-Hawley Little Skipper (2 hp) first built by St. Clair, and Skipper (4 hp). In a 1911 ad, Monarch claimed that more than 4,000 Little Skippers were in use, but this would have included those sold by Gray-Hawley. Probably, the biggest sales for these engines had been made by the time Monarch took over. Monarch returned to its core business and remained in Cincinnati until 1956 when it moved across the Ohio River to Covington, Kentucky. It was still listed there until 1987.

Monitor Vapor Engine and Power Co.
Grand Rapids, MI
Trade name: Monitor, Mogul
In 1911 *The Gas Engine* referred to this company as "old timers in the marine field." The article made clear, however, that the company was also past tense. The company was begun in 1894. The Monitor, at least, was a two-cycle engine. In addition to motors, the company made a two-section boat that could be bolted together.

Morristown Boat & Engine Co.
Morristown, NY
Trade name: Morristown
In production by 1914, if not earlier, the Morristown engine line included both two- and four-cycle models. Comparatively high rpm (900) and light weight were key selling points of the three-port two-strokes. A 4-hp twin and 6-hp 3-cylinder were offered. Both had jump-spark ignition, and the company took pride in pointing out that the bronze bevel gears driving the vertical timer were cut rather than die cast. The suggestion, presumably, was that the motor's timing gears would last longer than less expensive versions on some competing models. The motors had detachable cylinder heads and an aluminum base. The four-cycle engine was a 4⅛" × 5" 4-cylinder, L-head that developed 25 hp. A phosphor bronze connecting rod, said to be "far superior construction to the ordinary die-cast rod, for the material is not subject to crystallization," was used. Ignition was by high-tension magneto.

Morro Motor Co.
Morro, CA
Trade name: Morro

Morton Motor Co.
Detroit, MI
Trade name: Morton
Originally founded in 1911 and associated with Kermath as a sales and marketing group, Morton remained a distinct entity after Kermath reorganized its sales strategy, an event that probably occurred part way through 1913. Morton seems to have sold its own version of the 3½" × 4" 12-hp Kermath. The Morton was equipped with a Holley carburetor rather than the Kermath's Kingston, and it was rated 13 rather than 12 hp. The usual jump-spark or magneto ignition systems were offered. The $180 price was identical to the Kermath 12's. Apparently, Kermath also supplied its new 20-hp model to Morton. The 20-hp Morton was listed in 1918 by the Peninsular Auto Parts Co. of Detroit, which may have been a successor to the Peninsular Motor Co.
See also Kermath Manufacturing Co.

Motor Vehicle Power Co.
Philadelphia, PA
Trade name: Motor de Luxe
In 1901, the company was advertising its "Reliable and Simple" four-cycle, 6½-hp motor for marine and automobile use. The following year, a 3-hp twin and an 8-hp four were available. This was, for its time, quite a forward-looking motor with en bloc cylinders. Ads claimed no gaskets were used. The motor was finished in baked enamel.

Mottville Machine Works
Mottville, NY
Trade name: N/A
The company was listed in a 1904 *Marine Buyers' Guide*.

Mullins, W. H. Co.
Salem, OH
Trade name: Mullins
In 1882, W. H. Mullins purchased an Ohio company named Thompson & Bakewell, a maker of metal statuary, ornaments, cornices, and ceilings for buildings. The firm then became Bakewell & Mullins, but Mullins became sole owner in 1890. In 1894, having survived the sinking in icy water of his duck boat, Mullins began experiments with an unsinkable steel boat. The first one was manufactured in 1896. These and other models became immensely popular, and Mullins began work on a single-cylinder–powered rowboat in 1904. Soon the company became an important builder of steel boats powered by engines that were supplied by Detroit Auto Marine, Loew, Roberts, and Ferro. During the period 1914–1917, other brands including Sterling, Red Wing, Kermath, and Trebert were tested. The 1931 Mullins Sea Eagle was Lycoming-powered. Depending upon the business arrangement, some of the supplied engines carried a Mullins tag, but usually they also carried the maker's name in some fashion, such as "W. H. Mullins designed by E. W. Roberts." In 1956, Mullins Mfg. Co. was merged with American Radiator & Standard Sanitary Corporation.

Murray & Tregurtha
Boston and N. Quincy, MA
Trade name: M. & T.
Murray & Tregurtha sometimes referred to itself as the second oldest marine engine maker in the country without suggesting who it considered to be first. The company originated in East Boston. There, a Charlestown native named Alfred T. Tregurtha, together with his brother, George, and their Nova Scotia–born partner, John A. Murray, were in the process of making the transition from steam engines and equipment and launches to internal combustion engines in 1892. Readers of the early issues of *The Rudder* would have been familiar with the company

1905

1905

1905

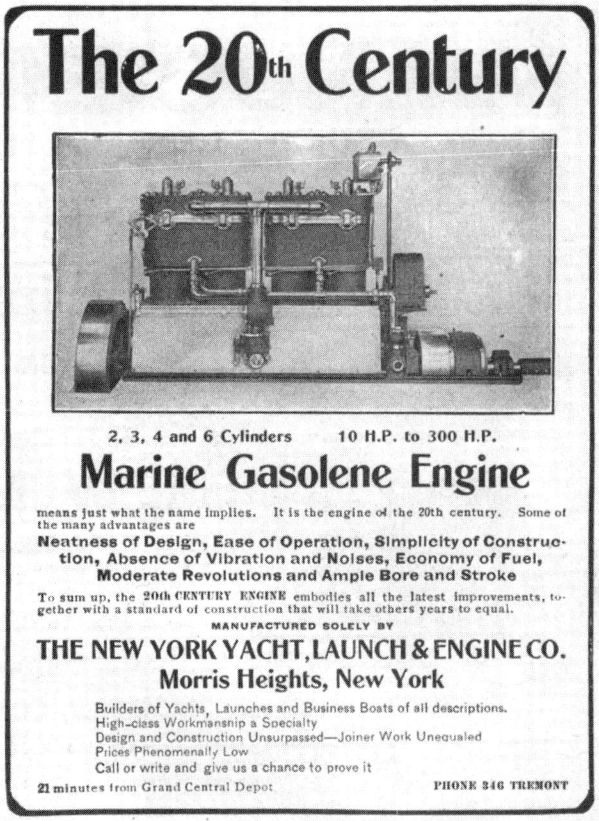

1911 New York Launch ad for the 20th Century

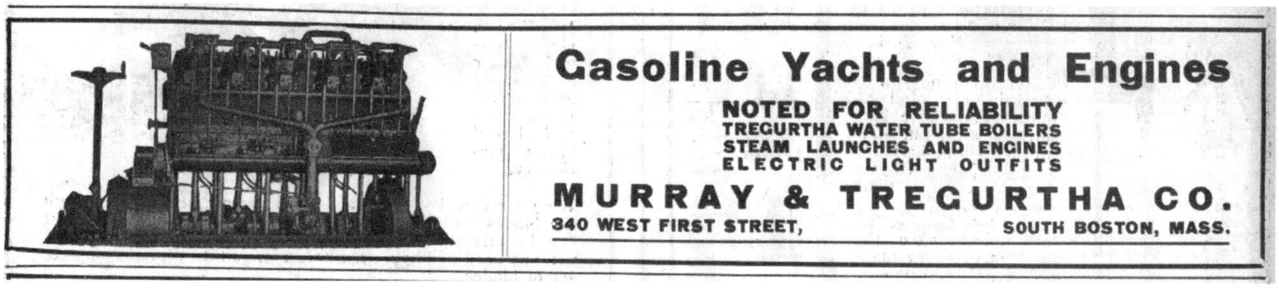

1913

for it was among the magazine's earliest advertisers, starting in 1896, which may be the year actual production began.

Heavy-duty, vaporizer-equipped, 1-, 2-, and 4-cylinder models were produced, together with steam engines and boilers, and the company kept that product mix for years. The company offered its engines in its own yacht tenders, launches, cruising sailboats, and power cruisers. Based on patents granted to George Tregurtha, those early engines were rather beautiful machines with a big, round crankcase that, for safety's sake, enclosed the flywheel. The cylinder had a polished brass water jacket. Ignition was electric via steel and platinum electrodes. A battery was used for starting, but then the engine was switched to its standard magneto. The engines included a built-in reversing gear, a piece of equipment considered mandatory to the advancement of gasoline engines if they were to equal this convenient aspect of a steam engine, and a muffler. The engines were initially available in 1½- to 32-hp versions, ranging in weight from 250 to 4,000 lbs. In 1916, the company introduced a magnificent overhead cam engine with four valves per cylinder. This 400 hp, 2,000-rpm six was among the most advanced marine engines of its time.

Murray & Tregurtha engines powered a wide variety of boats. Some examples include: a 3½-hp single that drove a 21½′ launch at six to eight mph; a 50′ × 10′4″ cruiser built by the Rice Brothers in East Boothbay with a 4-cylinder, 30- to 40-hp 6½″ × 8″ M & T that was good for a 10-mph cruise speed; and a Murray & Tregurtha-built 50′ cabin launch with a 30-hp engine of the company's own build.

During World War I, the company built Hickman Sea Sleds for the Navy, but following the war, in 1919, the firm was reorganized as the Murray & Tregurtha Corporation, whose focus was solely on marine engines. In 1920 Murray & Tregurtha introduced its beautiful 6-cylinder K-6, an engine whose double overhead camshafts were housed in aluminum tubes and operated the valves — four per cylinder — via lightweight roller push rods that acted on the valve stems. The cams were hollow and pressure-lubricated and driven by gear-operated vertical shafts off the crankshaft. A pair of en bloc steel castings formed the cylinder blocks.

Two distinct model lines were developed during the '20s. One had the cylinders mounted on steel columns similar to a steam engine, and the "crankpit" was enclosed by easily removed splash plates. These were the heavy-duty "G" engines, 4-cylinder models with 8½″ × 11″ bore/stroke dimensions that produced from 85 to 150 hp and weighed between 5,000 and 6,200 lbs. The other model, either initially or eventually included models called the "D," "E," "L," and "H," had a conventional cast-iron "box crank chamber." In M & T language, these 450- to 800-rpm engines were labeled medium duty and weighed between 1,200 and 6,000 lbs. Battery and magneto ignition was used.

A variety of invariably splendid machines followed. The 1928 lineup included 2-, 4-, 5-, and 6-cylinder engines. The company never stopped innovating and adapting. In 1934, the relatively high-speed K-6 (6¼″ × 7¾″) produced its maximum 325 hp at 1,650 rpm. The 6½″ × 8″ M-4 (65 to 90 hp) and M-6 (110 to 175 hp) were medium-speed 1,100-rpm engines. In 1936, an engine that started on gasoline but then ran on fuel oil proved popular. The company went on to supply an unusual heavy-duty outboard engine power unit to the Navy during World War II.

Muscatine Engine Co.
Muscatine, IA
Trade name: Muscatine
The company offered two-cycle, 2- to 40-hp models in 1909. A 4-cylinder, 50-hp model was available by 1913.

Mystic Motor Co.
Mystic, CT
Trade name: Mystic
In 1916, this company unveiled another of the many two-cycle engines that attempted to compress the incoming mixture somewhere other than in the base. In this case, the mixture was compressed in two separate air pumps and then passed into the combustion chamber through an automatic valve.

N and S Engine Co.
Seattle, WA
Trade name: N and S

Nadler Foundry and Machine Co., Inc.
Plaquemine, LA
Trade name: Nadler
The company was founded in 1888 by Henry Nadler, who by 1920 had bought the Sears Motor-Go engine (manufactured by Lockwood Motor Co.) for resale. Nadler acquired the rights and patterns of Lockwood-Ash circa 1924 and produced the engines for about 20 years. The engines were offered in 1-cylinder 2½- and 4-hp models and an 8-hp 2-cylinder. Most were used by loggers and fishermen in south Louisiana.
See also Lockwood-Ash

National Auto Engine
Trade name: N/A
The company advertised a four-cycle, 6-cylinder engine in 1912. This was probably an automobile engine.

National Boat & Engine Co.
Muskegon, MI
Trade name: N/A
This planned conglomerate was conceived as an umbrella organization that would be composed of the Racine Boat Manufacturing Company (the "leading company") in Muskegon; Truscott in St. Joseph; Pioneer Boat and Pattern Company in Bay City; Inland Lake Boat Company of Lake Geneva, Wisconsin; the Pope Boat Company in Fond du Lac; boat-builder Shell Lake; and the Pierce Engine Company in Racine. It is likely that the Outing Boat Company of Ashland, Wisconsin; the Western Boat and Engine Works in Michigan City; and the West Mystic Manufacturing Company in Connecticut were also part of the plan.

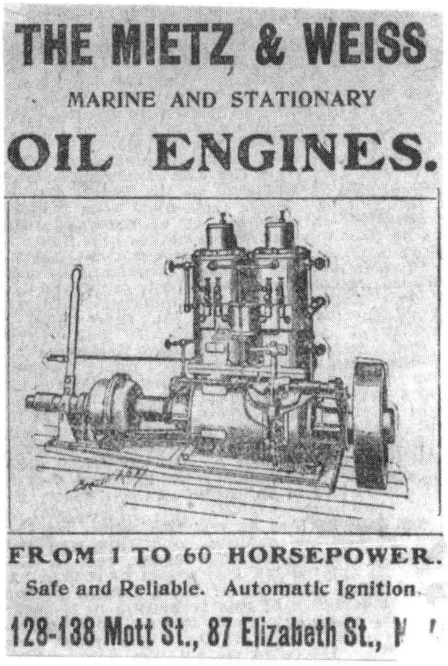

Circa 1911 Mietz & Weiss ad

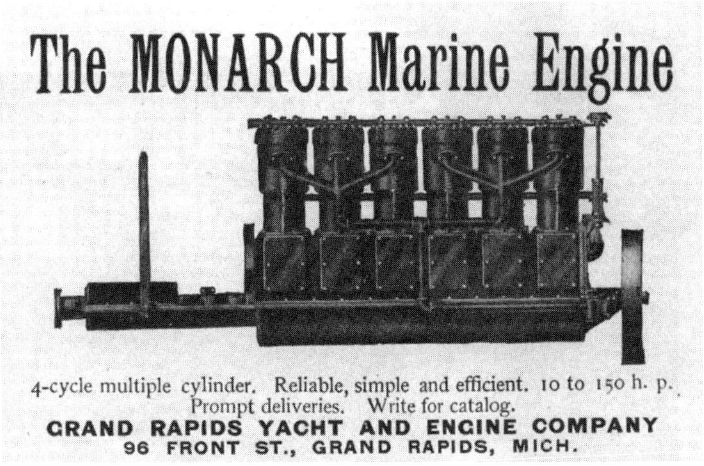

1906

1916 Miller

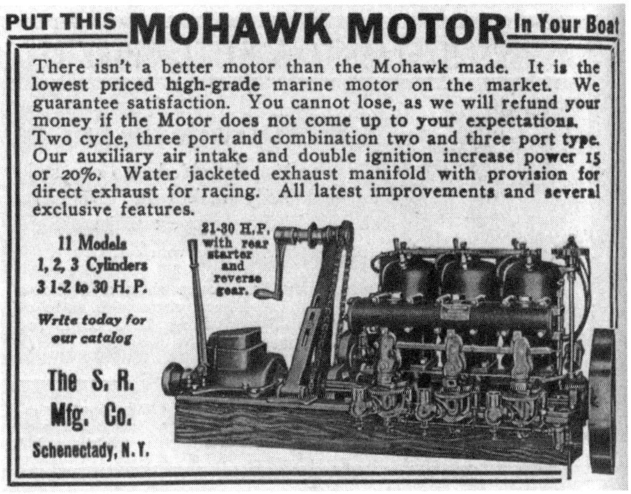

1913

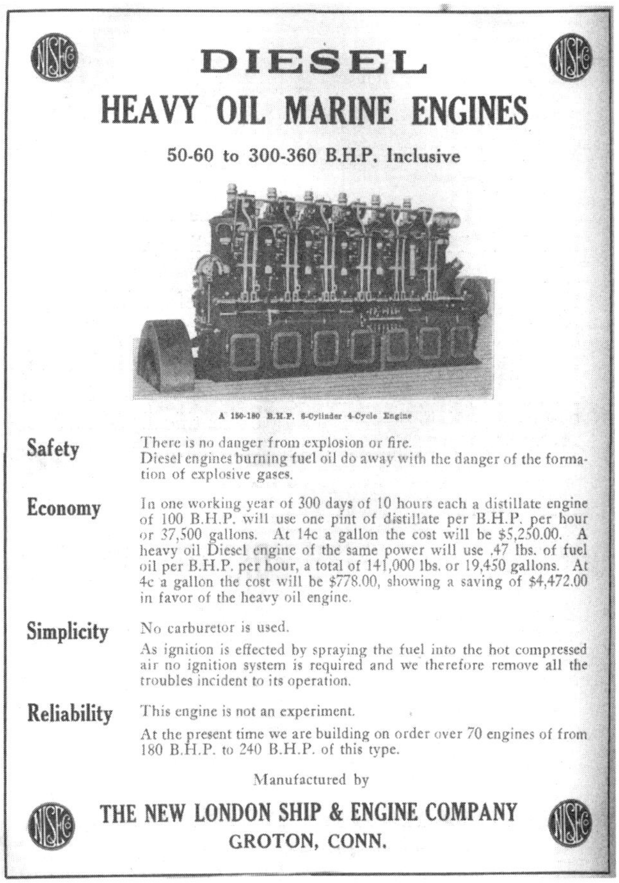

1915

This would-be General Motors-like assemblage of boat and engine builders was said to have $10 million in financing, but that was certainly hype. The company spoke of establishing showrooms in major cities "where the public may conveniently be shown our entire line of boats and engines." Also envisioned was the establishment of "supply stations" where prompt repairs could be made and supplies obtained.

Somebody among those involved calculated that all this would result in National controlling "90% of boat manufacturing from Maine to the Mississippi River." This was a very questionable claim! The costs of merging all the companies was to be covered by monies raised from a bond issue. After that failed in September 1911, National's principal property, the Racine Boat Manufacturing Company, was taken over in January 1912 by a receiver bank in Portland, Maine. By March, creditors were taking charges of fraud to the U.S. court in Grand Rapids and won the right to use the assets of the individual companies to pay off creditors and, perhaps, resume some sort of production. The business, such as it was, finally closed in early 1916.

National Machine Co.
See Daimler Motor Co.

National Motor Co.
Detroit, MI
Trade name: National

Notice of this engine appeared in *MotorBoat* in 1909. Samuel N. Rapp was said to have concerned himself with creating an engine that would produce exceptional power in a small package. "It is claimed to be the only engine upon which two explosions are obtained at each revolution ..." The engine was unconventional in all respects and seems never to have been a viable commercial endeavor.

National United Service Co.
Detroit, MI
Trade name: Simplex

Neponset Engine & Machine Works
Neponset (Boston), MA
Trade name: Rex

The Rex was a jump-spark engine said to be a "brand new combination two- and three-port engine." It was built to patents awarded Fredrick A. Thurston who lived not far from Boston in the city of Lynn. Thurston's patents had been awarded in 1908 and 1910 for a two-cycle engine in which the mixture was admitted through a pair of check valves, one in the cylinder wall, the other in the base. Ads for the Rex appeared in 1912 although it is possible that engines were built prior to that year. The engine was offered in 1-, 2-, 3-, and 4-cylinder models, all with 4½″ × 5″ dimensions. The engines did turn relatively fast, as they were rated at 500 to 700 rpm. Manufactured in Neponset, the Rex was sold in Boston by the Rex Motor Co. on Devonshire Street. (Another Rex engine, a 4-cylinder four-cycle that participated in motor boat racing in the Detroit area in 1907, is not known to have had a connection to this Massachusetts effort.)

Newport Engineering Works
Newport, R. I.
Trade Name: Newport

New England Marine Engine Co.
Noank, CT
Trade name: N/A

New Jersey Motors
Keyport, NJ
Trade name: N.J.M.

Charles A. Carlson had designed and built a 30-hp 4-cylinder, four-cycle marine engine in 1903 and later a heavy-duty engine, but none entered production. After World War I, Carlson redesigned the "Fordmarine" engine (based on Model T components) then being built by New Jersey Motors. New patterns, jigs, and production methods followed, and the resulting N.J.M. appealed to a dealer network seeking a reliable, low-priced marine engine.

New London Motor Co.
New London, CT
Trade name: Pequot

The two-cycle Pequot engine had cylinders that were cast in pairs. One-, 2-, and 4-cylinder models were offered in ratings of 4 to 16 hp. The four had jump-spark ignition; the others, make-and-break.

New London Ship & Engine Co.
New London, CT
Trade name: Nelseco

The New London Ship & Engine Company (Nelseco) began operation in the spring of 1911. The company was begun as a division of Electric Boat after that firm acquired the patent rights to the two-cycle diesel developed in Germany by M.A.N. Electric Boat had recognized that a significant market could develop from the Navy and another from commercial operators. A new factory was erected in Groton. Here, Nelseco produced a 6-cylinder, 300-hp machine that weighed 32,000 lbs. and was installed in a Standard Oil Company barge. This appears to have been the first installation of a commercial marine diesel in the United States. The engine was not a big success. It was afflicted with problems with its two-stage air compressor, various other mechanical ills, and high oil consumption. It was removed a few years later.

All these problems, together with the need to pay the license fee, seem to have been ample reason for Nelseco to abandon the two-cycle diesel. With the expiration of the Diesel patents, Nelseco's engineer, G. C. Davison, began work on a four-cycle engine. The engine was reasonably simple in its conception and featured horizontal overhead valves operated directly by big, vertical rocker arms that were themselves driven directly off the camshaft that ran along each side of the cylinders. According to the company, the valve layout resulted in locating the fuel injection valve "quite a little distance from the piston top and [the valve] does not receive so much of the intense flame on one spot as otherwise would be the case."

The first of these new engines was a 4-cylinder (9″ × 12″) that was rated at 120 horsepower. A Nelseco-patented fuel system with a single pump was used that was, according to the company, "made so durable and strong that there is little danger of disarrangement and no opportunity for trouble by operators changing adjustments, since there are no adjustments to change." The prototype easily developed 115 horsepower and actually produced a maximum 137 horsepower when pushed beyond its rated 350 rpm. A reversing gear was used rather than a direct-reversing mechanism.

The Nelseco must have been refreshing in its simplicity after the M.A.N.-style two cycles. The company reported that the engine's operation "is easily understood by men who were previously familiar only with gasoline and distillate engines, and is so very simple that there is little chance of its getting out of order in service." A demonstration engine was installed in an 84′ motor yacht, the *Idealia*, and ads reported a fuel economy of six gallons of fuel oil per hour. "It would take 15 gallons of gasoline to produce the same power, *at approximately ten times the cost.*"

The first engine was sold to exactly the sort of customer Nelseco intended. It was shipped to that region already buying semi-diesels in some numbers, the Pacific Northwest, where it went into the cannery tender *Warrior* in 1913. The following year, a 6-cylinder Nelseco powered what may have been the first U.S. diesel tug, the *Chicamauga*.

These big Nelsecos were joined in 1915 by a smaller displacement (7″ × 9″) 4-cylinder. It produced 60 hp at 450 rpm. In the period 1914–19, Nelseco reported that it sold two dozen of its four-cycle engines including a pair of 60-hp versions, 14 120-hp models, and eight more producing 180 or 240 hp. Nelseco had now established itself and would eventually find its real niche as a builder of big, heavy-duty marine diesels that by 1928 ranged in horsepower from 150 to 1,600 and in weight from 25,500 to 185,000 lbs. By 1936, the model line had been reduced to six MIR-series models. These included 6-, 8-, and 10-cylinder engines of from 250 to 2,000 hp. Operations ceased after World War II.

New Way Motor Co.
Lansing, MI
Trade name: New Way
A 5-hp, single-cylinder gasoline or kerosene model was listed in 1922.

New York Kerosene Oil Eng. Co.
New York, NY
Trade name: New York
New York was among the earliest marine engine makers to extol the safety and low cost of kerosene as a fuel. "The Safest and Cheapest Power known to man," said the company's ad in *The Rudder* in November 1900. In 1902, a 10-hp engine produced by this company powered the 38′ *Abiel Abbot Low* across the Atlantic, the first such crossing by a motor boat. The engine was started by heating a bulb for about six minutes. Compression was then released, and the engine cranked a couple of turns to draw air into the cylinders. Then, upon the compression stroke, the oil spray was injected as the piston approached top dead center and the motor was off and running.

New York Marine Power Co.
New York, NY
Trade name:
The company was in business by 1901 producing an open-base engine.

New York Safety Steam Power Co.
New York, NY
Trade name: Dock
Although a gasoline engine was noted in 1907, whether any were produced is uncertain.

New York Yacht, Launch & Engine Co.
Morris Heights, NY
Trade name: 20th Century
In 1896, New York Yacht established a plant on the banks of the Harlem River and soon advertised: "Design and Construction Unsurpassed — Joiner Work Unequalled." Management saw an opportunity in offering its own line of engines. To underscore the modern nature of the engine, it was named "20th Century." By the mid-teens, there were 2-, 3, 4-, and 6-cylinder models ranging from 10 to 300 hp. These fine, low-speed (400 rpm) engines had cylinders cast in pairs and were fully contemporary in design. The triple was especially impressive, based on a very streamlined-looking single casting. Until 1915, the valve mechanism — automatic inlet and mechanical exhaust valve — was exposed, but the 1915 models featured enclosed valve gear resulting in quieter operation. A fly-ball centrifugal governor was also new for 1915. The 20th Century had big main bearings, the one at the flywheel being 6″ long. Connecting rods were drop-forged steel with phosphor bronze bearings. Lubrication was force-feed. A fully enclosed reverse gear was standard.

The 20th Century was marketed as "strictly high grade four-cycle engines for heavy duty service." They found favor in many big power yachts of the time. Examples included a 6-cylinder, 100-hp 20th Century in an 82′ Gielow-designed motor yacht that could drive the vessel at 14 knots, and a 50- to 60-hp model in a fine 57′ × 13′ raised deck cruiser that cruised easily at 11 knots. The engines remained available into the late '30s. In 1936, an 1,128-cubic-inch 500 rpm, 20-60 hp, 4-cylinder and a 1,692-cubic-inch 75- to 100-hp six were listed.

Newton Crane Gas Engine Co.
Boston, MA
Trade name: Crane
The Crane was a horizontally opposed engine. As was usual for such configurations, it was advertised as being the "only absolutely balanced engine on the market." Each of the 4 cylinders contained two pistons, and they were driven by an "explosion between the pistons." The engine's supporting base was a manganese bronze casting supplied by the Cramp shipyard in Philadelphia.

1913 Niagara

1911 Niagara

1923 Niagara 9- to 15-hp "Special"

1913 Niagara

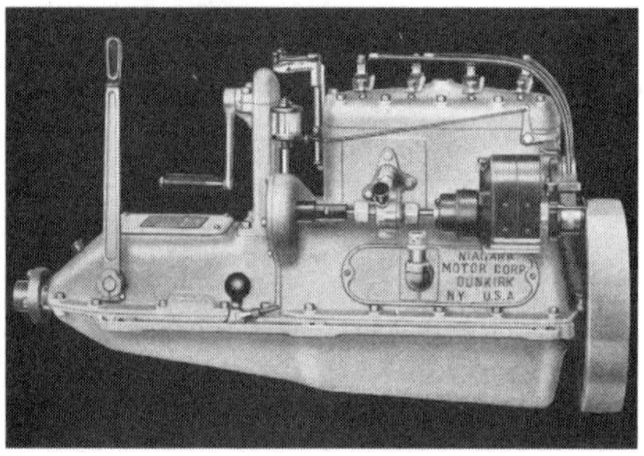

Circa 1930 Niagara 12- to 16-hp "Special"

1915 Northwestern

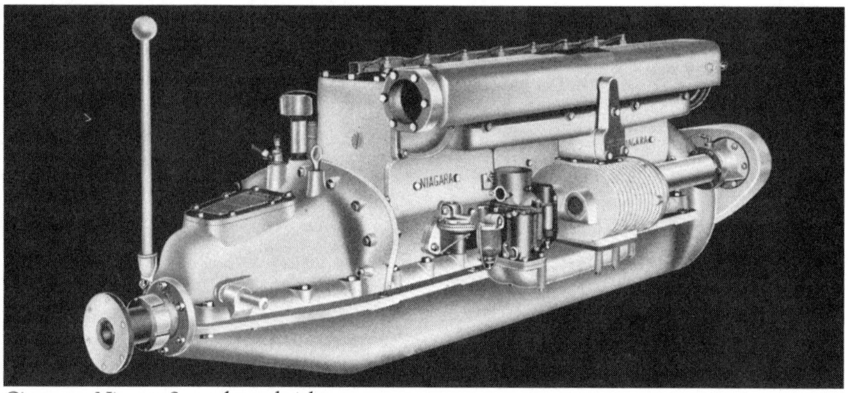

Circa 1930 Niagara Supercharged eight

In 1905, Crane advertised ratings of 25, 50, 125, 250, 500, and 1,000 hp. The company hoped to find customers in the stationary, automotive, and marine markets.

Niagara Gasoline Motor Co.
Buffalo, NY
Trade name: Niagara
One of four important marine engine manufacturers in Buffalo, this company was founded in 1904 by Louis Ducro and was known as the Ducro Manufacturing Company. When Ducro sold out to Charles Narraway sometime around 1912, the name was changed to the Niagara Gasoline Motor Company and a new plant was built in Dunkirk, New York. The early Niagaras were simple, functional T-head engines with separate cylinders cast without removable heads but with access plugs for the valves. "No intricate, complicated or superfluous parts," said ads, and they were accurate. Force-feed lubrication was standard and the company, throughout its existence, stressed that an owner had the "satisfaction of knowing that he can operate his motor with little or no anxiety in regard to lubrication as all working parts of the motor are at all times running in a veritable bath of oil."

The racing success of Ralph Sidway's *Arab III* in 1909 gave the engine's reputation a boost, although its primary benefit was probably to increase interest from those who planned to use the engine in a cruising or commercial application. After World War I, a line of affordable 9- to 15-hp engines was introduced known as the Niagara Special. So were D and E models of from 25 to 120 hp.

Niagara's emphasis on advanced metallurgy commenced with the first models and continued on through the '20s and '30s. The magnificent supercharged 370-cubic-inch 8L and 452-cubic-inch 8M of this later period had a crankcase cast of "special aluminum alloy" that weighed one-third less than cast iron and seven percent less than other aluminum alloys while being 50 percent stronger. Nickel-steel sleeves were pressed into the block. While the standard color of these engines is unknown, customers could obtain their engine in a color of their own choosing. Production ceased at some point in the '30s although the plant continued to supply connecting rods for other engines.

Nichoalds Co.
Detroit, MI
Trade name: Nichoalds
Nichoalds went out of business in 1912 and left little or no record of its production — or at least none that has surfaced.

Nichols Power Co.
Stamford, CT
Trade name: Perfection
The Perfection shown in 1910 was a 40-hp, four-cycle engine with 7″ × 6″ bore/stroke dimensions and make-and-break ignition. It was designed to operate on kerosene.

Nieland, J. E. Co.
San Francisco, CA
Trade name: Nieland
Nieland advertised as early as 1911 in *MotorBoat*. By 1918, a line of four-cycle engines that ranged from 1½ to 24 hp was in production. The company also was listed among American's marine engine makers in 1919.

Noank Motor Co.
Noank, CT
Trade name: Ideal

Norman, H., Co.
Oshkosh, WI
Trade name: N/A
The company was listed in a 1904 *Marine Buyers' Guide.*

North American Motors Co.
Pottstown, PA
Trade name: N. American

Northwestern Machine Co.
Detroit, MI
Trade name: Little Tiger

Northwestern Steel & Iron Works
Eau Claire, WI
Trade name: Northwestern
From a plant enlarged in 1910, the company shipped jump-spark 2-, 3-, 4-, 6-, and 10-hp two-port models throughout North America and to South America, Europe, and Australia. Engines were finished by brushed-on iron filler, sanding, paint, and varnish top coat.

Northwestern referred to its engines as being built of "semi-steel and not common cast iron." By 1914, a 2-cylinder, 18-hp motor was on the market, and the company advertised, "They are simplicity personified — nothing to get out of order and nothing to wear out except the main bearings which can be replaced at a nominal cost." In the spring of 1914, the company moved from 703 Spring St. to 772 and the name was shortened to Northwestern Motor Company.

Noye Mfg. Co.
Buffalo, NY
Trade name: Hopkins
Noye was making the two-cycle Hopkins in 1902.

Olds, P. F. & Son
Lansing, MI
Trade name: Olds
In July, 1911, *The Gas Engine* reported that Olds had "built a four-cycle marine engine, double cylinder in 2, 4 and 8 H.P. sizes. This engine was designed with the gasoline tank cast integral with the motor to avoid danger from leaky pipes and tank." Unfortunately, the article gives no specific date to the Olds marine engine. Pliny F. Olds and his son Wallace opened a shop to repair steam engines and farm equipment in about 1880. Later, Wallace's younger brother Ransom joined the company before moving on to become one of the automobile industry's great pioneers. It seems unlikely that the Olds marine engine found more than a very limited market although at least one has survived. Olds was known primarily for its stationary engines. The marine engines referred to by *The Gas Engine* may have been built circa 1897.

1913

1949 Packard marine engine ad

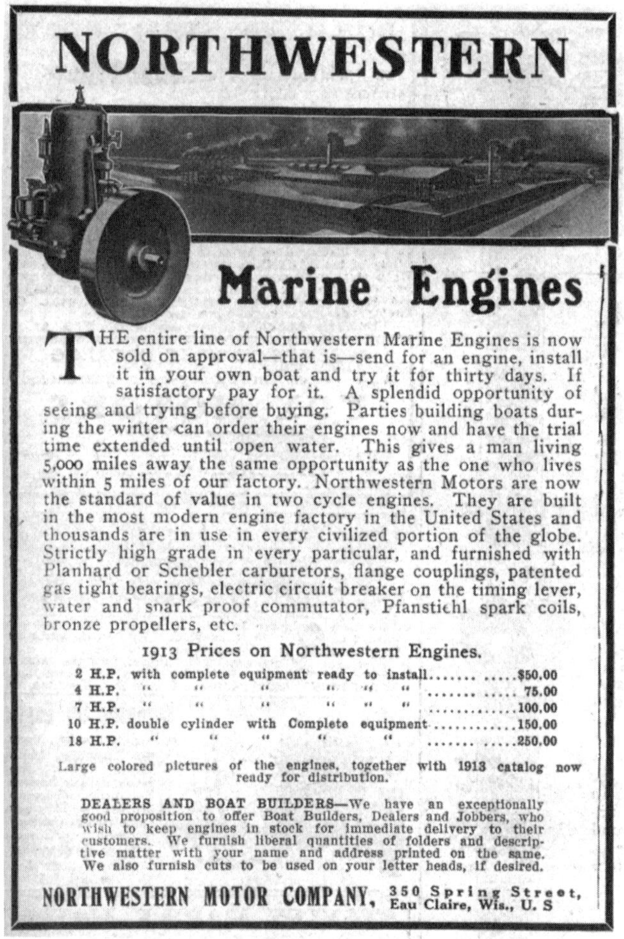

1913

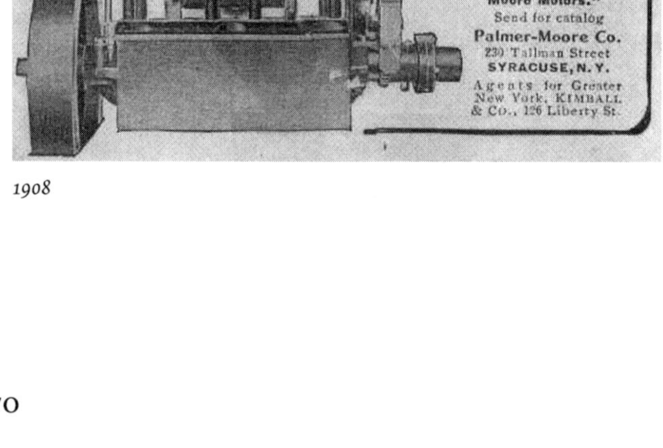

1908

Olmstead, A.E. & Son
Pulaski, NY
Trade name: Ontario

Ontario Iron Works
Pulaski, NY
Trade name: Ontario
By 1910, Ontario was building a two-cycle, 3-hp one-lunger with a 3½″ bore and 3½″ stroke that was rated at 900 rpm. That high rpm suggests the Ontario was a three-port motor with jump-spark ignition.

Original Gas Engine Co.
Lansing, MI
Trade name: N/A

Ormsbee, A.A. & Co.
Taunton, MA
Trade name: Pearle

Oswald Gas Motor Co.
San Francisco, CA
Trade name: Oswald

Oswald Motor Co.
Goshen, IN
Trade name: Oswald
The company was in business circa 1907.

Otis Machine Works
Rochester, NY
Trade name: Otis
An Otis ad of 1909 is one of the only documents that survives regarding this company. The motor was a two-cycle for marine or stationary applications. One- to 4-cylinder models were said to be available.

Otto Gas Engine Works
Philadelphia, PA
Trade name: Otto
The pioneer of pioneers, Nicolaus August Otto (1832–1891) was a German traveling salesman dealing in foodstuffs before he made the internal combustion engine his life's work. His story could fill a book. Otto received a patent on the four-cycle principle in 1877, by which time his factory in Deutz, Germany, was growing fast. The U.S. branch opened in 1880 with hopes of reaping the benefits of a vast market for stationary engines. With reorganization, it became the Otto Gas Engine Works in 1894. The marine engine market was also targeted. Otto sales agents found a good market for 15- to 80-hp engines on Missouri River stern-wheelers during the early days of conversion from steam to internal combustion. The Navy bought a 50-hp Otto to power its first submarine, the *USS Holland*, commissioned on Oct. 12, 1900. Larger 160-hp Otto models went into subsequent boats including *Adder, Grampus, Moccasin, Pike, Porpoise*, and *Shark*. These were 4-cylinder engines about 10′ long and 6′ high with a 5½′-diameter flywheel. The camshaft was mounted on the engine's starboard side and operated side-mounted air intake valves and overhead exhaust valves. Fuel entered through another valve. "Hammer break" ignition was used. These were 360-rpm engines.

The Otto found some success with pleasure boaters, too. In 1897, a 2-cylinder, 25-hp Otto powered a 53′ cruiser on a 75-mile run in 7 hr., 5 min. A single-cylinder model could be had in ratings of from 1 to 12 hp; *The Gas Engine* reported that the double was available in ratings of 17 to 100 hp.

Because of its German heritage and connection, the company was essentially taken by the U.S. government during World War I, and in 1923 the assets were sold to the Superior Gas Engine Company of Springfield, Ohio.

See also Superior Gas Engine Company

Oulton Motor & Mfg. Co.
Boston, MA
Trade name: Oulton
In 1909–10, Oulton became one of several companies that attempted to build and sell a practical, open-base, two-cycle engine that wouldn't rely on crankcase compression, with its possibility for leaks and explosions in the base. The Oulton's cylinder was closed at the bottom and the piston rod passed through into the cylinder. "The first compression of the gas is effected in this space between the bottom of the cylinder and the under side of the piston" is how the company described the engine's operation. Oulton said the engine would run smoothly between 100 and 1,800 rpm without missing or backfiring. Apparently, nobody got the message.

P. F. Olds & Son
See Olds, P.F. & Son

Pacific Diesel Engine Co.
Oakland, CA
Trade name: Werkspoor

Pacific Gas Engine Co.
San Francisco, CA
Trade name: Pacific
Based on patents awarded to Daniel Regan, who then assigned the rights to W. T. Garrett, Pacific was founded in April 1890 by Mora M. Bennett who acquired Garrett's rights to Regan's work. At least, that is what Mora believed. The result was that Pacific was competing with Regan in stationary engines and perhaps marine models as well. A lawsuit resulted. Ultimately, both Regan and Pacific were taken over by the Union Gas Engine Company in May of 1892. Like Regan, Pacific advertised both stationary and marine models. The Pacific was similar in general appearance to the Regan but possessed numerous details that differentiated it.

See also Regan Gas Engine Co. *and* Union Gas Engine Co.

Pacific Marine Engine Co.
Seattle, WA
Trade name: Pacific

Packard Motor Car Co.
Detroit, MI
Trade name: Packard
Among America's most fascinating automobile companies, Packard was also involved in aircraft engine design and manufacture (beginning with the Liberty engine during World War I) and marine engines. The first marine engine appeared in 1921: the 275-hp 1M-1551 (Model 1, Marine, 1551 cubic

1911

1951 Palmer ad from Yachting *magazine*

1898

1905

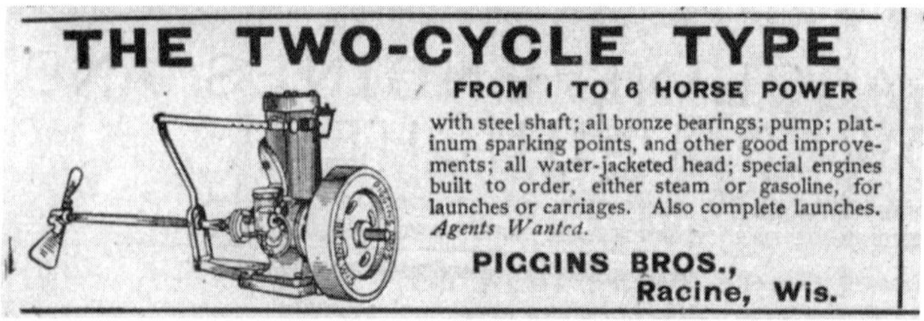

1902

inches) based on the engine that powered the airship *Shenandoah*. In ads, Packard called it the "Shenandoah Type." It was followed by the 6-cylinder, 45-hp 1M-268 and the 8-cylinder, 66-hp 1M-357, intended for runabouts and small cruisers. These models, with an aluminum crankcase and iron cylinders, remained in production until 1931. Also built were the 1M-618, a $5,000, 200-hp Gold Cup race engine and the $8,000, 400-hp "Sweepstakes Type" officially labeled the 1M-1237. The V-12 engine weighed 1,168 lbs.

"Supreme for every type of boat," read Packard's ad headline for these marine engines. Echoing the tagline that immortalized its automobile ads, Packard advised prospective marine engine customers to "Ask the Man Who Owns One."

In 1928 boat racer Gar Wood asked Packard for a marine version of its V-12 aero engine, as his Liberty-powered *Miss America* models were fast becoming outmoded. Packard produced four 1M-2500 engines for Wood between 1928 and 1929 and a production model, with transmission and reverse gear, the $15,625 2M-2500. This evolved into the 3M-2500 that powered early World War II PT boats. Packard had just completed its first 3M-2500 engines when it decided to modernize the design. By the time the production PT boats were ready for military use, the 3M-2500 was replaced by the 4M-2500. It powered all World War II PT boats. Between 1946 and 1950, the company built the 5M-2500 V-12 and a V-16 model — the 1M-3300. Packard also built smaller 6- and 8-cylinder engines for civilian use. From 1951 to 1956, Packard built 6- to 16-cylinder diesels of 300 to 800 hp, the last of which did not leave Navy service until 1990.

After the war Packard also introduced a pair of marine conversions of its 6- and 8-cylinder automobile engines. These were the 100-hp Packard Marine 6 (1MP-245) and the new-for-1947 150-hp Packard Marine 8 (1M-356).

Page Engineering Co.
Baltimore, MD
Trade name: Oriole, Straub

These engines, with their four-spoke flywheels, were in production as of 1902. Stated refinements in 1908 included a water-jacketed exhaust manifold. The flywheel now had six round holes. Ratings ranged from 4½ to 16 hp. In 1911, engines were available in ratings ranging from 5 to 17 hp.

Paige-Detroit Motor Car Co.
Detroit, MI
Trade name: Paige

This auto maker advertised its 16-hp, two-cycle engine for marine use in 1912.

Palmer & Rey
San Francisco, CA
Trade name: Palmer & Rey

By 1882, this company was a prominent manufacturer of printing presses and was the West Coast's leading supplier of type and other equipment needed for printing activities. Whether through acquisition of an engine or through its own development with an engineer consultant, the company was an early entrant in San Francisco's gasoline engine industry. It offered an open-base, vertical engine for stationary applications including, possibly, the powering of printing presses. However, Palmer & Rey soon divested itself of that portion of its business. It is included here because the company that acquired its engine became an important producer of marine engines.

See also Hercules

Palmer Brothers Co.
Cos Cob, CT
Trade name: Palmer

A great stalwart of the marine engine industry, Palmer Brothers traced its birth to the spring of 1894 when 37-year-old Frank Palmer became interested in having a small boat powered by one of the new internal combustion engines he'd read about. Unable to find an engine that suited him, Frank (1857–1945) and his brother Ralph (1860–1953) — known generally as Ray — decided to build one. What emerged was a single-cylinder, two-port, two-cycle with bore/stroke dimensions of 3½" × 3½".

If the Palmers couldn't find an engine that suited them, neither could they find a satisfactory make-and-break igniter. Testing two commercially available products, they encountered, in one case, problems with springs that lost their tension and, in another, an inability to adjust the device to achieve correct timing. So Ray Palmer designed an igniter with "a triple spring with thimble, rocker and latch, giving a short contact with a quick snap, drawing out the spark to its full capacity." A carburetor posed a similar problem, and that first motor did without. (See Chapter 2.)

At the time of these experiments, the Palmers owned a prosperous business called the Mianus Electric Company. There, they manufactured telephones and related electrical equipment. It was there, in 1895, that commercial production of the first Palmer motors was begun, and the Palmer marine engine was first advertised as a product of the Mianus Electric Co. The intricate cylinder was cast by a wagon maker in Bridgeport.

Looking back on Palmer's formative days in a brief mention in July 1911, *The Gas Engine* noted the brothers "were building the Amateur, a 1 H.P. two-cycle gas or gasoline engine for marine and stationary use. Its weight was 135 lbs. The castings and working drawings could be purchased if one wished to build his own engine." Approximately 100 motors were made during the first year of production, no more than three or four per week according to a report in *MotorBoat*. Frank oversaw the shop, and Ray methodically improved the motor's design.

In 1899, the line expanded to include both the original 1½-hp and a 3-hp version. Local oystermen became early converts, and this translated into what would become Palmer's long-standing good reputation among watermen. Like many early marine engine pioneers, the Palmers saw profit in offering both engines and boats. A plant dedicated exclusively to boat production was built on the Mianus River on a four-acre site two miles south of Mianus at Cos Cob in 1901. (Seventeen- to 30-foot launches were built there before production ceased in 1919.) That year, the Palmers exhibited two boats and five motors at the Pan American Exposition in Buffalo, New York, a venue that garnered worldwide publicity for the company.

As of 1900, the Palmer models were designated B, C, D, and E. The 1½-hp B and 3-hp C had non–water-cooled heads ("hotheads") for operation on kerosene while the 5-hp D and 7-hp E had water-cooled heads. (At some point, perhaps 1903, the C was made available with either head configuration.) With these two-port, 400-rpm models (500 rpm for the B), what would soon become an impressively diverse model line was established, each model carrying a different letter designation. The B would remain in production until 1912. Palmer experts are doubtful that the D and E lasted much past 1915. The C was still in production in 1930.

Not dogmatic about two-cycle *v.* four-cycle engines, and recognizing the practical limits of two-cycle motors in terms of size and fuel efficiency, the Palmers introduced their first heavy-duty four-cycle model in 1902. The 7½″ × 10″ K produced 16 hp at 350 rpm. It weighed 2,000 lbs. This engine went out of production in about 1907 with the introduction of the lighter, medium-duty M (5″ × 6″) and L (4½″ × 4½″) models. Both had closed crankcases for oil retention, a mechanical exhaust valve, atmospheric intake valve, and individual one-piece, L-head cylinders. By comparison to its predecessor 16-hp K, the 4-cylinder M-4 produced 18 hp but weighed just 700 lbs. Both the M and L had an external camshaft. They were lubricated by drip-feed oilers.

Seeing that two-port, two-cycle engines would not meet the requirements of pleasure boaters interested in lighter weight and higher speed, Ray Palmer developed a new line of three-port, dome-head, 1- and 2-cylinder engines for 1906, becoming one of the licensees for the Day patent that expired at the end of that year. These engines were rated at from 500 to 700 rpm depending on model. The models included the 2-hp Q and 4-hp Q2, the 3½-hp P and 7-hp P2, and the 5½-hp O and 11-hp O2. These were all jump-spark motors. For those who preferred make-and-break ignition, the same ratings were available but labeled U and U2, S and S2, and T and T2, respectively. All these Palmers were fitted with a carburetor rather than a vaporizer with check valve. While the weight savings was not dramatic in the smallest ratings, a 5½-hp O weighed 230 lbs. compared to the 400 lbs. of a 5-hp D.

In 1910, a new engine manufacturing plant was built on the same site as the boat-building operation. At that time, Palmer Brothers may have employed as many as 300 to 500 people. Now, what had become known as the Upper Works in Mianus — the original telephone equipment plant — was phased out and all production for the following years was at the Cos Cob plant. The new production location included a foundry and large machine shop with the most modern machine tools of the day driven by a steam engine up until the mid 1930s. Then the machine shop was converted to electric motor–driven machines.

By now, the Palmer lineup read like an alphabet soup of model designations. The first classic T-head Palmer, the Model R, joined the L and M around 1907. Both intake and exhaust valves were now mechanically operated, but all these engines had dome-head, water-jacketed cylinders to avoid the head gasket problems that had plagued earlier year engines with detachable heads. When an upgrade was unveiled at the end of 1912, it was labeled the NR (New R). Unlike the R, it had an enclosed camshaft and cam-drive gears. Like the R, the NR was a medium-duty, 5″ × 6″ engine rated at 500 to 600 rpm. It was offered in 1- (NR-1) to 4-cylinder (NR-4) versions of 5 to 20 hp. Also in 1912, the NL (New L) was introduced, basically a smaller version of the NR. Bronze oil pumps were used on the NR and NL until about 1915. Thereafter, they were built of cast iron. These engines became especially popular with fishermen and were widely accepted on the East and Gulf coasts, the Great Lakes, the Mississippi River, the Pacific Northwest and Alaska.

It's worth noting that Palmer offered stationary versions of its marine engines. The letter "S" after the marine-engine's designation indicated a twin-flywheel, stationary model on a steel base. The 1909 catalog, for example, showed a BS, CS, and DS. Later, stationary versions of the NR, YT-1, and ZR engines were produced. As of 1902, horizontal-cylinder stationary engines were also offered up to 15 hp. These are now among the rarest of all Palmer engines and are not to be confused with similar-looking engines from other makers. In addition to stationary engines, Palmer produced machinery. Perhaps the most widely sold piece of equipment was the Palmer dredge hoister for oyster boats. A Palmer wood-sawing machine and pumping outfit were also produced.

Following the introduction of the NR and NL, the NK (New K) models were introduced in 1913. They were heavy-duty 300- to 400-rpm models of from 2 to 6 cylinders with ratings of 20 to 50 hp. The F models, introduced in 1914 were rated at 400 rpm and built in 2-, 3-, and 4-cylinder versions of 16, 24, and 32 hp. A 60-hp F6 followed. By 1916, Palmer listed 38 models and displayed no fewer than 25 of them at that year's motor boat show in New York.

As good businessmen, the Palmers were certainly not immune to the potential of the automobile. In 1914, they began producing a 100-inch wheelbase vehicle with a 42-inch tread and a 4-cylinder engine. Such light cars, or "cycle cars," seemed to offer an exceptionally promising market, but the whole movement proved to be a fad

The 6-cylinder Packard Gold Cup engine developed 260 hp and weighed just 900 lbs. It dominated Gold Cup racing from 1922 through 1927.

1905 Palmer Model B, 1½ hp @ 500 rpm, 130 lbs.

1908 Palmer Model R (T-head)

1902 Palmer

Circa 1912 Palmer NL-3 or NR-3, 15 hp @ 400-600 rpm, 1,000 lbs.

1907 Palmer K, 16 hp, 2,000 lbs.

1913 Palmer NL-4 or NR-4

1913 Palmer T-head

that quickly ended, as did Palmer's car production in 1915.

Recognizing the need for a four-cycle one-lunger suitable for smaller pleasure and commercial boats, Palmer introduced the YT model in Dec. 1921. This 130-lb., 2-hp, single-cylinder was another winner that would prove long-lived. A 2-cylinder YT-2 was also available from 1924 to 1930. Also new for 1922 was a quite advanced engine, the 14- to 16-hp, 4-cylinder VH. Not only did the 1,200-rpm VH have an en bloc cylinder casting, but also its overhead valves were operated by an overhead camshaft. It was an interesting concept that did not work out entirely well in practice, thanks in large part to problems with the bevel gear camshaft drive mechanism.

By 1924 Palmer ads celebrated the fact that 75,000 of its "Fisherman's Friends" had been sold. The NL remained in production until 1923 when the ZR was introduced. A big step forward, the ZR had detachable heads with enlarged valves, a counter-balanced crankshaft that reduced stress on the hull, a water-heated intake manifold, electric starting, and a much-improved reverse gear.

In 1928 the classic dome-head NR was modified with new cylinders with removable heads as used on the 5½"-bore ZR. The NR was now designated as the PNR (Palmer New R). It remained in production until the mid-'40s. In 1929 the 251-cubic-inch, 25- to 40-hp Palmer Pal was introduced. This 4-cylinder en bloc engine (4" × 5") was aimed right at a big part of the market that clamored for a compact, high-speed (1,600 rpm) motor.

In 1931, another need was met when Palmer introduced a marine conversion of the Ford Model A and B 4-cylinder engines — "A Ford Engine Converted for Marine Use by Palmer Bros. Engines, Inc." (The company name had by now been slightly modified.) The 200.5-cubic-inch 4-cylinder Ford, which produced 23 hp at 1,000 rpm and 39 hp at 2,200, received a number of components to adapt it properly for marine use. It sold for $375. Palmer equipped the Fords with a full-length, cast-iron crankcase "to insure against the possibility of loss of oil through concealed leaks cause by corrosion, to which pressed steel cases are liable." A cast-iron, water-jacketed exhaust manifold was included, as were copper water piping, a housing for the water pump gears, and a Palmer disc-clutch reverse gear. A $445 marine conversion of the Ford V-8 followed. Conversion kits for Ford engines were also sold. Such converted engines were less expensive than a comparable, purely marine motor, but if not fresh water–cooled, the engine blocks quickly rusted out. Chesapeake Bay watermen began to use simple, home-made keel coolers formed from copper water pipe to get around the then-high cost of heat exchangers.

The mid-'30s saw Palmer continue its gasoline engine models and introduce a fuel oil engine. In 1936, the company introduced a version of the 4-cylinder Ford engine that it called a "Stratified Fuel Oil Motor." The term "charge stratification" referred to "two charges [of mixture] admitted to the cylinder at two different air-fuel ratios: one lean, one rich stratum." The key to the process was the "Palmer Stratifier," the device that replaced a conventional carburetor. The engine was intended to offer the usual advantages of fuel oil: lower fuel cost, better fuel economy, and reduction of fire risks. The engine was started on gasoline and, after warm-up, was switched to oil. It was a grand idea that didn't work well in practice because the valves could not stand the high heat of the stratified charge.

Toward the end of the decade, a line of 1-, 2-, and 4-cylinder diesels was built under license from Russell Newbery Diesel in England. "This Diesel Has Everything," said a Palmer ad in 1940. "New two diameter piston design gives better fuel economy and less detonation." The 4-cylinder, 40-hp model weighed 2,500 lbs. Production had just commenced when the U.S. entered World War II. Palmer was then directed by the office of War Production to build approximately 3,000 LH4 Pal 4-cylinder engines for use in lifeboats on ships being built for the US Maritime Commission.

Palmer Bros. made a large number of products besides marine engines. They made cider mill presses from small farm sizes to those for producing industrial quantities of cider and vinegar. This activity ceased in 1935 at the height of the Depression. They also made hoists for various marine purposes including the oyster dredge hoists. Old-time employees said that they would make anything that could be made from cast iron. That era ended when the Palmer foundry was shut down in 1937 and most of the casting work was done by Manufacturers Foundry Co. in Waterbury, Connecticut.

The company emerged from World War II in seemingly good shape. A small number of RND-1 diesels for use with air compressors and generators were sold. The RND design is still in use in some narrow boats plying England's canal system and is noted for its efficiency and low level of pollution. Still, the nature of Palmer's business, like that of the industry as a whole, had changed. The company was sold by the Palmer family to a New Jersey firm called Columbia Air Products. According to longtime Palmer employee Ray Bolling, poor management led to bankruptcy in only 18 months. Bolling and three partners were able to purchase the company assets with the aid of Carl Hathaway, one of Frank Palmer's sons-in-law. Twenty-five long-term employees were brought back and, by 1951, production resumed.

"We still had a line of good old heavy-duty, rugged gas engines of prewar design," said Bolling of those days. "We also had some conversions. They were very well thought-out. Palmer made a big full-length crankcase that supported the engine and its own reverse gear, a big and rugged water-jacketed manifold and various other parts. The carb was bronze instead of pot metal."

By 1951, the model line ranged from the 6-hp BH (the last Palmer-designed marine engine, introduced in 1946), and the 2-cylinder, prewar 10-hp HH, to the 4- and 6-cylinder PH models in ratings up to 125 hp. Later, a line of converted International Harvester Model 264 truck engines was offered. Those 6-cylinder engines were, according to Bolling, "very nice and rugged." An attempt by Bolling to sell the company to International failed and he left Palmer. Operations ceased for good in 1973.

Thanks to the enthusiasm and expertise of life-long Palmer enthusiast Richard Day, Palmer lore and technical information has survived, and Palmer's legacy has been better served than perhaps any other of America's pioneer engine companies. Those restoring a Palmer can identify the year of manufacture by the last two digits of the serial number. Paint color up to, it is thought, 1911 was a dark green. A 1913 catalog indicated that both dark green and a Navy-like gray were then in use. Thereafter, the color was a blue-gray that, Dick Day's research suggests, can be duplicated by color chips #25237 and #35237. Day also noted that, unless a customer specified otherwise, the Schebler Model D carburetor was used on all carburetor-equipped Palmers except the VH, the 1929 ZR-3, and 1936 models, all of which had a Stromberg M-2.

Palmer-Moore Co.
Syracuse, NY
Trade name: Moore
Although a two-cycle engine, this one had an open base. As always with such technology, the stated advantage was improved efficiency because the main bearings did not need to be sealed against loss of compression. However, oilers were required for the piston, connecting rod, and bearings. Ignition was jump-spark. Little is known of this machine or its level of production, if any. Ads for the Moore appeared from 1907 until 1910.

Parker, J. J. Co.
Fulton, NY
Trade name: Parker
Parker was advertising a conventional-looking, two-cycle, make-and-break engine in 1906. The company was succeeded by two others, apparently with little success.

See also Dutcher Machine Co. *and* G. J. Emeny

Parrott & Harter Engine Works
Olympia, WA
Trade name: Grizzly Bear
The Grizzly Bear was the work of David Garfield Parrott whose background and skills were typical of those possessed by the resourceful men who created small engine-building shops all across America. Born in Humiston, Iowa, in 1881, Parrott moved with his family to Olympia as a child and apprenticed as a machinist at the Pioneer Iron Works in Bremerton. In 1911 Parrott returned to Olympia and founded Parrott and Harter. The Grizzly Bear emerged in a Dec. 1913 ad in *Motor Boating.*

Paulsen
Geneva, NY
Trade name: Paulsen

Pausin Engineering Co.
Newark, NJ
Trade name: Arrow (Waterman)
Pausin's Arrow A-4 Special may have been acquired from Arrow Machine Works of Detroit. Pausin saw an increasing market for small, lightweight, inboard models and, in 1917, purchased the "patents, patterns, tools, goodwill, [and] stock on hand" of the Waterman Motor Co. Pausin continued to extol the virtues of the Waterman K-1 and K-2 with their corrosion-resistant, aluminum alloy crankcases. "In Norway alone there are hundreds of them in use in fishing boats in the open sea, where the failure of the engine might mean death." The K models were among the engines used by the Disappearing Propeller Boat Co. of Toronto. The heavier duty A-4 produced 4 hp at 600 rpm or 6 hp at 900 rpm and weighed 111 lbs. (compared to the 36-lb. K-1 and 60-lb. K-2). Pausin was still actively building the motors, fitted with Kingston or Schebler carburetors and Joes gears, as of 1928, but operations ceased sometime before 1936.

See also Waterman

Pease, Fred J. Co.
Grand Rapids, MI
Trade name: Pease
Four-cycle motors were advertised in 1915.

Peerless Marine Engine Co.
Detroit, MI
Trade name: Peerless
Another engine designed by Stanley R. Du Brie, this circa 1907 two-cycle had long, replaceable bearings ($1 each) sandwiched between machined top and bottom sections. Ignition was jump-spark and the engine developed 9 hp at 550 rpm. Finish was black with bronze trim. The company closed in July 1915.

Peerless Marine Motor Co.
Buffalo, NY
Trade name: Peerless
Peerless must have sounded like a fine name for an engine (or automobile) company. In the case of the Peerless of Buffalo, New York, the moniker fit reasonably well. Peerless started circa 1912 with a range of T-head engines positioned as "the ideal cruiser and workboat engines." In 1914, a 650-rpm 2-cylinder engine and an 850-rpm 4-cylinder became the line's feature models. These had cylinders cast in pairs and pressure lubrication.

In December 1916 Peerless, aware as was everybody of the increased price of gasoline, announced a kerosene engine for commercial service. It was available in 1- and 2-cylinder models producing 6 hp and 12 hp, respectively. Also new for the 1917 model year was an L-head, 4-cylinder engine with cylinders cast in pairs. A kerosene version of the engine was available in either 2- or 4-cylinder versions. A Peerless-built reverse gear was also introduced. Also that December, Peerless moved to a new factory at 2150 Niagara Street. It had 210 feet of frontage on the Erie Canal.

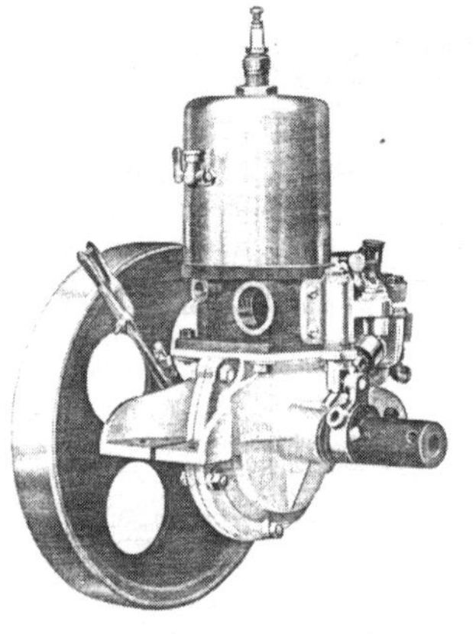

Circa 1917–1918 2-hp Arrow K-1 (Waterman) as built by the Pausin Engineering Co. The copper water jacket rested on a flange on the aluminum cylinder casting and was secured by a steel ring that produced "an absolutely tight joint."

1915 Globe (built by Pennsylvania Ironworks)

1911 Ralaco

1922 Pierce-Budd

1915 Penrose twin

1906 Regal

Circa 1916 – 1923 Pivert (Courtesy of Pierre McGraw)

1915 Penrose single

1911 Regal

In February 1918 Peerless unveiled new models. The classic T-head was made a thing of the past, and all models were now L-head designs. The Peerless-built clutch was now entirely enclosed. All-new models appeared, too: a 6-hp single-cylinder, a 12-hp twin, and a 16- to 20-hp four. The range grew methodically and by 1922 offered ratings of 5 to 50 hp.

In 1923 Peerless broke the mold by introducing a line of high-speed, lightweight engines. These included an overhead cam V-8 (5″ × 7″) with an aluminum base and separate cast-iron cylinders. It weighed 1,200 lbs. and developed 250 hp. A lovely 618.56-cubic-inch straight-six powered *Miss Mary*, a 1923 Gold Cup boat owned by George C. Hall of Buffalo. In 1924 a 151-cubic-inch racing engine with overhead valves and separate cylinders was unveiled. It developed 90 hp at 3,800 rpm. These new engines were produced in a new plant that the company acquired. Peerless survived through the 1930s.

Peerless Motor Co.
Lansing, MI
Trade name: Peerless

Pelton, T.G. & Sons
Lyons, IA
Trade name: Pelton
The company offered engines of 2 to 22 hp, and also built launches.

Peninsular Motor Co.
Grand Rapids, MI
Trade name: Peninsular
The company was in business in 1907.

Pennsylvania Iron Works
Eddystone (Philadelphia), PA
Trade name: Globe or *Pacific*
The Iron Works had built steam engines prior to making gasoline marine engines for the Globe Gas Engine Company, which it apparently absorbed circa 1895. With Globe came the license to Union's patents, and the Iron Works, like Union, wanted strongly to be associated in the public mind as being *first*. It stated its claim as follows: "The Pennsylvania Iron Works Co. were the first manufacturers in the Eastern states to take up and perfect the gasoline marine engine as a successful commercial product."

The Iron Works proceeded to name its horizontal models "Union" and the verticals "Pacific." Each, however, had a nameplate that read "Globe Gas Engine Co., Philadelphia." The horizontal Union model Globe pictured in an 1895 ad is identical in all ways to that year's Union produced in San Francisco. And the vertical models may have been equally identical to their West Coast counterparts. While the two companies' engines would soon begin to differ in appearance, it seems likely that the Iron Works' initial interest was prompted by the functional electric ignition, vaporizer, and governor patents controlled by Union.

Sales of Globe engines doubled twice in the period just before the turn of the century. By 1910, the engines had taken on new designs distinct from their predecessors. These remained heavy-duty engines but were now T-head designs (labeled Type "T") that were somewhat lighter than the standard Type "H" models. In 1911, the line included three 2-cylinder models (16, 20, and 25 hp), two 3-cylinder (30 and 60 hp), and four 4-cylinder engines (35, 50, 60, and 100 hp.) that ranged in weight from 2,075 to 14,500 lbs. Also in 1911, the company introduced an open-base, two-cycle model, the Eddystone-Globe. It was available in five ratings of 5 to 30 hp.

Although some Globes were installed in the company's own line of launches — the 35-foot *Luella*, with a double-cylinder 10-hp Globe, voyaged from Canada to Key West and covered 100,000 miles without a failure — and some yachts, commercial vessels were the key market. In 1898, the 25-hp Globe installed in the *Helen Miller Gould* became the first engine in the Gloucester fishing schooner fleet.

The Pennsylvania Iron Works had this to say about its engine's appearance in 1910: "After each engine is tested, it is removed to our paint shop and beautifully finished in black and gold. We particularly pride ourselves on the handsome appearance of our engines and they are a credit to the finest yacht." This was an aspect of the marine engine that had largely disappeared after World War I. It is one of the things that makes the rare early engine, still clad in its original paint, as evocative as an antique of any sort that has survived in its original condition.

This stalwart pioneer of a company seems to have been unable to adapt to the changing market. Published accounts of the company's products ceased after 1911.

Penrose Motor, Inc.
Philadelphia, PA
Trade name: Penrose
These motors with an "outside connecting rod" were said to have been built in a factory in Woodbury, New Jersey, as of 1914. The Penrose was yet another attempt to build a two-cycle engine that did not rely on the base for compression of the mixture before it was transferred to the cylinder. Although the company claimed the engine was manufactured in two types and four sizes — a 7-hp 800-rpm model and 10- to 20-hp 500-rpm models — there is little evidence suggesting many, or any, were sold.

Perkins Bros.
Havana, IL
Trade name: S & H

Perkins Launch & Motor Co.
Gloucester, MA
Trade name: RoNoMor

Peru Model Engine Co.
Peru, IN
Trade name: Model
The "Model" marine engine was a 4-cylinder that Peru claimed it had first built in 1895. This seems unlikely. These four-cycle engines were intended for truck and tractor applications until the company announced a marine version in *Motor Boating* in 1925.

Petoskey Iron Works
Petoskey, MI
Trade name: Petoskey
Born in Amherstburg, Essex County, Canada, in 1856, Phillip Middleditch moved to Saginaw, Michigan, where he became a naturalized citizen in 1877. In 1885 Phillip, with the help of his inventive younger brother Ben, founded the business that became the Petoskey Iron Works. Beginning about 1907, the Works began building a two-cycle "launch engine" designed by Benjamin. The company ads noted "Gasoline Marine Engines" together with general repairing, auto repairs, and the fact that the Works was an agent for the Alamo stationary engine. Marine engine production continued, it is believed, through 1911. After Phillip Middleditch died in 1914, the Iron Works was sold. It continued in business as an auto repair shop.
See also Middleditch, Benjamin

Phillips Gasolene Engine & Motor Works
Chicago, IL
Trade name: Phillips
Phillips was among several companies that attempted to popularize a horizontal marine engine by touting its smoothness. It also stressed the motor's ability to run reliably at any speed from 2 to 1,200 rpm without stalling. "You can run your launch just as you would with a steam engine, any speed, forward or reverse," claimed an ad in Feb. 1906. The company was still at it as of 1909 —"strong enough for a fisherman, good enough for a king." The twin "Duplex" was then available in ratings of 3 to 10 hp and the "Quadraplex" in ratings of 4 to 20 hp.

Pierce Engine Co.
Racine, WI
Trade name: Pierce
Begun in 1892 by Andrew Jackson Pierce, the Pierce Engine Company initially offered a horizontal-cylinder engine and then an unusual, inverted V-type twin. However, it was a line of headless, two-cycle marine engines that were most characteristic of the company's production. By 1910, jump-spark ignition was used, and the engines were equipped with a Schebler carburetor. That June, *MotorBoat* showed photos of two Pierce engines, the 3″ × 3″ 1½-hp model and the 4″ × 4″ 3-hp model. A 6-hp model was also available. The 1½-hp version was available in Pierce's 14′ × 4′ "dory" that was clinker-planked in cypress. These boats were finished in "lead gray" while the interior was a "dead grass color" (perhaps a tan color) and the gunwales, sheer strakes, and deck were varnished. In addition to completed launches, Pierce offered skiffs and canoes. The smaller boats were, it is believed, built by a subcontractor, the Racine Boat Company.

Parallel to his marine engine and boat offerings, Andrew Pierce pursued a dream of automobile building, and he sold 150 single-cylinder Pierce-Racine cars in 1904. New models with a 4-cylinder engine soon followed and production continued through to 1911. Marine engine production was still continuing in 1909 when the company was reorganized as the Pierce Motor Company. In 1910, Pierce Motor Company was acquired by J. I. Case. The sale did not include the marine engines. In fact, in 1910 the Pierce Engine Co. was among the entities slated to become part of the planned National Boat and Engine Company, a conglomerate of at least 10 boat and engine companies. That effort failed, and, sometime during the period 1914 to 1915, Pierce became part of Truscott-Pierce in St. Joseph, Michigan. The product was the Pierce marine engine.
See also National Boat *and* Engine Company *and* Truscott-Pierce

Pierce-Boutin Motors, Inc.
Merrill, MI
Trade name: Pierce-Boutin
See Pierce-Budd

Pierce-Budd
Bay City, MI
Trade name: Pierce-Budd
This company was a 1910 outgrowth of Pierce, Nye & Budd. After one of the founding trio, James H. Nye, dropped out, the company, located at 205 Saginaw St., was known as Pierce-Budd. Initial capitalization was $5,000. The technical genius behind the engines was James H. Pierce, who was born in Chester County, Pennsylvania, in 1864 and moved with his parents to Bay City in 1883. A year later, Pierce was working as assistant engineer on a Great Lakes steamer and, at age 21, became the youngest chief engineer on the Great Lakes. In 1889, Pierce was named superintendent of the Smalley Motor Company, a job he left in 1909 to start Pierce, Budd & Nye. (Nye also was a Smalley alumnus.) James H. Budd, who listed his residence as Wilmington, Delaware, may have been the principal investor.

The Pierce-Budd's polished brass water jackets, pressed on to form a watertight casing after the cylinder had been machined, gave them a distinctive look. "You can be certain of a fast and successful boat, a real race winner, if you install a Pierce-Budd," said a company ad. Pierce-Budd was responsible for what may have been the first published suggestion regarding mixing oil in gasoline. It appeared in the April 1909 issue of *Power Boating*.

In 1912, a three-port, 151-cubic-inch, 3-cylinder (4″ × 4″) engine developed 25 hp at 1,500 rpm at a weight of 217 lbs. The "high-speed" models were largely built of aluminum, while the "heavy-duty" model was said to share the same design but with "all parts grey iron." All engines were guaranteed to develop more than rated horsepower. Needless to say, in an engine of such comparatively high rpm, jump-spark ignition was used, fired either by a Bosch magneto or Atwater Kent electrical system. These 151-cubic-inch engines held a number of important speed records. The company offered models in various bore/stroke dimensions and in configurations of up to 6 or even 8 cylinders.

In 1916 the company showed what were probably its most powerful models, and likely among the most powerful two-cycle marine gasoline engines ever built. These high-speed Pierce-Budds were described as a "combination of the two and three-port systems." In this system, the upward stroke of the piston created the usual crankcase vacuum that caused air to be drawn into the crankcase from the carburetor through a check valve. Then, as the piston reached the top of its intake stroke, it uncovered an intake port through which mixture again entered, essentially maximizing the volume of intake mixture that would be compressed by the piston's downward stroke. All this resulted in really impressive power output. In 1916 a 6-cylinder, 6½″ × 7″ twin-ignition engine pictured in *The Rudder* was said to develop 335 hp at 1,500 rpm. At a weight of 1,200 lbs. without a reverse gear, this engine had a 3½:1 weight-to-power ratio. This rivaled that of an aviation motor, and the total output was competitive with the most advanced four-cycle engines of the day. No doubt, the engine could be finicky and required careful preparation and handling to avoid a mechanical failure.

Although one rightly associates Pierce-Budd engines with performance-oriented boats — they were still winning races in places as diverse as Peoria, Chicago, New Orleans, and Jamaica Bay into the early '20s — the heavy-duty versions were equally successful. A triple that had been installed in a 38′ × 10′ South African ferry proved both more fuel efficient and more powerful than a 35-hp four-cycle engine in a similar boat. By 1922, two-port heavy-duty 300- to 1,200-rpm models were also available.

The post–World War I economy and market challenged Pierce-Budd as it did all others, but a maker best known for exotic two-cycle engines must have been in an even poorer position than most. The company's organizational and, perhaps, financial difficulties were reflected in name changes and, finally, a move. When Pierce-Budd ceased operations, the company became J. H. Pierce, with a shop at 205 South Raymond Street. In 1920, an ad for the Pierce-Budd appeared in *MotorBoat*, but the company name was now Pierce Motor Co., "successor to William Goldie, Jr." By 1922 Pierce was doing business as Pierce-Boutin Motors, Inc., in Merrill, Michigan. That year, the motors were offered in four models with ratings of 6, 15, 25, 40, and 50 hp and carried the Pierce-Boutin trade name. In 1923 the company ad was for Pierce-Boutin Motors Inc. in Merrill, "formerly J. H. Pierce." The motors were then referred to again as Pierce-Budd. James H. Pierce died at age 79 in November 1943. Today, a Pierce-Budd would certainly be a choice find for any collector.

Pierce Motor Co.
See Pierce-Budd

Pierce, Nye, & Budd
See Pierce-Budd

Pierce-Scott Motor Co.
Detroit, MI
Trade name: Pierce-Scott
The company was in business in 1926.

Piggins Brothers
Racine, WI
Trade name: Piggins
The Piggins brothers were skilled machinists whose manufacturing interests ranged from marine engines to automobiles and trucks. A Piggins ad in 1902 showed a single-cylinder, two-cycle marine motor but noted that both two- and four-cycle types of ½ to 6 hp were available. What's more, Piggins claimed to offer the motors in either air- or water-cooled form. As always, when a company didn't focus exclusively on marine engines, success in the field was seldom achieved. The T-head automobile engine and the vehicle itself had a number of innovations but not enough to guarantee long-term business success. The Auto Parts Co. of Chicago was still advertising the Piggins for marine use in 1916.

Pittsburgh Filter & Engineering Co.
Oil City, PA
Trade name: Pittsburgh
An ad for this heavy-oil engine appeared in *MotorBoat* on Feb. 25, 1920.

Pivert Motor Co.
New Orleans, LA
Trade name: Pivert
Oliver Pivert was a skilled mechanic and machinist who founded the Pivert Motor Works on North Rampart Street when he just 18 years old in 1916. Oliver Pivert not only designed his engine, but also made all the casting patterns and did the machining and all the assembly. About the only thing he did not do, it seems, was produce the iron castings, which were supplied by a small, local foundry, the Shaft Foundry. Pivert did, however, make his own castings for brass parts, and he was said to have made his own crankshaft, pistons, and rings, a formidable accomplishment, indeed.

According to Pierre McGraw, a collector who, like others in the area, has long been interested in Pivert's work, these Piverts were substantial machines that probably produced about 25 hp. They had pushrod-driven overhead valves and magneto ignition and were painted red. Pivert's customers were local oyster and shrimp fishermen who used the motors in their 20- to 25-foot boats. As was typical with fishermen from Maine to California, an engine's reputation was spread by word of mouth, and Oliver Pivert and his engine earned a reputation good enough that some 300 were sold before Pivert ceased making them in the early '20s and went to work in the scrap metal business. Pivert's descendants recall that, although he had given up making engines, he continued to involve himself in things mechanical and applied for a patent on a mechanical governor and a tracked tractor. Oliver Pivert was working as an instructor at the Higgins shipyards, teaching machining and mechanics, when he died in 1941 at age 43.

Pocahontas Marine Motor Co.
Detroit, MI
Trade name: Pocahontas
In 1909, the company was listed at 1121 River.

Pope-Hartford
Hartford, CT
Trade name: Portola

Pope Motor Car Co.
Toledo, OH
Trade name: Pope-Toledo
The car maker (1904–1909) made an effort to position its copper water-jacketed engine for marine use in ads in *The Rudder* in 1906. The 6″ × 6″ engine had 6 cylinders and developed 100 hp at 1,000 rpm.

Portage Boat & Engine Co.
Portage, WI
Trade name: Portage
The company offered 4- to 24-hp models in 1914: "The Old Reliable."

Porter Engine & Machine Co.
York, PA
Trade name: N/A

Powell Engine Corp.
New York, NY
Trade name: Powell
In 1908 Powell presented to the boating public the virtues of its open-base, two-cycle marine engine. The key point was that, in theory, base compression would not be lost through the bearings as it could on a typical two-cycle motor. In the Powell, a "cross head" driven by the crankshaft was attached to the piston's connecting rod by a steel wrist pin. What sealed the section of the cylinder through which the piston rod passed? Powell described this component as a "compression plate [that] is cone-shaped and has a deep recess in its upper end which contains three packing rings, which make a gas-tight joint around the piston rod." The individual cylinders had removable heads. The Powell apparently worked better than similar motors produced by other makers. By 1910, Powell offered models of 5 to 20 hp. By 1912, the line had grown to include 1-, 2-, 3-, and 4-cylinder motors of up to 40 hp. The Powells were 500-rpm engines with jump-spark ignition and a Schebler carburetor.

Premier Motor Manufacturing Co.
Indianapolis, IN
Trade name: Premier
In 1910, according to *The Rudder*, the Herreshoff Manufacturing Company of Bristol, Rhode Island, first participated in the Boston Boat Show. The boat was a slender, lightly built mahogany launch about 30′ long. It was powered by a 6-cylinder, 60-hp four-cycle engine made by Premier.

Priestman & Co.
Philadelphia, PA
Trade name: Priestman
This engine was the work of two brothers, William and Samuel Priestman, in England. The American market was to be catered to by a Philadelphia branch with production occurring at the Aultman Company shops in Canton, Ohio. The 1893 marine version Priestman was an en bloc double-cylinder with an open base. It is unknown if any sales were made.

Progressive Mfg. Co.
Torrington, CT
Trade name: Eagle
An outgrowth of the Eagle Bicycle Co. (1888), Progressive began engine production of Eagle motors by about 1904 in Connecticut. These two-port, two-cycle engines were, like most of their counterparts, medium-speed engines, but the check valve in the base was of a unique design. The valve itself was comparatively large, flat, and light. It could be adjusted with the engine running to improve performance. The finish was glossy, baked black enamel.
See also The Eagle Co.

Prouty Mfg. Co.
Boston, MA
Trade name: N/A
The company was listed in 1909–1910.

Puritan Engine & Boat Co.
Boston, MA
Trade name: Puritan
This builder of 550-rpm, two-cycle engines was in business during the period circa 1910–1914. In 1912, the line consisted of 1-, 2-, and 3-cylinder models of up to 18 hp. Cylinders were individual castings. Ignition was Orwell jump-spark. The carburetor was a Schebler.

Putnam Motor Mfg. Co.
Springfield, MA
Trade name: Putnam
Putnam was in business by 1909.

R. N. Walker
See Walker, R. N.

Racine Boat Co.
Racine, WI
Trade name: Racine
In 1906, a boat builder named Peter Godske left the Racine Boat Manufacturing Co. in Muskegon and returned to Racine where he founded the Racine Boat Co. Company brochures intimated that the firm's boats were available with its own motors "bench-tested at its factory." There is no evidence, however, to suggest that motors were built here. In 1911, *MotorBoat* included the Racine Boat Co. on its list of two-cycle motor builders and listed 12 models. These ranged from a 3″ × 3½″ 2-hp single to a 4-cylinder, 6⅛″ × 6″ 60-hp model. The majority of the engines shared bore/stroke dimensions and weight with Termaat & Monahan models, and it is likely that T&M supplied Racine Boat Co. as it did Toppan and others. The *MotorBoat* list also included a pair of Racines in the four-cycle section, a 12-hp double and a 24-hp, 4-cylinder version, each with 6″ × 7″ bore/stroke dimensions. The origin of those engines is currently unknown. The company ceased operation in 1928.

Racine Boat Mfg. Co.
Racine, WI; later Muskegon, MI
Trade name: Racine
Founded in 1893, this boatbuilding company prospered with a wide-ranging

line of steam engines, canoes, boats, and, probably starting in 1896, gasoline engines. "It is quite refreshing to read in a catalogue of a manufacturer of both steam and gasoline launches that the latter are 'in fact the most complete, simple, economical, powerful, safest and effective outfit ever offered to the boating public,'" said *The Gas Engine* in March 1902. By then, Racine's "Electro-Vapor Motors" were available in ratings of ¾ to 6 hp in a single, 8 to 16 hp in a double, and 18 and 21 hp in a triple. Like many motors of the time, the ratings were noted by the company as being conservative but guaranteed. These were two-cycle motors.

Fire destroyed the plant in 1903, and the company moved to Muskegon, Michigan. The line then included both two-cycle and four-cycle engines. Details of these motors' general features appeared in both the 1911 and 1912 listings of American marine engines published by *MotorBoat* and *Power Boating*, respectively. The two-cycle models were offered in 1- and 2-cylinder models. Bore/stroke ranged from the 3½" × 3½" of the 3- and 6-hp models to the 5" × 6" of the 10-hp and 5½" × 6½" of the 12-hp version, both of which were doubles. A 2- and 4-cylinder four-cycle was available with 6" × 7" dimensions and ratings of 12 and 24 hp. Ignition was jump-spark on all engines. The four-cycle models had Splitdorf magnetos. Schebler carburetors were used.

At the Boston Boat Show in 1910, the company showed a 16′ dory equipped with a 3-hp Racine motor; an 18′ "speedboat" with a twin-cylinder, 6-hp Racine; and a 30′ Loew-Victor–powered cruiser. By now, the Racine Boat Manufacturing Co. was a big enterprise for the time and locale, employing more than 300 people. That is when apparently bad management caught up with the firm. That March, it was reported that $1 million would be invested in the company, but this was likely a murky cover for what was really going on. In November 1910, when a planned conglomerate of boat and engine builders called the National Boat and Engine Company was announced, the Racine Boat Manufacturing Company was declared its "leading company." Leading or not, the firm went the way of most others who sought relief under the new arrangement. The official dissolution date was January 1912.

A new entity, the Racine-Truscott-Shell Lake Boat Co. was formed in Nov. 1914 with offices in Madison, WI. It closed in 1915.

With its exposed bronze parts, the Racine makes a wonderful collectible.

See also National Boat and Engine Company *and* Truscott-Pierce Engine Co.

Racine Gas Engine Co.
Racine, WI
Trade name: Racine

The company was in business as of 1899, producing 1- and 2-hp engines. Production continued until about 1910.

Ramsey Manufacturing Co.
Philadelphia, PA
Trade name: Ramsey

In 1907 Ramsey claimed its little two-cycle engine produced 20 to 40 percent less cylinder friction than others.

Raspberry Island Boat & Engine
St. Paul, MN
Trade name: Fish

The Fish was available in 1911.

Rathbun-Jones Engine Co.
Toledo, OH
Trade name: N/A

This firm, in business in 1913, was successor to Rathbun-Lacy.

Rathbun-Lacy Co.
Toledo, OH
Trade name: Rathbun-Lacy

Incorporated in 1905 by engineer Edward Rathbun and V. E. Lacy, Jr., the company was, in 1906, placing ads that emphasized the simplicity of its four-cycle overhead valve engines with "inlet and exhaust pipes ... carried down within the water jacket." Rathbun had previously been a design engineer for S. M. Jones, makers of the Ralaco while Lacy had worked as a designer at the Pope Motor Car Co. The company had a capital stock of $30,000 and offered 3- to 15-hp two-cycle engines and four-cycle models of 20 hp and higher.

Ray Motor Co.
Detroit, MI
Trade name: Ray

Listed in the 1922–23 *Motor Boat Trade Directory*, this firm built an engine that was based on Ford Model T parts.

Raymond Engineering Co.
Boston, MA
Trade name: R-V

By 1913 Raymond was offering an unusual lineup of two-cycle motors ranging from a 10-hp single to 160-hp V-12 that listed at $2,750 including magneto. The motors featured an aluminum crankcase and ball bearings. The R-V had comparatively large ports, and Raymond stressed both their high power output and favorable weight-to-hp ratios (as low as 5 lbs. per hp). The water pump and magneto were driven by a silent chain. Such costly and exotic engines apparently found only a very small market.

Red Wing Motor Co.
Red Wing, MN
Trade name: Red Wing (later Thorobred)

What would become the Red Wing Motor Company had its origins, according to *MotorBoat* magazine, in the summer of 1900. That is when an intense-looking 24-year-old boat builder named John R. Trautner and a partner named Charles Gardiner started Trautner & Gardiner to build gasoline engines and launches in Winona, Minnesota. In late 1902, Gardiner sold his share to a Red Wing businessman, and Trautner and his new partner promptly moved the operation to Red Wing, a city of fewer than 10,000 residents of whom more than 400 belonged to the local yacht club. The Red Wing Boat Manufacturing Company was formed in 1903.

Trautner recalled that his first gasoline-powered boat was a 16′ fishing skiff powered by a two-cycle, 2-hp motor. The first engine actually sold by Red Wing was a two-cycle that was installed in a flat-bottom 26′ boat used to tow logs. It was still in service 20 years later. Trautner foresaw great potential in the market for motor boats and motors but believed the only way to be successful would be to build standardized products at a reasonable price. The first two-cycle motors followed that philosophy as did subsequent models. In 1910, the firm's name was changed to become the Red Wing Motor Company. Red Wing methodically expanded its line of boats and the power of its engines as dictated by the market.

In 1913 a 4-cylinder, four-cycle engine was introduced that would "make" the company. This was the Thorobred, destined to continue in series production for many years and become a staple among pleasure boaters for some four decades. Eventually, Thorobred would become the company trade name applied to a variety of models. With the future direction established, Red Wing began phasing out boat building, but two-cycle production was said to have continued into the late '20s although the motors were not regularly listed in ads or annual listings of marine engine products. Instead, they found their primary market as stationary engines on farms where many were mounted on grain binders.

A limited product line saw Red Wing through World War I and the immediate postwar years. By 1922, the company was identifying models according to bore/stroke dimensions. The 10- to 14-hp "D" was 4″ × 2½″; the 14- to 20-hp "A" was 3½″ × 4½″; the 18- to 24-hp "AA" was 3¾″ × 4¾″; the 23- to 30-hp "F" was 4¹⁄₁₆″ × 5″; the 32- to 40-hp "D" was 4½″ × 5″. By the late '20s, the model line had expanded considerably and included 1- to 6-cylinder models of 4 to 150 hp. Most models were described as medium duty (1,000 rpm), but high-speed (1,800 rpm) engines were also available bearing their own names: Big Chief, Warrior, Courier, and Arrow. In 1936 the 282-cubic-inch, 85-hp Hiawatha was introduced. A handsome 6-cylinder, L-head motor, the Hiawatha had a seven-main-bearing crankshaft, and press releases boasted it had powered a runabout to a 40-mph top speed.

Supporting all these engines was a sales force that stretched from Minnesota throughout the U.S., together with agents in Canada, Europe, and Asia. Throughout production, Red Wing refined its engines in detail. What's more, recognizing changing market demands, Red Wing introduced its Hesselman line of medium-duty fuel-oil engines during the '30s. These 4- and 6-cylinder engines were, by 1936, available in ratings of 35 to 180 hp. A compact, compression-ignition diesel was also available by then, the Comet.

Red Wing was among the rare marine engine companies that remained in business from the industry's early days right through and beyond World War II, when it produced a little 4-cylinder engine for use in lifeboats together with larger models. When Trautner retired in 1953, the company was acquired by a group of businessmen from St. Paul who changed the name to the Red Wing Marine Corporation. Steel-hulled houseboats powered by the 6-cylinder Meteor engine were produced. Red Wing ceased operation in 1965, but the name was kept alive by local businessmen, who continued selling parts and even some engines. The last owners, Harry Munson and Virgil Mischke, also preserved the company records, a service to both Red Wing owners and industry historians. Only a few marine engine companies' records fell into such good hands.

Regal Gasoline Engine Co.
Coldwater, MI
Trade name: Regal

Founded in 1901 by Arthur Robinson with three workmen and a small shop on the east side of Coldwater, this company grew steadily to become a consistent producer of long-lasting, serviceable four-cycle engines. In fact, Robinson had been building engines even before officially incorporating his company. Those first models were open-base engines with automatic intake valves. The production Regals were not only simple and long-lasting, but also easily serviced. All the connecting rod and crankshaft bearings were die cast and interchangeable. The top end of the rod had a bronze-bushing bearing that could be taken up as wear occurred. All but the one-lunger had pressure lubrication.

The Regal's popularity prompted a move to a fine brick building on the southwest corner of Pearl and Monroe Streets. By 1912 Regal's extensive lineup was divided into lightweight, medium-duty, and heavy-duty models; horsepower ranged from 3 to 45. Cylinders were cast individually except for a new 4-cylinder "Auto marine" model with cylinders cast in pairs. The 1,000 rpm, 32-hp engine weighed just 600 lbs. (By comparison, the Regal 28-hp Heavy Duty turned 400 rpm and weighed 2,500 lbs.)

Such a lineup had wide appeal to engine dealers, and Regal was well represented both domestically and overseas. Once, a dealer from the Fiji Islands, on a visit to America, made a special trip to Coldwater to shake hands with Robinson. Robinson himself went to Finland in 1914 to successfully bring a patent infringement suit. In 1914, Regal introduced a new 32-hp, high-speed 6-cylinder (4½″ × 5½″) motor. It had a weight-saving aluminum crankcase and came equipped with electric starting and a generator.

In 1936 Regal introduced a 2-cylinder 28- to 35-hp diesel with solid injection and a Bosch fuel pump. This was not enough to save the business, however, and in 1939 the company was sold and the name was changed to Regal Manufacturing Company. It produced bomb components during World War II, but there was no real engine know-how left after the war. An attempt to manufacture wheelchairs failed, and the old building was razed and replaced by a bank. A local businessman named Charlie Stokes acquired most of Regal's

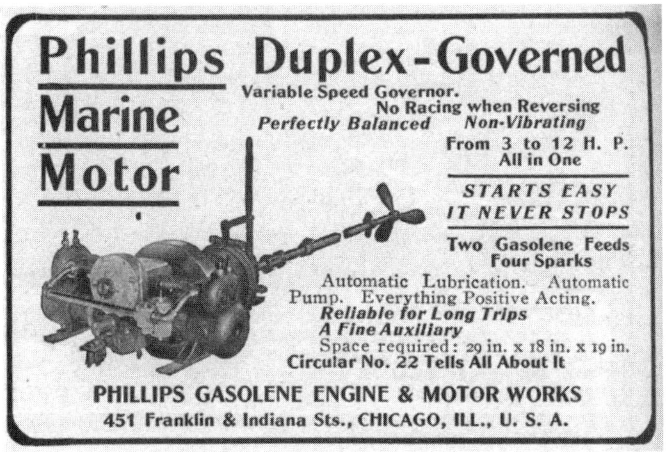

1906

1915

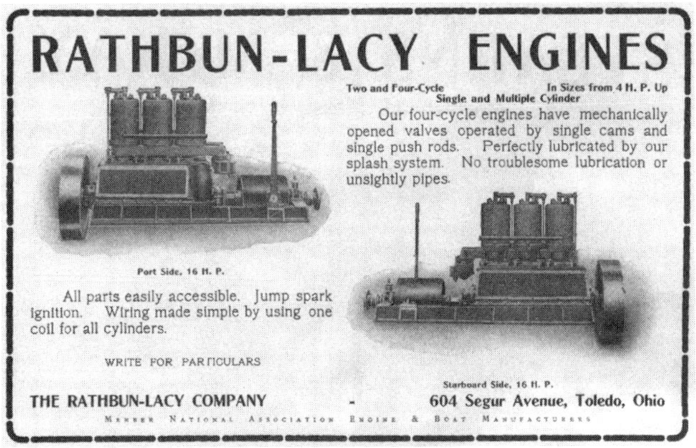

1906

1905

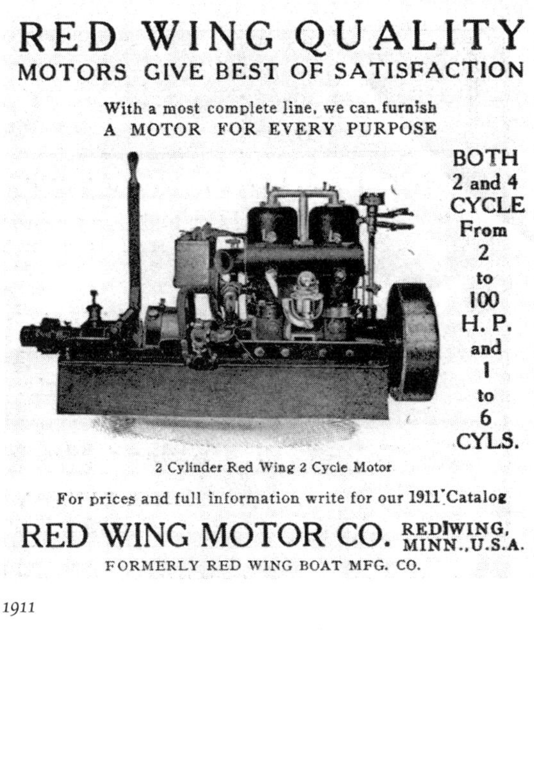

1911

patterns and, while production never resumed, Stokes sold parts and marinized AMC engines for many years.

Regan Vapor Engine Co.
San Francisco, CA
Trade name: Regan
A prophet without honor, an inventor without reward, Daniel S. Regan was born in San Francisco in 1855. By the time he was 38 years old in 1893, Regan had been granted 18 patents and was known in the city as a gifted mechanical engineer. His inventions included not only a variety of engines for marine and stationary use but also a street railway locomotive. How Regan developed his ideas and turned them into reality is a question that may never be answered unless some primary source material about him is unearthed. Little has survived about Daniel Regan other than a few articles demonstrating that he was a prescient and gifted mechanical engineer. His was a tragic story, and even in Regan's own time, the men who knew the tale, and knew what Regan had lost, spoke of him with a certain discomfort.

Addressing a banquet of the California Metal Trades Association in 1911, Willis C. Dodd, then general manager of the Union Iron Works in San Francisco and a man with a broad background in technical innovations in the city, posed the inevitable question that is still discussed today: "Who made and put in operation the first successful gasoline engine that you now see everywhere?" Dodd answered the question himself: "A poor fellow by the name of Regan, who originated the make-and-break type of electric ignition.... Mr. Regan had built gasoline engines in 1883 and 1884. However, the patent for the make-and-break type of electric ignition was not granted until June 16, 1885." Regan's brother recalled that he had personally seen Daimler launches (*Violett* and *Pass-Partout* — after the Jules Verne character in *Around the World in 80 Days*) at the Paris Exposition in 1889, and that Daniel's first engine had been built *five years earlier*.

There seems no reason whatsoever to doubt that Dan Regan emerged with a practical, electric-ignition marine engine in 1885. Regan was doing business under his own name from 1885 or '86 until 1892. The first gasoline-powered boat on San Francisco Bay was a Regan-powered launch named *Amy*, purchased, perhaps, from Dan Regan himself, by a commercial waterman named Henry C. Peterson in 1884. Regan envisioned building and selling a line of elegant launches, and his catalog listed models ranging from 18′ to 35′.

"There is but one Regan Vapor Engine built," said the company catalog issued in 1891, "the distinctive features of which are using no fire or flame. . . ." The firm was then offering a line of horizontal and upright engines intended for marine or stationary applications including boot and shoe machinery, hoisting machines, well-drilling equipment, mill machinery, polishing machines, and more.

The term "vapor" referred to the gasoline vapors piped to the engine from what Regan called a "carburetor or oil tank" that could be located, as he said, "five feet or fifty feet from the engine." The engine operated on the four-cycle principle with make-and-break electric ignition. Regan brochures described the engine's operation as follows:

"The charge is drawn into the cylinder as the piston is drawn out by the crank and flywheel; on the return stroke it is compressed in the head of the cylinder and just before the piston arrives at its highest point or 'upper dead center,' a shoulder or point projecting from it (which is an electrode) comes in contact with a flexible piece (also an electrode), the two comprising the poles of the small battery, thus completing an electric circuit which is broken just as the piston has passed its 'center' and commences to descend. At the instant the electrodes separate the spark is emitted, firing the charge. On every other stroke of the piston this is repeated automatically, save when the supply is 'cut off' by the governor."

By the time his 1891 catalog was published, showing launches with horizontal engines, Regan was already immersed in the nightmare that would smother his dreams. Regan and a business partner had assigned his valuable patents, for reasons now unknown, to others. Among them was William T. Garrett who was granted rights in the western states. The patents were used by his Pacific Gas Engine Company located in San Francisco on Fremont Street. The engine in Pacific's ads was unmistakably similar in design to the Regan and included the make-and-break ignition system. The ads even listed the patents protecting the engine. Regan's case went to court in the summer of 1891 and an appeal in his favor was handed down in January 1892. But by then, there was no recovery, and Regan's engine and patents were taken over by the newly created firm, Union.

See also Union Gas Engine Co.

Reliance Mfg. Co.
City Island, NY
Trade name: Reliance
This company was listed in the 1904 *Marine Buyers' Guide.*

Reliance Motor Boat Co.
New York, NY
Trade name: Reliance-Continental
In 1912, Reliance was among the featured companies displaying at the National Motor Boat Show, where it showed a 40′ cruiser named *Peter Pan Sr.* powered by twin 250-hp Reliance-Continental motors. A 21′ tender was also shown, equipped with a 4-cylinder engine. These engines were presumably marine-converted Continentals, perhaps carrying a Reliance-Continental nameplate.

Remington Automobile and Motor Co.
Utica, NY
Trade name: Remington Standard
"A Good Motor with a Good Name" is how this company advertised its 2-cylinder motor in *The Rudder* in July 1901. Positioned for either automobiles

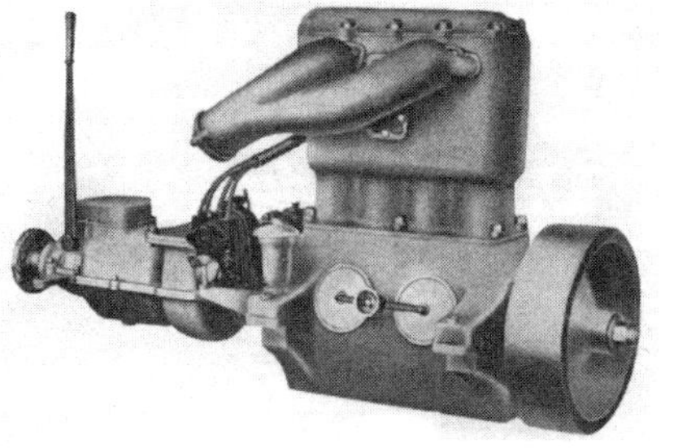

1911 Reynolds

1906 Roberts with rear starting handle

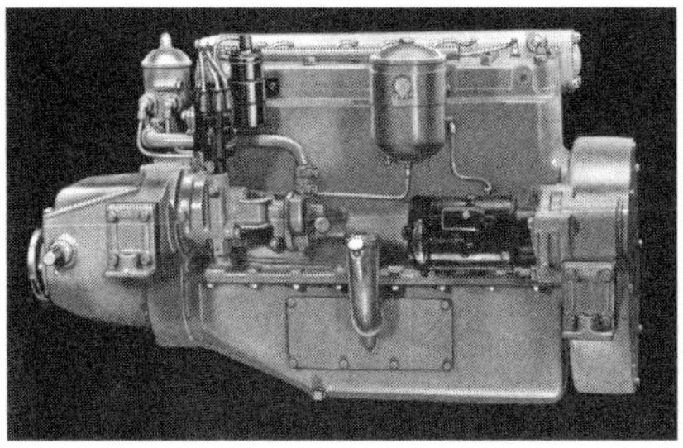

The 6-cylinder, 221-cubic-inch Scripps Series 90

1905 Sittman & Pitt

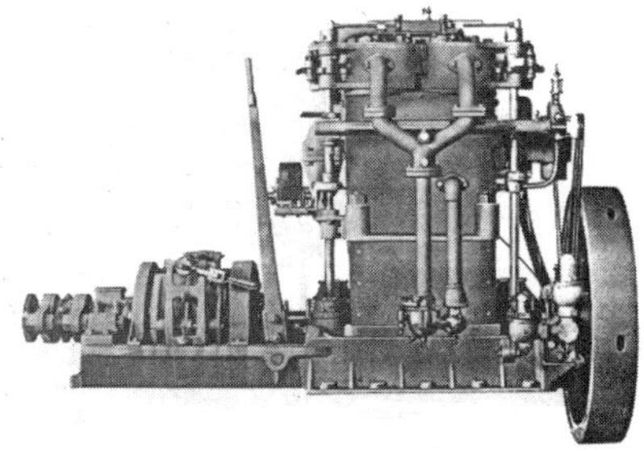

Circa 1910 'Frisco Standard twin

This is the Scripps conversion of Ford's flat-head V-8. The 95-hp engine could be had with cast iron or aluminum heads and a steel or cast iron base.

1906 Sparks

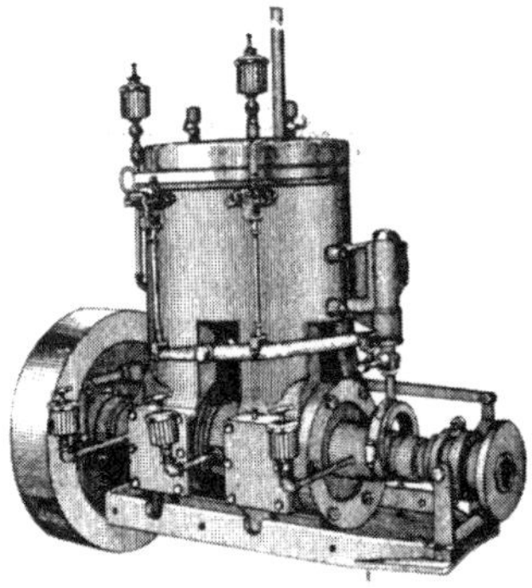

1902 Superior Engine Co.

1893 Sintz make-and-break with drip-oiler

A 4-cylinder 1908 Standard Motor Co. of New Jersey engine undergoing testing

or launches, the engine appears to have been a four-cycle with overhead valves and camshaft, the latter chain-driven off the crankshaft.

Remington Oil Engine Co.
Keyport, NJ (also Stamford, CT)
Trade name: Remington

Rex Motor Co.
See Neponset Engine & Machine Works

Reynolds Detroit Co.
Detroit, MI
Trade name: Reynolds
See Reynolds Motor Co.

Reynolds Motor Co.
Detroit, MI
Trade name: Reynolds
The brainchild of engineer Robert Albert Reynolds (1882–1937), the 20-hp Reynolds was a four-cycle, rotary-valve engine. Said an ad: "A four-cycle motor with not a spring of any kind, nor a cam shaft nor a push rod in its construction has arrived." An example was displayed at the1911 Boston Power-Boat Show, and a magazine ad that year showed the engine together with a modern-looking factory at 100 Hilger Ave.

It is unclear whether the engine ever entered production. The company was incorporated in November 1909 and was still listed in 1913.

Rhode Island Motor Co.
Providence, RI
Trade name: Rhode Island
The company was listed in the 1904 *Marine Buyers' Guide.* By 1906 the two-cycle Rhode Island was touted for its water-cooled exhaust and the interchangeability of all its parts.

Rialto Mfg. Co.
Toledo, OH
Trade name: Rialto
The Rialto was another example of that curious two-cycle species that didn't rely on a sealed crankcase for compression. The 2 cylinders were cast en bloc. Hidden within would have been some method of enclosing the bottom of the cylinder and then providing a means of sealing the hole through which a cross-head passed, joining the piston to the connecting rod. At least that is the most likely scenario. The jump-spark engine was rated at 12 hp. Introduced in 1914, the Rialto appeared at a time when the two-cycle engine was dwindling rapidly in popularity. There is no indication that it enjoyed any success.

Rice Brothers Co.
See Boothbay Engine Co.

Rice Gas Engine Co.
See Rice, John V., Jr., & Co.

Rice, John V., Jr., & Co. (later Rice Gas Engine Co.)
Bordentown, NJ
Trade name: Rice
John V. Rice, Jr., was known locally as a highly inventive machinist who received a number of patents on marine steam and gasoline engines. One of the latter, described as a "two-cycle, double-acting gasoline engine," was adapted by Rice to power a rock drill. This device apparently earned Rice a tidy sum, and the inventor purchased a landmark Italianate residence (long ago destroyed by fire) in Bordentown, among his other properties.

Rice was advertising two-cycle marine engines, jump-spark or make-and-break in 1905 but production may have begun earlier. The ad also claimed that four-cycle engines were available and that horsepower ratings ranged from 1 to 150 hp. Built "in the best equipped shop in the United States" is what the company had to say about its engines. While Rice's machine shop impressed all who saw it, it is now unclear whether the engine's castings — at least those for early models — were poured there. Six miles up the Delaware from Bordentown, the Trenton Malleable Iron Co. was placing its own ads for Rice Gasolene Engines, and it likely was involved in their production. The Rice pictured in the 1905 ad had a distinctive, domed, removable cylinder head. It was built in Canada as the St. Lawrence.

In 1906 Rice claimed to have developed a scavenging system that permitted "the enormous speed of 3,200 revolutions per minute in a two-cycle engine." Details of this 1½-hp, 80-lb. engine have yet to surface. In 1912 Rice's company became the Rice Gas Engine Co. It was said the business was capitalized at $1 million, then a very impressive sum. Adjacent to the machine works was a boat-building facility capable of constructing up to 60-footers.

See also St. Lawrence Engine Co. (Canada)

Richards Iron Works
Manitowoc, WI
Trade name: Manitou

Riotte, C. C. Co.
New York, NY
Trade name: Empire
Carl C. Riotte had begun experiments with engines in 1890 and introduced the Empire in 1898. The engine was offered in 1- and 2-cylinder versions of from 1 to 20 hp and a 40-hp 4-cylinder model. Mixture was supplied from a vaporizer, and ignition was electric make-and-break, the current generated by a magneto. A reversible propeller was used. According to naval architect William J. Deed, writing in *The Rudder* in 1930, the Empire "looked very nearly like the later perfected Standard." The Standard Motor Construction Co. succeeded C. C. Riotte Co.

See also Standard Motor Construction Co.

Robbins, L.D. Co.
Lynn, MA
Trade name: Oxford
Robbins claimed to have been influenced by steam engine design. In 1913 the company offered 6- to 30-hp models of unique design in which the base held lube oil and was not used to compress the mixture before transferring it to the cylinder. Robbins was a short-lived effort, in business from 1910–1913.

Roberts Motor Co. (et al.)
Sandusky, OH
Trade name: Roberts, Aero-Lite
The son of a physician who had served in the Civil War, Edmund Wilson Roberts was a Cornell-trained mechanical engineer. After graduation in 1895, Roberts went to England, where he worked with Maine-born aviation pioneer and machine-gun inventor Hiram Maxim. In England, Roberts designed one of the first wind tunnels. After his return to America, Roberts began designing engines for aviation as well as marine, automotive, and stationary applications. It is said that the first parachute jump in the U.S. was made from a Roberts-powered airplane. During World War I, Roberts headed the aeronautics bureau in Washington, D.C., a post that involved him in the Liberty engine's production process.

Roberts was always a prolific writer, and his articles on engine design and operation appeared regularly in boating magazines and other publications. But Roberts did more than write. As a consulting engineer, Roberts was responsible for some 90 engines between 1903 and 1908 alone. Among these engines was one for the Elmore Manufacturing Co., for which Roberts was listed as chief engineer for a period around 1903. The Elmore car had a three-port, two-cycle engine. Later, with the assistance of family members — his brother Benjamin became company president — Roberts founded the Roberts Motor Co. in Clyde, Ohio, in 1905. The factory was located at 132 E. Maple. Toward the end of 1909, the company moved to Sandusky, Ohio. The factory, at 1501 Columbus Ave., was served by the B&O RR.

The Roberts two-cycle marine engine had an integral cylinder/head casting, with an aluminum base (cast iron bases were also used), and jump-spark ignition. After much analysis, Roberts concluded that the three-port, two-cycle engine, thanks to its absence of a check valve, was the most practical for a majority of customers. In 1910, Roberts patented screened ("cellular") transfer ports and thereafter stressed that his were "the motors that never backfire." That year, Roberts advertised that "the W. H. Mullins Co. (the world's largest boat builders), recently placed an order for 2,000 specially designed ROBERTS MOTORS."

The Roberts line was extensive and by 1912 included engines of 1 to 6 cylinders, all with jump-spark ignition and Krice carburetors. Mechanical oilers were still being used on engines above 6 hp. One original engine that survives from the period is finished in maroon paint. Joe's gears were used. The Roberts line of lightweight two-strokes was labeled Aero-Lite and used alloys stronger and lighter than aluminum.

Like many other engine makers, the Roberts Motor Co. felt the attraction of the automobile industry. In 1915 the Roberts Motor Manufacturing Co. offered an automobile that seems to have lasted one year. The focus then returned to engine manufacture. In 1915 Roberts and his family moved to Cincinnati where he became editor of *The Gas Engine* magazine. Back in Sandusky, engine production continued. The 1916 Model "X" (4-cylinder 4½" × 5") was typical of the most advanced Roberts two-strokes. It had cylinders with polished water jackets, produced 50 hp at 1,150 rpm, and weighed just 215 lbs. The 5" × 5" "6X" version produced 100 hp. By then, Roberts had converted to oil in gasoline lubrication.

Four-cycle models — the 2-cylinder 8-hp Model H and 4-cylinder 16-hp Model R — followed by 1922, presumably Model T based. The Model "J," introduced in 1924, was a 4-cylinder 16-hp engine with "all working parts interchangeable with those of Ford motors." By about 1928, the "J" was joined by four more 4-cylinder models including the 40-hp "A"; the 35-hp "RS"; the cruiser-oriented, 50-hp "OX"; and the 30 hp at 4,000 rpm "HS" ($2^{11}/16$" × 3½"). The basis of these engines — the CL and OX were also referred to as "Johnston Type" — although what this meant is presently unclear. Roberts production continued into the mid-'30s by which time an 80-hp 8-cylinder had been added to the line. Paragon or Joe's gears were used.

During his career, Roberts claimed to have designed 131 engines for boats, airplanes, automobiles, tractors, and for stationary applications. He died at age 80 in Cincinnati, Ohio, on March 14, 1947.

Rochester Gas Engine Co.
Rochester, NY
Trade name: Rochester
Company president James S. Graham claimed that Rochester had built its first engine or engines in 1897 although the official start-up date for the firm was given as 1899. Either way, Rochester was an early producer. The company offered 3½- to 60-hp, make-and-break or jump-spark two-cycle models with reverse gears. The engines were equipped with forged and balanced crankshafts, the latter atypical at the time. The 1912 models had Detroit oilers, Schebler carburetors, and Atwater-Kent ignition. Like many early engine companies — Rochester was advertising in *The Rudder* by 1900 — the company offered both engines and launches. It is likely that, as demand for two-cycle marine engines began a steep decline around 1912, Rochester saw its business fall. That said, the company didn't disappear from the City Directory until 1921.

Rockaway Motor Co.
Rockaway Beach, NY
Trade name: Rockaway
This Long Island company was producing 4- to 30- hp engines with make-and-break ignition in 1906 and claimed to have been in business since 1896, although what business that was is not stated.

Rockingham Machine Works
Kittery, ME
Trade name: Nancy Lee
The Nancy Lee was advertised in 1912.

Rome Gasoline Engine Co.
Rome, NY
Trade name: M.A.F.

1905

1898

1915 Roberts ad showing the company's breadth of offerings

1913

1906

1911

This was a two-cycle, jump-spark, 3-hp motor with a 13" flywheel.

Root & Van Dervoort
East Moline, IL
Trade name: R&V
A renowned maker of stationary engines widely employed in farm applications, R&V also built a two-cycle marine model during the heyday of such engines around 1912. The engine featured a separate head casting, make-and-break ignition, a Schebler carburetor, and a water-cooled exhaust manifold. Such features would have made it familiar to an owner of a two-cycle marine engine from any of the New England or Midwestern companies.

Royal Engine Co.
West Mystic, CT
Trade name: Royal
Royal built L-head engines during the period of about 1910–1920.

Royal Engine Equipment Co.
Bridgeport, CT
Trade name: Royal
Founded in 1904 as the Royal Equipment Co., Royal was one of several engine makers in the industrial center that was Bridgeport. A year later, it established a plant at 165 Housatonic Ave. and, it is believed, offices at 1378 Main. The "Equipment" in the company name referred to now unknown automobile components. As for the engines, these were two-cycle, two-port engines with make-and-break ignition. The water-jacketed exhaust manifold had a rust-proof bronze valve to control flow. Royal's 3½" × 4" single-cylinder developed 2½ hp at a claimed rpm of 1,200, at least twice the rpm one would normally expect of such an engine. When a boating writer asked how this speed was achieved given the engine's apparently ordinary check valve, no answer was forthcoming.

In 1908, Royal fell prey to the apparently irresistible urge to develop a two-cycle engine with an open crankcase to improve scavenging and eliminate pressure loss through a normal two-stroke's bearings. It was claimed that the first time the idea was tried in a two-cycle marine engine was in Royal's motor. At present, there is no evidence to suggest the engine was ever successfully marketed.

Royal had more success with its original, conventional two-port, two-cycle engine and changed its name to the Royal Engine Co. in 1910. The line then consisted of six medium-speed (500-650 rpm) models of 2½ to 15 hp. The business continued through 1916.

Runnings, J. R.
Detroit, MI
Trade name: N/A
The company, located at 94 Brush St., was listed during 1907–1908.

Ryan, W. M. Co.
Chicago, IL
Trade name: Shortt
In 1910, Ryan claimed that its 2½-hp single-cylinder Shortt with an unusual rectangular, headless cylinder casting, possessed notably efficient scavenging properties. But no detail was provided. There may well have been something to the claim, for the Shortt was rated at 1,200 rpm.

S. M. Jones Co.
See Jones, S. M. Co.

S-R Manuf. Co.
Schenectady, NY
Trade name: Mohawk
S-R made the sort of fascinating engines that keep collectors on the lookout. The company's "combination Two and Three Port Motor" had a slotted piston and twin spark-plug ignition, and a Schebler carburetor for each cylinder (1, 2, or 3). The motor could develop up to 1,200 rpm. A removable plug enabled a straight exhaust to be fitted for racing. A medium-speed (700-rpm), two-port model, the K-33, could be run on kerosene or gasoline and had water-cooled main bearings to address a perennial problem of kerosene-fueled two-cycle motors. Make-and-break or jump-spark ignition could be had. A heavy-duty, 500-rpm, two-port motor with make-and-break ignition was also offered. The engines were finished in black enamel. The 1916 line, presumably little different from its predecessors, included 1-, 2-, and 3-cylinder models of 3½ to 18 hp. S-R was in production from circa 1912 until 1922 when 16 models were listed, the most powerful being the 3-cylinder, 21- to 30-hp S27. As a two-cycle builder, S-R did well to survive as long as it did.

Safford, W. F. Manuf. Co.
Westerly, RI
Trade name: Narragansett
The company built power dories and, possibly, gasoline engines circa 1907.

Sagamore Engine Co.
West Lynn, MA
Trade name: Sagamore
Sagamore appears in the Lynn City Directory in 1910.

Salter Motor Mfg. Co.
Kansas City, MO
Trade name: Salter
The Salter was a four-cycle engine with the exhaust valve mounted atop the cylinder head. In 1912 the company's advertising suggested that the Salter combined four-cycle economy with two-cycle power.

Samson Iron Works
Stockton, CA
Trade name: Samson
The Samson Motor Works was begun in 1897 by Danish immigrant John Kroyer. A machinist who gained extensive experience with foundry work, Kroyer started his company with a stationary pumping engine that immediately gained acceptance in the San Joaquin River delta. Marine engines were introduced in 1902. These were all heavy-duty, four-cycle engines with make-and-break ignition. They were used in commercial vessels like tugs and schooners. In 1912, it was noted that the Samsons were finished in dark-blue enamel with nickel-plated fittings. Later, a lighter color may have been

1905

1915

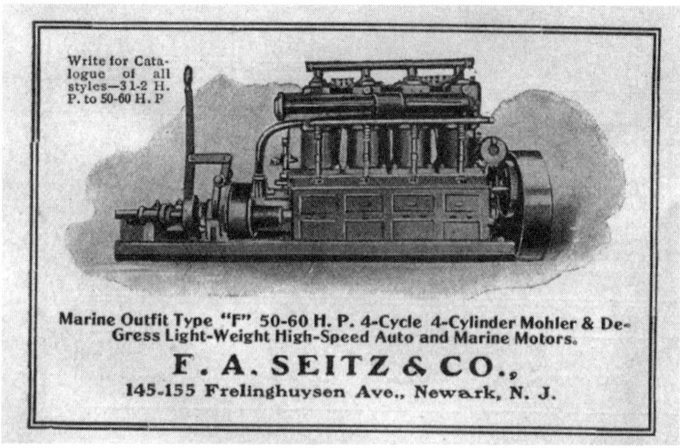

1906

1905

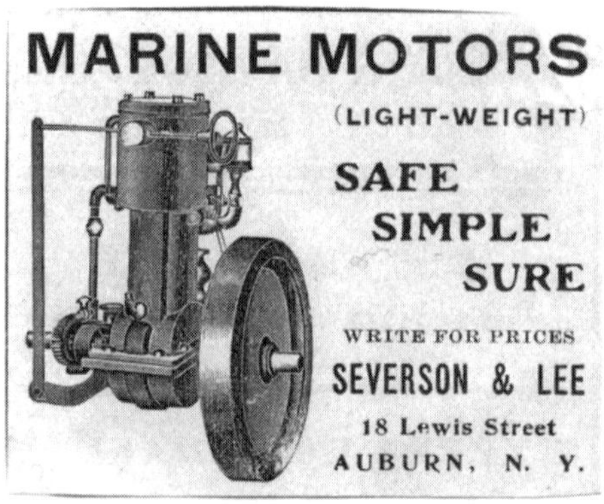

1905

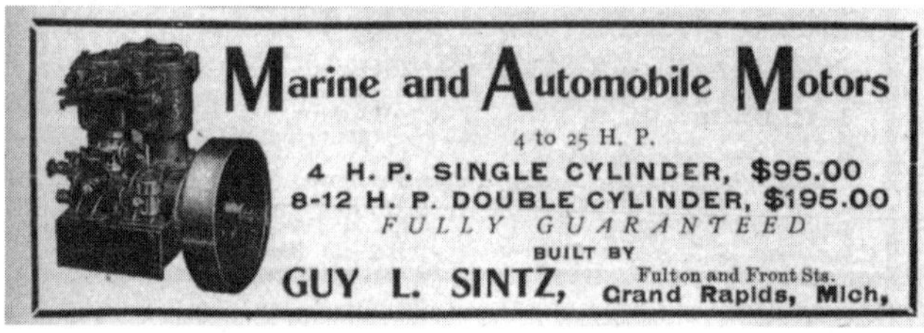

1905

used. When General Motors expressed interest in the tractors that Kroyer had developed, he made an outstanding decision and sold out in 1918.

San Francisco Standard
San Francisco, CA
Trade name: Campbell, Essex

Sanderson-Cyclone Drill Co.
Orrville (later Cincinnati), OH
Trade name: W-S-M
In 1925 or '26, this company purchased the rights to the Wellman-Seaver-Morgan, a 4-cylinder, 425-cubic-inch engine originally developed by Joe Van Blerck and first sold as the J.V.B.
See also Wellman-Seaver-Morgan

Sarvent Marine Engine Works
Chicago, IL
Trade name: Sarvent
Initially built under the auspices of Marine Iron Works, Gas Engine Department, by 1905, if not earlier, the Sarvent was an open-base, four-cycle engine. By 1906, models ranged from a 2-cylinder, 10-hp engine to an 8-cylinder, 500-hp model. Air starting was typical.

Schaefer Mfg. Co.
Berlin, WI
Trade name: Schaefer
Given its proximity in central Wisconsin to W. E. Schaefer, it is possible that the Schaefer Mfg. Co. and W. E. Schaefer were related. The company had an ad in *Motor Boating* in Dec. 1911.

Schaefer, W. E.
Ripon, WI
Trade name: Schaefer
William Schaefer was a well-known steam-engine specialist in Ripon who had entered the marine-engine business by or before 1906 and had an ad in *MotorBoat* in July 1908. The company offered 1-, 2-, 3-, and 4-cylinder engines. It is plausible that when he closed the company, Schaeffer became involved in another venture in nearby Berlin, Wisconsin.
See also Schaefer Mfg. Co.

Scheppele, Charles A. (*also* **Scheppele Machine Works**)
Dubuque, IA
Trade name: Scheppele
Machinist Charles A. Scheppele (1866–1941) developed a two-cycle engine that he began manufacturing, according to his daughter's best knowledge, around 1900. These engines found a market with commercial boatmen, including those in the freshwater clam fishery that supplied Iowa's pearl-button industry with shells. Pleasure- and racing-boaters also became customers. According to the family, Scheppele's production was some 2,000 engines before 1912, which, if accurate, suggests a well-organized company. The shop employed eight to ten machinists and was located on Central Ave.

Schlosser Mfg. Co.
New York, NY
Trade name: Schlosser
In 1909, Schlosser claimed to offer 4-cylinder engines with an aluminum crankcase that were suited to automobiles trucks, and boats.

Scripps Motor Co.
Detroit, MI
Trade name: Scripps
Founded in 1906 by boat racing and engine connoisseur William E. Scripps, a son of the founder of the *Detroit News*, the company produced high-grade engines with a reputation for quality. Great publicity was earned when, in 1912, a 16-hp, 2-cylinder Scripps was used in a gasoline-powered, transatlantic crossing — New Rochelle to Russia — in the 35′ 10″ *Detroit*. The journey was recounted by the boat's skipper, *Rudder* magazine editor Thomas Fleming Day, in *The Voyage of the Detroit*. For years thereafter, "The Motor that crossed the Atlantic" was a hallmark phrase of Scripps ads and company letterhead.

The 1912 Model L Scripps of the *Detroit* was part of an extensive engine line that ranged from the single-cylinder, 6-hp Model A to the 6-cylinder models R (6″ x 6½″, 100 hp at 1,000 rpm, and weighing 1,510 lbs.) and D (7½″ x 9″, 96 hp at 450 rpm, and weighing 3,300 lbs.). These Scripps were side-valve engines with individually cast cylinders and Bosch magneto/Delco battery ignition. In 1913 Scripps unveiled the Midget — "the smallest four-cylinder, four-cycle marine motor ever offered for practical use." The L-head cylinders were cast in pairs and had Bosch dual ignition and even an air starter. The 7-hp Midget was an elegant engine intended for tenders up to 20′ long. Boat builder Scott Matthews instigated the engine, and it was designed in-house by engineer Walter Meyer. It was painted deep cream color, and the upper crankcase had an engine-turned finish.

The core of the Scripps line by 1914 included medium-speed (600- to 800-rpm) engines available in eight models of 1, 2, 4, and 6 cylinders. Dual Bosch/Delco battery/magneto ignition was standard on all. Semi-speed (1,000-rpm) 4- and 6-cylinder models were also offered. In 1916 the Scripps HB Series "B" was selected to power the Hacker Hydrocar, a $2,000 boat that presaged Chris-Craft's more successful effort to build on a production basis boats affordable to increasingly wealthy middle-class buyers. En bloc cylinders had arrived by 1917. Among these models was the beautiful Scripps Model C "Greyhound," a long-stroke racing engine.

After World War I, the company continued development while maintaining the popular 4-cylinder D and E models that ranged from 15 to 100 hp. In 1924, the F-series of 4- and 6-cylinder models was introduced. As always, these were L-head engines with cylinders cast in pairs. The engine featured electric starting and a new Schebler marine-type carburetor. *Motor Boating* observed a dynamometer test in January 1924, in which the 330-cubic-inch, 700-pound F-6 produced 133 hp at 2,800 rpm.

By 1928 the 2-cylinder 12- to 18-hp "E" anchored the smaller end of the line while the F-4 and F-6 covered 25- to 125-hp market needs. By then, the company had also introduced the H line

1906 Claude Sintz ad for the horizontally opposed model

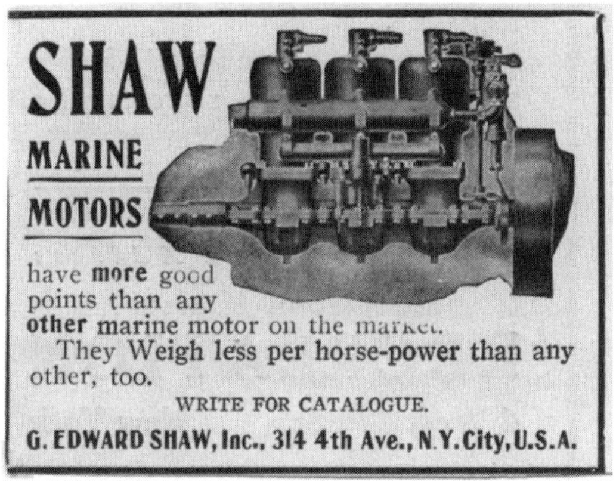

Circa 1905

1906

1905

with numerical model designations: the 100-hp "60," 140-hp "62," 125-hp "70," 160-hp "72," 115-hp "80," 150-hp "82," and 200-hp "200." All these engines shared a 5¾" stroke dimension and a 4¾" bore. (The 100-hp "60" and 140-hp "62" models were 4½" x 5¾".) In 1929 these engines were refined with a Holley downdraft carburetor, self-priming fuel pump, oil cooler and Purolator filter as well as a new model Joe's reverse gear. The engines were now designated by power rating: 160, 162, 170, and 172. The higher-horsepower versions were rated at higher rpm than the medium-duty versions.

Scripps remained profitable even during the Depression. In 1932 the popular "150" series 6-cylinder models were introduced in 110 hp at 1,800 rpm and 165 hp at 3,000 rpm versions. The 5" x 5¾" "202" (210 hp at 2,300 rpm) became especially popular with Gar Wood and Hacker. In 1933 the big news was a new L-head, 894-cubic-inch, dual ignition V-12 series based off the Model 152. Models included a "high-speed" 330 hp at 2,800 rpm and a "medium-speed" 225 hp at 1,800 rpm version. The engine was directed at the market for fast runabouts over 25′ and cruisers up to 90′. Boat builder Greavette was among the first customers. In 1936 a compact, 95-hp version of the Ford V-8 was unveiled, the Model 43. Scripps designed an intake manifold that permitted the engine's installation at an angle without affecting performance, a stamped steel oil pan that permitted operating angles up to 18 degrees, and its own cast-iron or aluminum cylinder heads. Scripps enjoyed great success with the V-8, which found a market in work as well as pleasure boats. One company sales letter pictured a Scripps-modified V-8 with reduction gear in a 42′ Seattle salmon trawler. Throughout the '30s, Scripps never hesitated to refine an engine. Lubrication systems were improved, carburetors refined, reduction gears made available, and as Scripps liked to point out, performance gains made without having to resort to premium fuel.

If the Depression had been a challenge to Scripps and other engine builders, the post–World War II era proved, if possible, even worse. On Lincoln Avenue in Detroit, at the Scripps Motor Company, business had already undergone a fundamental change by the time company founder William E. Scripps died of respiratory failure in the early morning hours of June 12, 1952. The Scripps sons-in-law continued the business, selling spare parts and getting the occasional windfall order such as one for 14 Scripps V-12s for use in Canadian patrol boats. But the business could no longer justify remaining in the old Scripps factory. Scripps was moved first to a building that had housed a car dealership and finally to a former supermarket in Walled Lake, Michigan, about an hour from Detroit. That is where an engineer and motor boat enthusiast named Pete Henkel purchased the assets and parts of the grand old company in order to meet the needs of Scripps owners for years to come.

Seabury, Charles L. Co.
Morris Heights, NY
Trade name: Speedway
Founded as a boatbuilding company in 1885 at Nyack, New York, this firm merged with the Gas Engine & Power Company to form Consolidated in Morris Heights (Bronx).
See also Consolidated Shipbuilding Corp.

Sears, Roebuck & Co.
Chicago, IL
Trade name: Kenwood, Harvard
By 1908, the Sears catalog included the four-cycle Kenwood in ratings of 1½ to 10 hp and the two-cycle Harvard in ratings of 2 to 10 hp. Each could be had with fittings for either fresh water or salt water. Both models had jump-spark ignition and were finished in green enamel. While the manufacturers were, of course, not named, the Kenwood was shipped from "southern Michigan," the Harvard from "southeastern Michigan."

Seattle Standard Engine Mfg. Co.
Seattle, WA
Trade name: Seattle Standard

Seitz, F. A. & Co.
Newark, NJ
Trade name: Mohler & DeGress
The company was begun 1902, was bought by Mohler & DeGress in 1905, and was sold to Jencick Motor Corp. in 1912. The classic Seitz-built motor was a T-head with individual cylinders and an unenclosed reverse gear. Ratings ranged from 6 to 50 hp.

Seneca Boat & Engine Co.
See American Gasoline Motor Co.

Seneca Falls Engine & Supply Co.
Seneca Falls, NY
Trade name: K & D
The company was in business by 1911.

Servis
Trade name: Servis
The brand was sold in Chicago by the Auto Parts Co.

Severson & Lee
Auburn, NY
Trade name: N/A
The company built "Safe Simple Sure" two-cycle engines with removable heads and a solid flywheel. Precise company dates remain unknown, but the engines were in production circa 1905 at 18 Lewis St.

Shadegg Engine Co.
Minneapolis, MN
Trade name: N/A

Shaw, G. Edward Inc.
New York, NY
Trade name: Shaw
This was a lightweight, two-cycle engine that was available in 1906 in 1- to 3-cylinder models.

Sheriffs Mfg. Co.
Milwaukee, WI
Trade name: N/A

Sherman, F.W.
Buffalo, NY
Trade name: Sherman

Sibley-Houfley Machine Co.
Detroit, MI
Trade name: Peerless
The company had an ad in *MotorBoat* in July 1908.

Simplex Engine Co.
Detroit, MI
Trade name: Simplex
The company offered 2- and 4-cylinder, two-cycle engines in the early to mid-teens.

Sinclair Motors Inc.
New Orleans, LA
Trade name: Sinclair
This engine had rotary valves and was advertised in *Motor Boating* in Feb. 1920.

Sintz, Clark Gas Engine Co.
Springfield, OH
Trade name: Sintz
See Sintz Gas Engine Co.

Sintz, Claude
Grand Rapids, MI
Trade name: The Leader
Born in 1876, Claude Sintz was the eldest son of engine pioneer Clark Sintz. He was involved with his father and younger brother, Guy, in the creation of Wolverine in 1894. After service in the Spanish-American War, Claude returned to Grand Rapids, Michigan, and continued his involvement in various Sintz enterprises. The Sintzes had first launched an unsuccessful car-related effort in 1897, and Clark Sintz tried again unsuccessfully in 1904 after he left Wolverine. At the same time as this effort was under way, a return was made to marine and stationary engine production by Claude. This presumably occurred sometime in 1906 after Claude decided not to move east with the Wolverine Motor Co. The first Leader engine (the name would soon reappear on a totally different model) was a 1½-hp two-stroke said to have been mass-produced using "special jigs and templates" that permitted a $75 price. A 2-cylinder, 3-hp model followed. The company was located at the corner of South Front and West Fulton streets. Like the other Sintz family efforts, this company had only a brief life and ceased operations in late 1906 or early 1907. By then, Claude Sintz was focused on a new engine entirely.

See also Sintz-Wallin Co.

Sintz Gas Engine Co.
Grand Rapids, MI
Trade name: Sintz
The Sintz Gas Engine Company was among America's first and most influential makers of marine engines, all thanks to the genius of Clark Sintz. In 1888, Sintz, together with engine pioneer John Foos, started the Gas Engine Company in Springfield, Ohio. Within less than two years, however, Sintz became more interested in the potential of marine engines than stationary models and he sold his interest to Foos in 1890 and then established his own company. Both stationary and marine models were offered, but it was the latter that made the greatest impression, both on customers and on others interested in the potential of the marine market. In December 1892, at the suggestion of a wealthy customer named Addison A. Barber, Sintz moved to Grand Rapids and opened the Sintz Gas Engine Company. Barber was president. The Sintz engine and the launches they powered soon became highly visible examples of internal-combustion technology.

In 1893 Sintz introduced a mechanical fuel pump that delivered the charge of gasoline into the intake chamber in a process vaguely similar to what we would now call fuel injection. The 1893 Sintz was an extraordinarily successful two-cycle marine engine. It prompted one of Sintz's customers, Detroit advertising executive Ora J. Mulford, to buy into the business under the auspices of his Michigan Yacht and Power Company. Capitalization was increased from $30,000 to $100,000 in December 1894, when "a considerable business," according to *Michigan Tradesman*, developed without any special advertising efforts. It was at this time, current research suggests, that Clark Sintz separated from the company. Two years later, in January 1896, *Michigan Tradesman* reported that more than 100 orders were on the books with more anticipated, thus requiring either plant expansion or a move. By 1898 stand-alone Sintz ads were largely justified in claiming, "We are the oldest builders of Marine Gas Engines in the United States." (They probably weren't thinking of the Union Engine Co., whose antecedents predated their own efforts.) Sales branches then existed in New York City, Boston, Philadelphia, and Detroit.

An "improved model" available by 1900 was a 2-cylinder based on a single casting, most unusual for that pre–en bloc period. In January 1901, *The Gas Engine* reported that "the Sintz gas-engine plant of Grand Rapids, Mich., has been sold to Detroit capitalists." The magazine reported the plan was to move the company to Detroit later in the year. The lead "capitalist" was Ora Mulford who, at some point in 1901 or early 1902, bought Sintz Gas Engine in its entirety, moved it to Detroit, and integrated it fully with his Michigan Yacht and Power Company, located at 1564 Jefferson Ave. Ads then read Michigan Yacht and Power Company/Sintz Gas Engine Co. (Consolidated).

By now, the Sintz engine had opened the eyes of many Detroit entrepreneurs and created a new breed of boater, the motor boat enthusiast. The following year, and probably looking ahead to a new venture, Mulford took a first step by either selling a portion of Michigan Yacht to W. A. Pungs or bringing in Pungs as president while he, Mulford, took the title of secretary. For his part, Pungs brought in his son-in-law, engineer Edward B. Finch, as treasurer in 1903. In 1904, the two men converted the Michigan Yacht facility to be the home of the Pungs-Finch automobile company (1904–1910). Mulford's son John reported that Mulford sold his last interest in Michigan Yacht (and Sintz) in late 1905. An advertising genius who worked with Henry Ford, Packard, and others, Mulford would be among those who founded the Gray Motor Company in 1906.

As for the Sintz marine engine, having influenced many other engine designers, it disappeared from the scene while its creator went on to a fascinating, always convoluted, and financially unrewarding career, first with the newly organized Wolverine Motor Co. and later with a succession of other engine and automobile-building efforts and related inventions. His sons Claude and Guy did much to support their father in his later years. Clark Sintz's inventive tinkering was brought to a sudden end when he was struck and killed by an automobile in July 1922. Today, a complete Sintz marine engine in original condition would be an object worthy of prominent display and interpretation at any credible museum dedicated to the history of American invention or nautical history.

See also Wolverine Motor Works

Sintz, Guy L. Co.
Grand Rapids, MI
Trade Name: Sintz-Wallin

Guy L. Sintz's company filed for incorporation in June 1908 with Guy Sintz (Claude Sintz's younger brother) as president and three investors as principals. It is possible that the company was begun as an attempt to produce or market a steel alloy developed in France and first introduced in Grand Rapids by the New Process Steel Company in 1906. Soon, however, the Sintz company appears to have become a sort of parallel or alternative to Sintz-Wallin. The Sintz-Wallin 1,500-rpm, horizontally opposed Leader engines were touted as being "absolutely free from vibration." Their inherently low center of gravity was potentially an aid to a boat's stability. The company also advertised a "special 12 to 14 horse power, 2-cylinder opposed engine for replacing the Cadillac single. Price $150 complete." A notice in *Michigan Investor,* Vol. 6, no. 35, of June 6, 1908, spoke of "the horizontal opposed automobile and marine engines [to be] manufactured on contract." The same article did nothing to distinguish Guy from his brother, the creator of the engine. Apparently, the hoped-for contracts were not forthcoming. In 1910 the company moved to the Chesher Mill Building on Exchange Street. According to *A History of Marshall,* the company declared bankruptcy in 1910. That was the year in which Sintz-Wallin ceased to exist.

See also Sintz-Wallin Co.

Sintz-Wallin Co.
Grand Rapids, MI
Trade name: Leader, Sintz-Wallin

The first mention of this venture appeared in March of 1907 when *Michigan Tradesman* announced that Claude Sintz had "merged his business into a stock company under the style of the Sintz-Wallin Co. with an authorized capital stock of $40,000 common and $20,000 preferred." Besides Sintz, who was treasurer, stockholders included F. B. and W. C. Wallin and William Andrews. A 1907 company brochure proclaimed Sintz-Wallin as "successors to Claude Sintz."

The genesis of this particular Leader engine dated to 1904 when Claude Sintz first developed a 2-cylinder, horizontally opposed, 6-hp marine engine. According to Sintz, that first version had been inspired by the work of Feodor Hirsch, a Daimler shop superintendent who had been brought to America by William Steinway to oversee production of Steinway's Daimler engine venture. Unlike Hirsch's engine, Sintz's had the valves on the top side of the cylinders where they wouldn't be exposed to bilge water.

When ads began appearing, the engine was said to be available in ratings of 4 to 65 hp. "I am the original builder of a successful Horizontal Opposed Four Cycle Marine Gasoline Engine," claimed Sintz in a *MotorBoat* ad in April 1908. The ad listed Claude Sintz as the company general manager and treasurer. According to Claude Sintz, the Sintz-Wallin was popular with the Navy for installation in 36′ battleship tenders. Even assuming that was the case, sales volumes were not enough to support the business as competition increased. The company closed in 1910. Claude Sintz then busied himself with development of a farm tractor, and, in 1912, he moved to Detroit to build his Steel King motor plow.

Sittmann & Pitt
Brooklyn, NY
Trade name: S & P

Already well established by 1905, this boat- and engine-building company offered two-cycle engines of from 5 to 20 hp. "You can reverse it with two fingers," said the ads, doubtless referring to the potential to change direction by manipulating the timing just so.

Skipper Marine Engine Co.
Cincinnati, OH
Trade name: Little Skipper

Listings also appeared for the Little Skipper Marine Engine Co. in Cincinnati. The company was advertising in *Motor Boating* in December 1911. It seems most likely that, whatever the listing, the engines were built by Monarch.

See also Monarch Tool & Engine Co,

Slitz, F. A.
Newark, NJ
Trade name: N/A

Sloane-Daniel
New York, NY
Trade name: Sloane-Daniel

Sloane Motor Co.
Chicago, IL
Trade name: Sloane

Smalley Bros. Co. (Smalley Motor Co.)
Bay City, MI
Trade name: Smalley

Smalley was founded in 1889 although it is unclear when engine production commenced. Ads appeared by 1902 for single- and double-cylinder two-cycle engines of 1½ to 18 hp. The acceptance of its engines prompted local announcements in 1904 that an expanded factory would be needed to meet increased demand. The company was said to have 200 employees working in this larger plant in 1906. By 1907, however, the company had entered receivership and was reorganized as the General Machinery Co.

See also General Machinery Co.

Smith & Baldridge Machine Co.
Detroit, MI
Trade name: S & B
A 1908 ad suggested, "Territory open to responsible agents alive to a quality proposition."

Smith & Mabley
New York, NY
Trade name: Simplex
This company was founded by A. D. Proctor Smith and Carlton R. Mabley to import high-grade automobiles from Europe for sale in the wealthy New York market. In 1904 the company introduced its own automobile, the chain-drive Simplex. Its engine was a T-head, 4-cylinder with cylinders cast in pairs. In 1904 one of these Simplex motors powered *Vingt-et-Un II* to victory in that year's second Gold Cup race. The 6½″ × 6½″ motor produced 60 hp at 800 rpm and 75 hp at 1,400 rpm. An 8-cylinder version went into a 1904 Harmsworth Trophy racer called *Challenger.* These engines, however, remained essentially automotive designs, and Smith & Mabley was not a factor in the marine engine industry.

Smith Electrical Engineering Co.
Watertown, NY
Trade name: Boynton

Smith-Laingmaid
Portland, ME
Trade name: Casco
The two-cycle Casco may have been based on a cylinder and base casting purchased from Gray & Prior, makers of the two-cycle Hartford, according to an experienced collector familiar with both brands. Smith & Laingmaid was founded in 1905 and was located at 5 Central Wharf, just off Portland's busy Commercial Street. When the company ceased building the Casco, it became an agent for a variety of other marine engine companies. In 1926, these included Gray, Red Wing, Sterling, Kermath, and Hubbard. They also handled Evinrude outboards. The company appeared in the City Directory through 1966.

Smith Mfg. Co.
Chicago, IL
Trade name: Great Western

Snecker Motor Co.
Stamford, CT
Trade name: Snecker
See Stamford Motor Co.

Southern Motors
Brunswick, GA
Trade name: Southern
In 1923 Southern advertised in *MotorBoat* a 3¼″ × 4″ engine with 4 cylinders producing 12 to 16 hp. This 177-cubic-inch engine was almost certainly a Ford Model T–based block derated for marine applications. Atwater-Kent ignition was used, and a reverse gear was an option ($225 so equipped *v.* $190 without).

Southwark Foundry & Machine
Philadelphia, PA
Trade name: Southwark-Harriss

Sparks, C. F. Machine Co.
Alton, IL
Trade name: Sparks
By 1906 Sparks was advertising a small but interesting line of engines. Included were a pair of 4-cylinder models of 20 and 40 hp, a 10-hp double, and a 5-hp single. In addition, 7-hp and 15-hp opposed twins were offered. These were four-cycle engines.

Spaulding Gas Engine Works
St. Joseph, MI
Trade name: Spaulding
In operation in 1904, by 1912 Spaulding built a line of 1- and 2-cylinder two-cycle engines of 5 to 20 hp. In 1912, six models were offered: the 5-hp B, 7½-hp C, 10-hp D, and three twins including the 10-hp BB, 15-hp CC, and 20-hp DD. They featured make-and-break ignition and a Schebler carburetor. While most companies had by 1912 switched to a gasoline/oil mixture, Spaulding still listed gravity sight-feed lubrication.

Specialty Manufacturing. Co.
South Norwalk, CT
Trade name: Toquet Special
Probably at some time during 1910, Specialty acquired the rights and patterns to what it named the Toquet Special. However, this engine bore no resemblance to the Toquet that preceded it. It was a conventional-looking, headless, two-cycle engine available in single- and double-cylinder models of from 2½ to 10 hp. Specialty labeled the single-cylinder engines as the H (2½ hp), I (3½ hp), and K (5 hp), while the doubles were the R (7 hp) and S (10 hp). A Krice carburetor was used. Ignition was make-and-break. Specialty's shop was located on Water Street in South Norwalk. The engine was still being made in 1912, but production may have ceased thereafter.

See also Toquet Launch and Motor Co. *and* Stanley Company

Speed Changing Pulley Co.
Anderson, IN
Trade name: N/A

Sphinx Motor Co.
New York, NY
Trade name: N/A
The company was in business in 1911.

Springfield Motor Co.
Springfield, MA
Trade name: Springfield
By 1910, Springfield was in production on a line of two-cycle motors of from 1 to 4 cylinders and 3 to 20 hp.

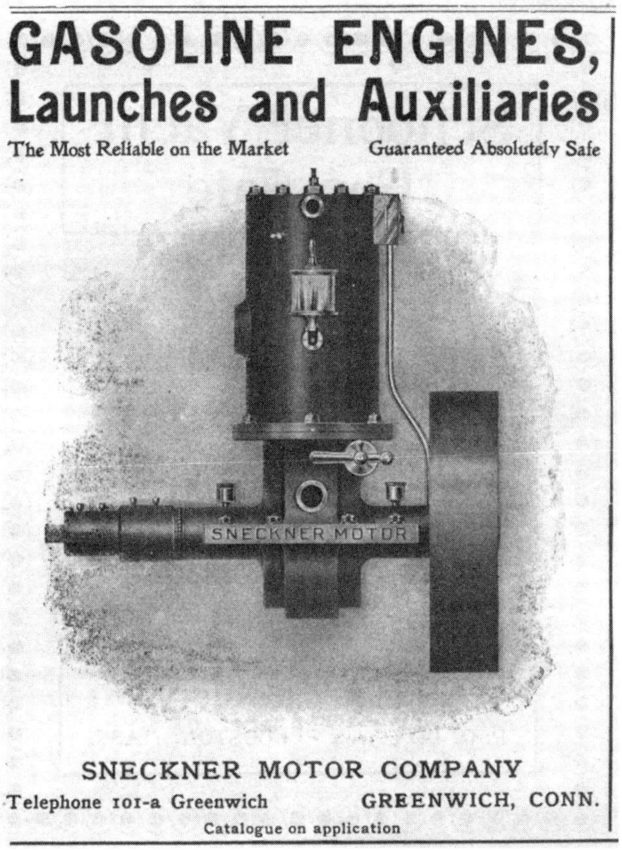

1901

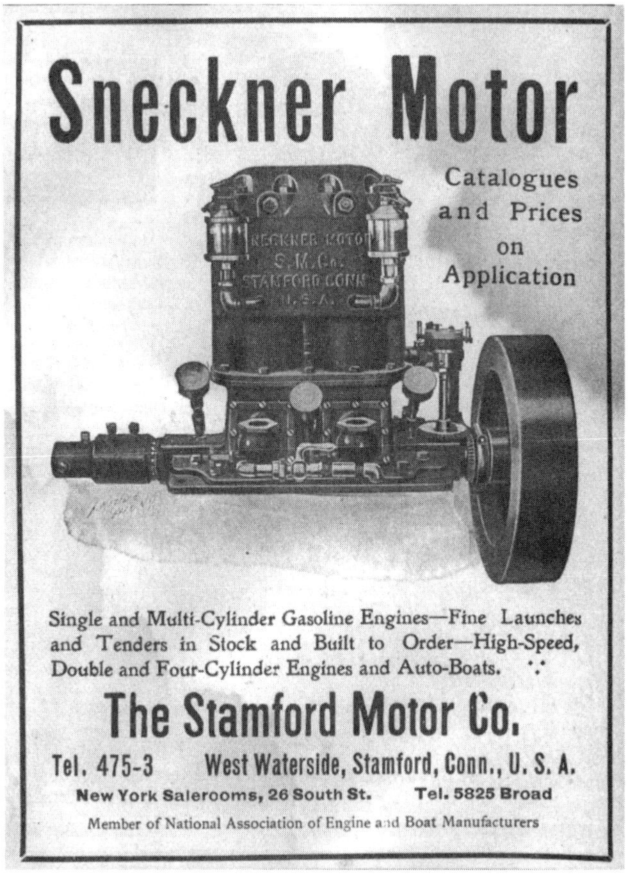

1905

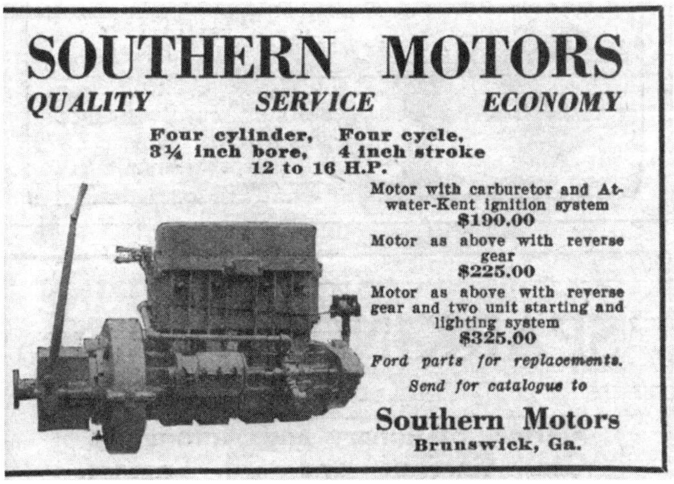

1923

1915

The 2- and 4-cylinder motors were unusual for the period in that they had en bloc cylinder castings. Jump-spark was standard, make-and-break an option. Schebler carburetors were used. A gravity-feed lubrication system directed oil from a cylindrical tank to the bearings where the splash then lubricated interior parts.

St. Clair Motor Co.
Detroit, MI
Trade name: Little Skipper
See also Monarch Tool Co.

St. Joseph Motor Co.
St. Joseph, MI
Trade name: Miss Simplicity
"The two-port engine as manufactured by us possesses every desirable feature of the three-port, with the added advantage of being easier to start, and developing a greater power at lower speed," read the elaborately worded St. Joseph catalog. What the company was referring to was an intake manifold with "an adjustable gas inlet valve which takes the place of the usual check valve on the two-port engine, and the third port in the three port." Mounted on the inlet manifold was a float-feed carburetor "because there is no known form of mixing device which makes gas from gasoline that is better adapted to this type of motor."

The Miss Simplicity had unusual over-square bore/stroke dimensions (3¾" × 3½") to reduce piston travel and wear. The cylinders were dome-head because, according to the company, "there is no vegetable substance from which a gasket can be made that will withstand the action of the heat, oil and water to which it is subjected." These motors were listed in the brochure as being available in single-cylinder models of 3, 5, and 8 hp, doubles of 6, 10, and 16 hp, and 4-cylinder versions of 12, 20, and 30 hp. All were jump-spark and equipped with Champion A. C. spark plugs.

With that as background, it must be admitted that these motors currently remain mysterious. *MotorBoat* reported that the "full line of Miss Simplicity motors manufactured by the St. Joseph Motor Co, of St. Joseph, Mich. were on exhibition" at the 1910 Boston Show. They reported that the only change over 1909 models was an increase in the length of the main bearings. The Miss Simplicity was pictured in a 1910 list of marine motors but made no further appearance in the boating press.

While a St. Joseph Motor Supply was in business in the city as of at least 1925, no listing has yet been found for a St. Joseph Motor Co. Presumably, the word "Supply" was not included in the brochure since it implied the company may have been a dealer or distributor rather than a manufacturer. A motor called Miss Simplicity appeared in none of the major lists of marine engine builders of the 1912–1928 period and, although the motor's general appearance is evocative of Lockwood-Ash, none of the key bore/stroke dimensions match. There are also differences in equipment. As for the St. Joseph Motor Supply Co., it disappeared from the city directory after 1929.

St. Lawrence River Motor & Machine
Clayton, NY
Trade name: St. Lawrence
The company was in business as of 1925.

St. Louis Boat & Engine Co.
St. Louis, MO
Trade name: St. Louis
The company was in business by 1905, but no further information has yet surfaced.

Stamford Foundry Co.
Stamford, CT
Trade name: Shipmate, Snecker

Stamford Motor Co.
Stamford, CT
Trade name: Snecker
The Snecker was a two-cycle motor first advertised as early as 1901 by the Snecker Motor Company of Stamford. It was produced at the Stamford Foundry.

Standard Company
Torrington, CT
Trade name: Eagle
Sometime between 1913 and 1917, Standard took over production of Eagle two-cycle motors from the Progressive Manufacturing Company in Torrington. A 1916 ad pointed out that the cylinder, intake manifold, and water-jacketed exhaust manifold were all a one-piece casting, much reducing the number of gaskets. As for the detachable cylinder head, it was mounted atop a modern "double-jacketed copper-asbestos gasket." By 1918, the line consisted of seven 1- and 2-cylinder models ranging from 1½ to 12 hp. The single-cylinder 1½-hp Eagle was a two-port, 600-rpm engine with make-and-break ignition. The 3½-hp one-lunger was a three-port, 1,000-rpm model with jump-spark ignition. Both two- and three-port 6- and 7-hp models were offered. The most powerful model was a 12-hp double. This three-port engine was rated at 1,300 rpm, and was fitted with a standard-equipment reverse gear. All this must have seemed confusing to customers of that era. Sometime in 1920 or 1921, production ceased.
See also Progressive

Standard Gas Engine Co.
San Francisco (after 1906 earthquake, Oakland), CA
Trade name: Standard, 'Frisco Standard
Sometime in 1900, a strike idled the Union Gas Engine Company. This shutdown was the impetus for the founding of the Standard Gas Engine Company by a group of former Union engineers and machinists. They were led by Peter Mohrdieck, who had worked in Germany with Rudolf Diesel before emigrating to the United States. While Standard maintained a salesroom on Market Street in San Francisco, the factory itself occupied a seven-acre site in Oakland where there was access to a deep-water dock and a railway spur.

Thanks to solid construction based on gray-iron blocks and cylinders, and an efficient, overhead-valve design, Standard engines enjoyed a rapid success. The installation of a 3-cylinder, 120-horsepower Standard in the San Francisco whaling schooner *Monterey* was cause for favorable comment in the press. The early reputation that Standard gained for quality laid a solid foundation. The company had agents in New York, New Orleans, Seattle and Portland, Australia and New Zealand, Hawaii, Fiji and Papeete, and Manila.

Standard focused on 1- to 3-cylinder engines that ranged in output from 4 to 110 hp. Applications were primarily commercial: fishing schooners, seiners, tenders, and workboats. (The 3-hp Kid was popular for small workboats.) Although Standard built a number of 400- and 500-horsepower Southwark-Harris two-cycle diesels under license beginning in 1913, the company's real market was the smaller gasoline engine so well adapted for a majority of marine uses, as well as for farm and hoisting applications. The 1916 line was typical. It included 16 models. These included a trio of one-lungers (4, 5, and 7 hp), six twins (8, 10, 12, 16, 20, and 30 hp), and six triples, the most powerful being a 175-hp model rated at 270 rpm. The company offered larger sizes on special order. While make-and-break was the standard ignition system, jump-spark was available on special order as well. These engines had en bloc cylinders with separate heads cast in the company's own foundry. The cylinder heads were pressed into the top of the cylinder so that the lip protected the head gasket from direct exposure to flame.

Good as its engines were, Standard did relatively little in the way of product development. By 1930 this failing combined with the Depression to put Standard into a precarious position. A line of lighter weight DF and GK models was added alongside the old 'Frisco Standards, but the new line had little impact. Standard continued production at a much diminished rate. The post–World War II market finally put an end to Standard as it did to all the engines born in the "iron age." By the time Standard ceased production, its engines were well on their way to becoming wonderful collectors' items rather than a viable source of power.

Standard Gear Co.
Detroit, MI
Trade name: N/A

Standard Motor Construction Co.
Jersey City, NJ
Trade name: Standard

Not to be confused with the Standard Gas Engine Co. of San Francisco, this pioneer firm grew out of C. C. Riotte and Co. in New York (1898), builder of the Empire engine designed by Carl C. Riotte. The Standard Motor Construction Company was formed in 1902 and two years later rolled out a 6-cylinder, direct-reversing (sliding the camshaft could permit rotation in either direction) gasoline engine. It was that engine, according to Riotte's brother, Eugene, that "eventually became the Standard."

Two particular events put the Riottes on the map. In June 1904 Eugene Riotte drove a 60′ launch to victory in the first Gold Cup race, garnering extensive press coverage. The boat was powered by the brothers' 6-cylinder engine, a 3,200-lb., open-base design that produced 110 hp. The launch was named, appropriately, *Standard*. Then, in July, *Standard* defeated John Brown Herreshoff's fast, steam-powered *Swift Sure* in a boat-for-boat competition. Given Herreshoff's stature, this event resulted in more publicity. Standards were among the brands in use by the Navy in 1906 for yard ferries and in the Presidential yacht *Sylph*'s tender and gig.

Comparatively large 6-cylinder engines were staple fare at Standard, but the range had grown by 1912 to include 2- and 4-cylinder models as well. In 1913, the company introduced flanged exhaust manifold piping to ensure leak-proof fits and easy replacement thanks to standardized threads. Most models were heavy-duty with cylinders cast in pairs although most 100-plus–hp engines had separately cast cylinders. Ignition was then make-and-break. The 100-plus–hp engines had reversing camshafts.

Standards were popular with a great many boat builders, especially for big cruising yachts. Two were installed in a 75-foot Mathis built around 1914, two more in a Cox & Stevens 99-footer with six staterooms and three bathrooms. That said, smaller Standards could be found in a variety of launches and auxiliary sailboats. A Norman Skene–designed 32-footer designed in 1911, for example, was equipped with an 18-hp Standard. In 1917 Standard received the biggest order in its history, and probably one of the biggest ever for a marine engine maker. The order was for more than 1,100, 220 hp at 400 rpm engines for installation in submarine chasers to be built by Elco.

Although it is extraordinarily rare to find a first-hand description of living with such an engine, British Air Commodore F. R. Banks shared his memories of those Standards in his book *I Kept No Diary*. As a youngster of 18 or 19, Banks spent time aboard the Elcos, known to the British as an "ML," in the North Sea: "... [T]hey were open crankcase jobs with drip feed lubrication via slinger rings, to plain bronze main bearings and big ends. They had automatic inlets and cam and pushrod-operated exhaust valves with a sliding camshaft for directly reversing the engine, initiated by compressed air ... Sometimes, due to hurried wartime workmanship, the keying of the exhaust cams to the shaft was not always well done and an order to reverse the engines was rendered ineffective owing to the cams sliding off their keys!

"Great care had to be taken to avoid pieces of cotton waste getting caught in the adjustable drip feeds to the big end slingers, since, if one feed was diverted, this would quickly cause a complete seizure of the particular big end on its pin bringing the whole engine to a dead stop."

Diesel models of 40 to 400 hp entered the line in the mid-'20s, but Standard had to be reorganized after it

1913

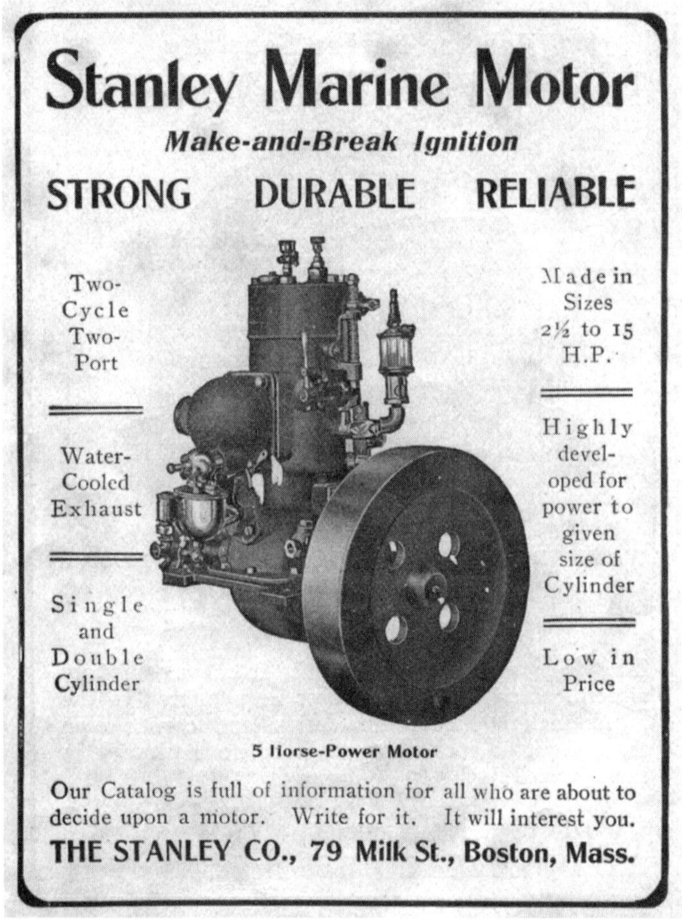

1911

1905

1906 Sterling ad for an early four-stroke model

1911

entered receivership in April 1929. The company was actively listing its engines in 1934 but succumbed in 1937.

Stang Engine Company
Harvey, IL
Trade name: Stang
The company was in business by 1906 at 140 Dearborn St., offering gas, gasoline, and kerosene engines. Records show it was still in business in 1912.

Stanley Company
Salem, MA
Trade name: Stanley
The Stanley Company is a good example of a marine engine maker that, like many others, left behind some fine-looking engines and a few knotty, organization-related mysteries to keep researchers guessing. The first Stanley ad seems to have appeared in 1908 and listed the company's address as 79 Milk St. in Boston, with the factory located in Lawrence, Massachusetts. Research suggests that Frank Stanley's principal business was shoe manufacturing, for the Milk Street address was listed in the City Directory as home to the Stanley Shoe Machinery Co. By 1911 the Stanley Company office, at least, had relocated to 48 Loring Ave. in Salem, which was closer to the expansive Swampscott summer home of president Frank F. Stanley. Lyman Stanley was the company's treasurer.

Stanley produced a line of two-port, two-cycle motors. Single-cylinder models typically had make-and-break ignition, while the doubles came standard with jump-spark ignition. It is possible, however, that both systems were available on at least some models. Stanley's ratings evolved. New for 1910 were a 7-hp single, a 14-hp double, and an 18-hp triple, all with jump-spark ignition. In 1911 the single-cylinder engines were listed at 3, 5, 7½-, and 1 -hp. The doubles were rated at 6, 7, 10, 14, and 15 hp. The following year, the 12-hp single was dropped and so was the 14-hp double. Like most two-port motors, the Stanley was rated at 425 to 500 rpm depending on model.

The Stanley's float-feed carburetor was mounted on the intake manifold and fed directly into a check valve. The inlet/exhaust manifold was a single casting to warm the incoming mixture. This "special feature," claimed the company, avoided "excessive condensation of water in and about the Carburetor and its connections, which trouble happens when the Carburetor is set out on a pipe containing a check valve of the usual construction." In 1910, the check valve was redesigned so that the valve seat was horizontal and the valve opened directly into the crankcase. A vertical orientation was more common.

In 1913 a 3-hp kerosene model was introduced, and other models were refined with revised ports, intake manifold, and higher compression to improve fuel economy as gasoline prices rose. A Stanley brochure circa 1911 noted the engines were finished in three coats of dark blue enamel and all brass fittings were highly polished. An oil-tight reverse gear of Stanley's own build was included.

Stanley advertised, as early as 1908, that in addition to its own motors, "We also manufacture the well-known Toquet Motors which can be purchased direct from us."

Details of the Toquet-Stanley arrangement are currently unknown, but there seems to be no doubt, based on bore/stroke dimensions and the few available pictures, that the engines were different. Stanley disappeared from the Salem City Directory after 1917 but did appear on the first postwar list of two-cycle engine manufacturers published by *Motor Boating* in December 1918. Boston was again listed as the city of record. Stanley did not appear in subsequent directories of marine engine makers and it is likely that few, if any engines were produced after 1917.

Star Marine Engine Works
Elmhurst, CA
Trade name: Star
By 1936 the conversion specialist offered a line that included engines based on the Ford Model A and V-8, three Chrysler models of 70, 72, and 75 hp, and a 55-hp Flint. Joe's reverse gears were fitted and Joe's 2:1 reduction gears were available as options.

Staten Island Shipbuilding Corp.
Port Richmond, NY
Trade name: N/A

Stationary & Marine Supply Co.
Port Washington, NY
Trade name: Uncle Sam

Stearns Motor Mfg. Co.
Ludington, MI
Trade name: Stearns Extra Reserve
What Stearns meant by its engine's name was that its exceptionally large main and crankpin bearings, combined with a forged, heat-treated crankshaft, gave owners an extra reserve of dependability and long service life. "Money Never Bought More in a Marine Engine" was the slogan. Stearns may have entered the business with the purchase of Gile Boat & Engine Co. By the '20s, a line of 40- to 150-hp, four-cycle engines was offered. These were impressive products. The 1925 Super Power Six was typical of Stearns output. The side-valve motor had an en bloc cylinder casting to which the three separate heads were secured with 17 big nickel-steel bolts. An automatically tensioned silent chain drove the camshaft, water pump, and accessories. An "oil rectifier" was included in the lubrication system to extend oil life and strain out contaminants. Ignition was provided by a 12-volt American Bosch firing two plugs per cylinder. A fully enclosed Joe's reverse gear was fitted with the Stearns logo cast into the cover plate. This fine engine developed 120 hp at 1,200 rpm and 160 hp at 1,600 rpm at a weight of 2,100 lbs. for a model with aluminum equipment and 2,480 lbs. for the model with aluminum parts.

Sterling Engine Co.
Buffalo, NY
Trade name: Sterling
One of the most innovative, successful, and longest-lived of all the marine-

engine companies, Sterling began modestly enough when a local businessman and boater named Charles A. Criqui became interested in purchasing a two-cycle engine designed by Albert B. Shultz and built by the Sterling Engine Company. According to Criqui himself, the company's then-owner decided at this time to quit the business. Criqui and several colleagues then bought Sterling, probably in 1903. Sterling's model line was made up of two-cycle 1-, 2-, and 3-cylinder motors of from 1½ to 24 horsepower. They were three-port engines built under license from Joseph Day.

Apparently, plans were soon under way to expand into four-cycle engine production. It is possible that the first such engines were available in 1905. The company's 1907 catalog shows both a two-cycle lineup and a four-cycle. Already, the company was talking about power-to-weight ratios and listed its 60-horsepower engine at 850 lbs., which was then a *very* light weight. Did Shultz adapt a European automobile engine? It's possible. The same brochure refers to cylinder manufacture as being "similar to Panhard," the great French company that pioneered the automobile and whose engines were frequently found in racing motor boats in the early years of the century. Right from the start, apparently, the four-cycle T-head Sterlings had cylinders cast in pairs to reduce the chance of leaks and improve overall engine stiffness.

Sterling constructed a new factory on the corner of Niagara and Auburn Streets and outfitted it with the most modern machinery. Rough castings entered the factory on the bottommost floor, where cams, piston pins, gears, bolts, and other components were heat-treated. On the next floor up, bases and cylinders and brass castings were ground and smoothed before being passed on to the machine shop. Other floors held the drill-press room, paint shop, and testing room. On the topmost floor were the offices and a drafting room that kept four men busy full time.

Criqui endeavored to offer a Sterling model to meet the needs of different customers. By 1910 Sterling was producing the 2-cylinder Model C and the 2-, 4-, and 6-cylinder Model D heavy-duty engines. In 1910 designer George Crouch described one of these Model Ds as "a handsome piece of engine building." He noted that the splash lubrication for the crankshaft was based on "inclined troughs" that led from one to the other, directing oil splashed up the connecting rods. Also offered were the 4-, 6-, and 8-cylinder Model B. These were available as a "semi-heavy duty" or "speed" model.

C and D models might be used for lower-speed, heavier-duty applications such as pleasure launches or naval launches, and the Navy equipped many tenders with the 40-horsepower Model D. The launch for President Taft's yacht *Mayflower* had a 4-cylinder Sterling Model B of 25 to 40 horsepower. B models also were used in a variety of fast runabouts and cruising boats. Six- and 8-cylinder Model Bs were popular with owners of fast pleasure and racing craft. Such engines were the basis for the glory that became the Sterling Engine Company. Adequate capitalization, disciplined management, and a model line diverse enough to generate a significant volume of orders combined to form the foundation for a company that always would be recognized for quality, reliability, innovation, and performance.

In 1915, the new Model F appeared. Initially, it was an 8-cylinder, 200-horsepower model. This was an enclosed flywheel design with cylinders cast in pairs and mounted to a one-piece, manganese-bronze base. The lower base was iron, with aluminum available as an option. The Model F was intended to combine high rpm (1,500) with the durability necessary not only for 55′ to 60′ pleasure boats but also for naval dispatch and patrol vessels. Ford's Childe Harold Wills, always a staunch supporter of high-quality engine and boat companies, ordered *three* for his 100-foot yacht. A 5½″ × 6¾″ 8-cylinder model was also offered; this 200-horsepower eight with an aluminum base weighed in at 2,100 lbs., just 10½ lbs. per horsepower. A larger-displacement model with 6¾″ bore and 9″ stroke was available. The motors found an immediate naval market and the Italian navy, among others, ordered them for patrol boats and submarine chasers.

Just six months after introducing the Model F, Sterling unveiled another innovation. It was "An improvement which the Sterling Engine Company are incorporating in their new Model F, 6-cylinder machine, and which it is stated has eliminated certain disagreeable features of marine engines so totally as to surprise even the manufacturers themselves." *It* was a counterbalanced crankshaft. Both Criqui and the Sterling engineers were well aware of advances made by automobile engineers with counterweighted cranks. When such a crank was introduced to the Model F, the company spoke of it in automotive terms. "There is no jarring or jerking as the motor speeds up under load from slow to full speed, but an evenness of acceleration such as is found in the Hudson Super-Six, the 8-cylinder Cadillac and other high-grade cars."

By now, Sterling was in the midst of its domination of Gold Cup racing. In 1915 when the race was held on Long Island Sound, the first five boats were Sterling-powered. In fact, the only non–Sterling-powered boat did not finish. The engine used was the big Model R (Racing) straight-eight which, with its 5½″ bore and 6¾″ stroke, produced an easy 250 horsepower. This engine in 1916 powered the Chris Smith–designed *Miss Minneapolis* to complete domination of that season's Gold Cup. In 1916 Sterling introduced its 6- and 8-cylinder racing engine, the GR. It was an overwhelming advance. That year, Sterling again powered the Gold Cup winner, *Miss Minneapolis*. The following year, a Sterling GR-powered boat also set the record for fastest complete heat and fastest complete race for all Gold Cup events held between

1916 Sterling F6

1913 Sterling 6-cylinder heavy-duty

1911 Sterling six, port side

Circa 1923 Sturtevant

1911 T&M twin

1913 Sterling "Kid"

1905 Strelinger

1905 Strelinger

1915 T&M "Special"

1904 and 1919. This was achieved by Gar Wood in his *Miss Detroit II* at Minneapolis in 1917. Sterling's domination would continue through 1917 and then be challenged and overwhelmed by race boats powered by converted Liberty or other airplane engines and, finally, by Packard.

While such glamorous engines and the boats they powered obviously generated lots of attention, Sterling had also introduced the 40-hp "Kid" so that the model line was basically capable of meeting the needs of any boater. Sterling was very well placed at the outset of the United States' entry into World War I. It received government orders for engines to power everything from admirals' barges to patrol boats and Coast Guard cutters. After the war, the variety of Sterling-powered boats continued to grow. Commercial boats of all types, tugs, offshore oil-rig supply boats, the toll-collector's boat at the Cape Cod Canal, and a tourist boat on Lake Chelan in Washington are but a few examples. These craft were equipped with the Model F heavy duty, an engine that also saw service in big power and sailing yachts. An 8-cylinder, heavy-duty Model F was put into the 136′ schooner yacht *Finback* built in Essex, Massachusetts, by the famous Story yard for a wealthy Bostonian named Charles Henry Wheelwright Foster whose great-grandfather had once been sailing master on the *U.S.S. Constitution.* Whether in the finest yachts of the era or working boats, Sterling engines were used from coast to coast, and they were exported, too.

After the war, Sterling continued development of the Model F. In January 1920, it introduced "a most pronounced departure from the usual" when it unveiled its 4-, 6-, and 8-cylinder GR models. With bore/stroke dimensions of 5¾″ × 6¾″, the engine was similar in its basic construction to the F but included what Sterling called a "dual valve-in-head" configuration — what today would be called a 4-valve engine. Sterling recognized, as had some others, that "the principle admits of greater valve area and the use of smaller valves, which cool more rapidly and retain their shape and effective seating. It also gives great flexibility of operation." The new engine had a pressure-lubricated, hollow crankshaft and, an elegant touch, was delivered with a can of touch-up paint.

In 1922 the company dropped designations involving letters of the alphabet and substituted names based on elements from the company's coat of arms. The crest included three seagulls, a Viking helmet, a petrel, a trident, and a dolphin. While all were marvelous engines, the Sea Gull was, and would remain, extraordinary. Development work on the 6-cylinder Sea Gull must have begun by 1919 or 1920 under the direction of Sterling's intense-looking chief engineer, E. T. Larkin. Bolted atop the aluminum base were black-enameled alloy cylinder castings with steel sleeves. The cylinders were topped by an aluminum cam cover, for the Sea Gull had an overhead camshaft driven by bevel gears off the crankshaft. With bore/stroke dimensions of 4¹¹⁄₁₆″ × 6″, it was rated at 150 horsepower. The Sea Gull sold in 1922 for $3,450, about $500 less than a Cadillac touring car. The Viking, a larger companion model intended for heavy cruising boats, shared the Sea Gull's design features but added four valves per cylinder. With a 7½″ bore and 8½″ stroke, it developed in excess of 300 horsepower.

In the mid-'20s, Sterling was approached by the Coast Guard to develop an engine suitable for use in boats that could pursue bootleggers. The Sterling "Coast Guard" model was the result, and the factory produced 550 engines so quickly between April 1924 and June 1925 that most sat in storage until the boats could be completed. Criqui and Sterling were awarded a special citation by the Coast Guard for this performance.

In 1927, the 779-cubic-inch Petrel was introduced. It was similar to the Viking but had an L-head, making it less expensive to produce. The Petrel was a versatile, 150- to 250-hp engine intended for big cruisers or as an auxiliary in a sailing yacht. The Dolphin and Admiral (based on a design by Vimalert) models rounded out the line in the '30s and were produced in volume for the government during World War II.

Charles Criqui retired from Sterling in 1939 and died in April 1948. By the time of Criqui's passing, Sterling was in trouble. The company entered receivership in 1948 and was reorganized under new management by 1950. At that time it was announced that diesel engines would be Sterling's focus. Sterling's diesel lineup was acquired from the National Supply Company in 1946 with the acquisition of the tooling to produce that company's Superior Models A and D. The Model A had been developed during the mid-'30s and was then a quite modern concept, with a patented combustion chamber and en bloc cylinders. It was a 4½″ × 5¾″ engine rated at 1,600 rpm. A larger displacement version, the Model D, with a 7½″ bore and 8″ stroke, was introduced in 1937. Sterling sold these engines as the Viking Diesel, and a supercharged version of each engine was available, the 8-cylinder supercharged Viking Diesel producing 650 horsepower at 1,200 rpm, and weighing some 13,000 lbs. Stationary, industrial, and marine versions of these engines were manufactured, finding application in generator sets and switching locomotives.

Whatever the virtues of these Sterling diesels, the domination of the diesel market by General Motors and by war-surplus "Jimmys" really made success unlikely. Even a 1946 order for $1,500,000 worth of engines from the Whitcomb Locomotive Company and others was not enough to prevent the need for a major restructuring and borrowing two years later in 1948. Graham-Paige, which had previously been an automobile manufacturer but had by then become an investment company, acquired 32 percent of Sterling's stock in October 1951. Finally, on February 8, 1955, it was announced that the manufacturing equipment of what was then known as the Sterling Precision Instrument Corporation would be acquired by the Prescott Company of Menominee, Michigan. Subsequent production of engines or parts was carried out at a facility in Manistee, Michigan, but it was not enough to keep Sterling a viable entity. Within a few years, operations ceased. In 1962, the Baldwin-Lima-Hamilton Corporation purchased the Viking parts business. That marked the end of a company that could rightly calls it products "The Masterpiece of Motordom."

Sterling Mfg. Co.
Fall River, MA
Trade name: Quesquechan
This two-cycle motor was said to be available in 2- to 16-hp models in 1907.

Stevenson & Leonard
Albany, NY
Trade name: N/A
The company was advertising marine engines and electric generating sets in 1909–1910.

Stilson Mfg. Co.
Pittsfield, MA
Trade name: N/A

Stoddard-Dayton
Dayton, OH
Trade name: Stoddard-Dayton
A builder of high-quality automobiles commencing in 1905, this company is listed here only because it ran an ad for its 50-hp, 4-cylinder engine in *Motor Boat* on March 10, 1915. There is no indication that the Stoddard-Dayton was ever used in any quantity, or in fact at all, in marine applications.

Stork Motor Company
Saginaw, MI
Trade name: Stork
Fred W. Stork and partner Fred L. Helfrecht started this company in 1905. Helfrecht sold his interest to Stork 18 months later to start his own machine shop. Stork introduced an extensive line of two-cycle motors by 1908, and these were replaced by four-cycle models of 8 to 75 hp. The 1916 line was typical. All the engines were then four-cycle. The smallest was a 2-cylinder (5″ × 6″) 10-hp engine that weighed 850 lbs. The 6-cylinder, 75-hp Stork was a 7½″ × 9″ engine that weighed 6,000 lbs. Among Stork's early customers was the Defoe Boat and Motor Works of Bay City, which had begun in 1905. Defoe was listed for many years as Stork's sales agent. By 1928, Stork listed more than two-dozen models. By then, the company was known as the Stork Engineering Co., and its primary focus had become machinery for oil field use. Another firm, Stork Oil Co., was formed in 1929 to operate 29 oil wells of its own. Stork was still operating in the early '60s.

Stover Motor Car Co.
Freeport, IL
Trade name: Stover

Strauble Machine Co.
Green Bay, WI
Trade name: Strauble

Strawbridge, J.B. Motor Co.
Baltimore, MD
Trade name: Peerless

Strelinger Marine Engine Co.
Detroit, MI
Trade name: Strelinger
Strelinger was in the marine engine business starting circa 1899, building its "old reliable 4-cycle Marine Motor." In 1903 it introduced the Little Giant two-cycle model. By 1907 customers could choose either jump-spark or mechanical ignition. In 1907 Strelinger acquired the two-cycle engines formerly built by the Detroit Gas Engine & Machinery Co., and that firm's predecessors. It is possible that Strelinger produced the 1½-hp version of that engine for resale with a Fairbanks-Morse badge.

Strickland
Bound Brook, NJ
Trade name: Strickland
Introduced in 1905 with ports located near the middle of the stroke and equipped with poppet inlet valve and cam-driven intermediate and exhaust valve, the Strickland had a water-jacketed head and finned cylinder. Jump-spark was used. It's not known whether any were sold.

Stroud-Humphrey Mfg. Co.
Hastings, MN
Trade name: Stroud Valveless

Stuart, John Co.
Wollaston, MA
Trade name: Stuart
In 1902 Stuart advertised single- and double-cylinder motors of 1½ to 40 hp. A few years later, models up to 100 hp were advertised.

Sturtevant, B. F. Co.
Boston, MA
Trade name: Sturtevant
Benjamin Franklin Sturtevant began his company in about 1863 with six employees. The focus was on ventilation of buildings, and centrifugal industrial fans became the staple product. Eventually, the company engaged in the design and manufacture of pumps, turbines, heat exchangers, and everything related to heating, cooling, and ventilation. Sturtevant fans would be used to ventilate the Holland Tunnel. At some point, probably during 1912, Sturtevant began manufacture of a high-grade, 4-cylinder (4½″ × 6″), 75-hp marine engine. The E4 had two inlet and two exhaust valves in each of its T-head cylinders. The base was of cast aluminum. Force-feed lubrication was standard, and a fine-mesh screen filtered impurities before oil reached the pump. The oil was cooled by a built-in pipe carrying the engine's cooling water. A Stromberg carburetor was used. Ignition was high-tension magneto with a separate battery ignition system and electric starter firing a second set of spark plugs. The engine's maximum rating was 90 hp, giving it a weight-to-hp ratio of 9.3 lbs. per hp — outstanding. It

1923

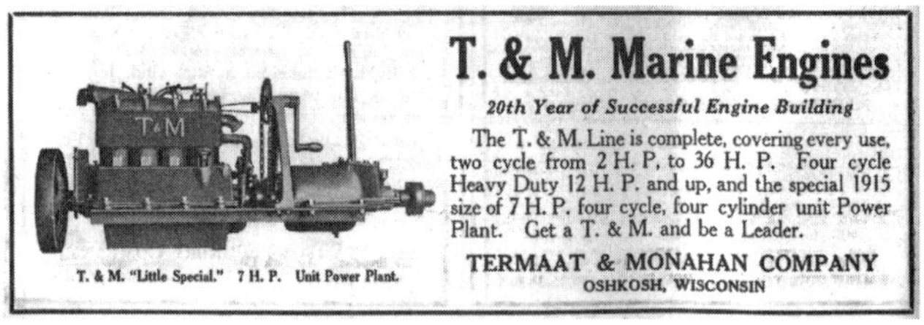

1915

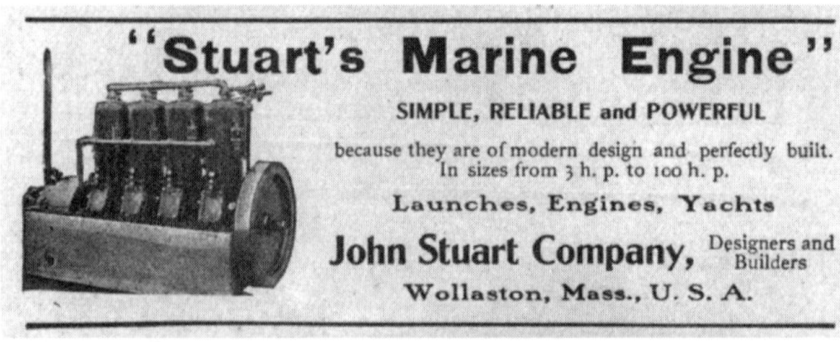

1906

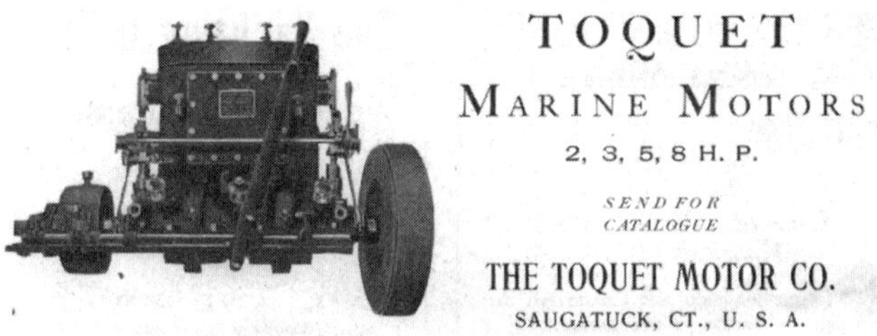

1906

1908

1913

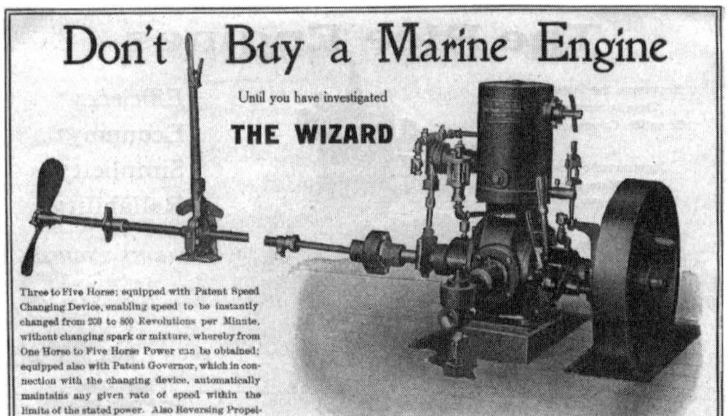

Circa 1905

was rated for continuous operation at 1,500 to 1,600 rpm.

The Sturtevant found some success in pleasure boats and, during World War I, was purchased by the Navy and used in the Hickman Sea Sled. After the war, a 40-hp four, and 60-hp six were offered by 1923. These were L-head, fully enclosed engines of very contemporary design. In 1923 it was reported that Sturtevant acquired the defunct 10-acre, Corliss, Wisconsin, plant of the Wisconsin Engine Co., which was then making Corliss pumping engines. By that time, Sturtevant had plants in San Francisco and Philadelphia. Marine engine production was apparently a small sideline for this impressive industrial company. Its marine models and an all-alloy, V-8 airplane engine, probably developed before World War I, were never more than sidelines. Sturtevant itself eventually became a division of Westinghouse, which sold it in 1985, the most recent owner being the Howden Fan Company.

Sumner, H. W. Co.
Seattle, WA
Trade name: Sumner

Superior Engine Co.
Springfield, OH
Trade name: Superior

Begun in Oct. 1889 as the Superior Gas Engine Company, this firm's original products were stationary engines. Production of such engines ceased in 1920, and Superior acquired the Otto Engine Works of Philadelphia in 1923. A few years later, a new plant was built in the Holmesburg section of Philadelphia, and the company name was changed to the Superior Engine Company. A new line of Superior-designed models was introduced in 1927, and the original Deutz-patent engines were continued until 1930. Marine diesels were certainly in production in the '30s. These were built both in Philadelphia — the high-speed — 1,400- to 1,800-rpm MA and MRA series — and in Springfield, under the name of The National-Superior Company. The latter were slow-speed, heavy-duty engines rated at from 325 to 700 rpm.

Superior Machine & Engineering Co.
Detroit, MI
Trade name: Yale

Superior Motor Works
Jackson, MI
Trade name: Superior

Sutter Bros.
New York, NY
Trade name: International 16

Syracuse Gas Engine Co.
Syracuse, NY
Trade name: Syracuse

Just as some makers of two-port, two-cycle engines eloquently described the superiority of their designs, Syracuse literature explained the virtues of the three-port motor. "...Valves are unnecessary to prevent the gases drawn into the engine from passing back in the carburetor. These valves depend upon the action of a spring to close them, and suction, or an impulse from a cam or otherwise, to operate them. If they are worn or improperly adjusted, the incoming gas does not reach the cylinder at the proper time or the outgoing gas is not allowed to discharge itself freely. Or, if not perfectly tight, a leak occurs seriously impairing or entirely destroying the efficiency of the motor."

The ports cast into the Syracuse cylinder walls were somewhat unusual. They were deliberately made longer rather than higher, in theory increasing the engine's functional stroke. (There would be slightly more mixture drawn in during the upstroke and exhaust scavenging would be complete before the transfer port was uncovered.) Syracuse also had a patented "water jacketed condensing exhaust manifold" that was cast with four internal passages, one for the exhaust in the center and one each, above and below, for the water circulation system. The fourth was an "air flue" that supplied the carburetor with hot dry air. In addition to improved engine operation, Syracuse claimed an advantage because cold water wasn't in contact with the "outside of the firing chamber [which] could cause a cylinder to go out of shape and cause leakage."

The Syracuse model line was extensive and divided into three different series, each with its own bore/stroke dimensions. The Type B models (4" × 4") were offered in 1- to 6-cylinder versions of 5 to 42 hp. The Type C models (6" × 7") were offered in 1- to 4-cylinder versions of 12 to 48 hp. The Type D models (5½" × 5½") were offered in 1- to 6-cylinder versions of 10 to 100 hp. All models had jump-spark ignition. The Syracuse was finished with "two coats of best black paint" that contrasted nicely with the aluminum parts. The heads were removable.

Syracuse appears to have started production before 1909, but nothing more was heard from the company after 1912 when it offered one of that era's typically well-written and nicely illustrated brochures. Like most other champions of the two-cycle marine engine, Syracuse had been fighting a losing battle.

Taylor Engine Works
Norfolk, VA
Trade name: Taylor

Taylor built three two-cycle engines in 1912. These included a 4-hp single and a pair of two-cylinder models of 8 hp. Ignition was jump-spark.

Teel Manufacturing Co.
Medford (Boston), MA
Trade name: Teel

In 1910 Teel was in production of four-cycle engines, but also was at work on a two-cycle in which "the crank-case compression is replaced by compression in an annular space about the larger diameter of a two-diameter piston." *MotorBoat* reported in February of that year that "the design is simple and should be effective." Little was heard thereafter about the Teel motor.

Temple Pump Co.
Chicago, IL
Trade name: Wizard

The circa 1905 Wizard was a two-port, two-cycle motor equipped with a patented "Speed Changing Device" said to permit instant throttling between 200 and 800 rpm. The motor was rated at 1 to 5 hp.

Termaat & Monahan Co.
Oshkosh, WI
Trade name: T & M

This pioneer company owed its beginning to John D. Termaat who was born to a wealthy family in Aalten, Holland, in 1867 and moved to Wisconsin in 1884. Among other things, Termaat was an avid fisherman who, when reminiscing about how he became involved in the engine business, said it was all because he had tired of rowing and sought a better way to get his fish home in time for dinner. According to Termaat, the "better way" had its basis in the work of inventor Louis J. Monahan, who, in 1896, received a patent on a vertical, two-cycle engine.

Because of the success achieved by what became Termaat & Monahan, and because of his local prominence, Termaat was the person usually turned to when the press sought to recount the company history. The results of complete reliance on Termaat were some discrepancies and, possibly, omissions, which leave open a few questions about dates, titles, and at least one possibly key figure. In *Wisconsin, Its Story and Biography*, published in 1914, Termaat indicated that the business was begun in 1892 and expanded "so rapidly that in 1895 it was found expedient to incorporate." The granting of Monahan's patent a year later would seem to fall neatly in place with this timeline. However, in a retrospective published in *MotorBoat* in April 1924, Termaat claimed that he met Monahan in 1898 and that they then began two years of engine experimentation. Termaat wrote that "after two years of experimenting, we developed an engine that we felt was really reliable. In 1901, we started to produce these engines . . . The first few engines met with favor with local users, and the demand for our engines increased." In 1902 the Termaat & Monahan Company was begun to build marine engines on a production basis.

Whether the real start date was 1895 or (more likely) 1902, Termaat & Monahan began in a small way in a backyard shed, this particular shed located in Winnebago, Wisconsin. The initial ½- to 4-hp single-cylinder models became popular with boaters on Lake Winnebago and the Fox River. Twin-cylinder models of 5 to 8 hp soon followed. These "first generation" T&M models had a removable head and a solid flywheel. At least until about 1904, they may have had proprietary make-and-break igniters. Still, a 1904 brochure shows a spark plug mounted in the center of the head, and jump-spark may by then have become standard equipment.

By 1906, as Termaat remembered, "we were going strong." That was when a new design was introduced. These were the T&Ms with the now familiar four-hole flywheel and the distinctive, rectangular, water-cooled exhaust manifold ("condenser" in the parlance of the day) for which Monahan had filed for a patent in Oct. 1905. The 1906 T&M brochure showed a cutaway drawing of the condenser, noting its dual function of reducing exhaust back pressure and permitting an underwater exhaust. Intake mixture entered a "walled off" portion of the water jacket where it contacted the outside of the warm cylinder wall casting. The latter, it was said, both did away with potential condensation and permitted operation on kerosene or alcohol.

By 1911 the line included ratings of 2 to 36 hp. The 8-, 12-, 18-, and 20-hp models were twins. There was also an 18-hp triple with a smaller (4¾″ × 4¾″) bore/stroke than the 18-hp twin; a 27-hp triple; and a 36-hp 4-cylinder. A 1911 4-cylinder, 36-hp model weighed just 550 lbs. By now, T&M had adapted its engines to run on either gasoline or kerosene and had added stationary models aimed at those in need of "small power outfits in country homes, farms, etc." In 1912 T&M's lineup offered 1-, 2-, 3-, and 4-cylinder models in two model lines, A and B. The smallest model was the 125 lb., 1-cylinder, 2-hp Model A. The largest was the 1,100 lb., 4-cylinder, 60-hp Model B, a 6⅛″ × 6″, 800-rpm engine. Cylinders were, as previously, cast separately, and ignition was jump-spark. The 2- and 3-hp Model A singles were equipped with a vaporizer but the others had a Schebler carburetor. The engines were painted a medium green color although color renderings of the period also suggested a darker shade of green.

When Termaat remembered sales as "going strong" by 1906, he was not exaggerating. The engines' good reputation and the company's ample production capacity resulted in the T&M's being re-badged for several boat builders who wanted to offer complete "outfits." By 1909, these included Toppan in Massachusetts. "A complete line of the T&M marine motors is carried in stock by the Toppan Company of Boston and New York," noted an entry in *MotorBoat* in June 1910. This was one of the very rare times that an engine builder gave away the "secret" that not all engines bearing a specific nameplate were actually the work of that company. The Racine Boat Company in Wisconsin was another T&M user. Current research suggests that a number of engine builders also re-badged T&M-built motors.

By 1913 T&M was a busy concern that employed some 100 people who handled every phase of manufacture from drawing board to foundry to assembly. Both marine and hopper-cooled engines were produced and sales were made throughout the country and to export markets. In 1913 Termaat was said to have always been the company president, but in the 1924 *MotorBoat* article he claimed that Monahan was president and he was vice president. Never did Termaat mention Emmery Homer Fahrney, the son of a wealthy Chicago doctor who had developed an engine of his own. Fahrney had a summer place in Oshkosh and was likely to have been an important investor in T&M.

According to Termaat, the company's 1906–1913 growth "attracted outside capital and likewise outside ideas." The investors saw in T&M's fine factory the opportunity to expand into the production of machinery as well as engines. Rather than engage in manufacture of what Termaat called "some foreign product," both Termaat and Monahan promptly resigned. By the time T&M entered receivership in 1917, the two partners and, it is quite likely, Fahrney, had long since started a new endeavor, which they named the Universal Motor Company. T&M had disappeared from the Oshkosh City Directory by 1922. The company's 1919 listing was as a manufacturer of farm engines.

See also Universal Motor Company *and* Fahrney, E. H.

Terry Engine Co.
Brooklyn, NY
Trade name: Terry

In 1908 this company's ads noted that it was the successor to the Geo. H. Terry Co. The slogan was "Built for Business at Business Prices." The Terry was a four-cycle T-head engine. The 5- to 30-hp models were said to be "medium weight and medium duty" while a 20-hp model and those from above 30 to 100 hp were a "heavy duty slow speed type."

The Twice-Two Cycle Engine Co.
See Twice-Two Cycle Engine Co., The

Thelma Engine Works
Detroit, MI
Trade name: Thelma

"The fastest motor boats for their size on the great lakes are equipped with Thelma engines, viz: *Thelma I, Echo,* and *Thelma II,*" said a company ad in 1912. The first Thelma was a 35-hp 4-cylinder built in 1905 for the 28′ × 4′ *Thelma I,* which achieved a speed of 26 mph. A similar engine was installed in the 40-foot cabin launch *Robin.* In 1910, a 6-cylinder version was installed in the 32′ × 4′ *Thelma II,* which sped over a Detroit River course at 29 mph. These were certainly competitive speeds in that era; the attention gained by the boats led to the formation of the Engine Works at some point in 1910 or 1911. The Thelma was built as a 35- to 45-hp, 4-cylinder and a 70-hp, 6-cylinder T-head engine for a couple years. Like many engines of the period, two-cycle or four-, the Thelma's separately cast cylinders did not have removable heads because of potential sealing difficulties with gasket materials of the day. Two locations in rather close proximity to each other were given for the company, 223 Jefferson Ave. and 223 Junction Ave. It seems doubtful that the effort survived beyond 1912.

Thomas & Grant
Ithaca, NY
Trade name: Thomas

Thomas, E. R. Motor Co.
Buffalo, NY
Trade name: Thomas

Having achieved great success with his Cleveland bicycle, Erwin Ross Thomas proceeded to build a motorcycle and then, in 1903, began the E. R. Thomas Motor Company. The Thomas Flyer became famous when it won the epic New York to Paris race in 1908. Located as he was in Buffalo, Thomas was well aware of the rapid growth of motor boat racing, and, at least during 1904, he advertised his car's engine in the *Marine Buyers' Guide.* Although Thomas didn't become a factor in the marine engine industry, he funded a perpetual Thomas Trophy that carried with it a substantial $2,500 prize. In 1913 the Trophy was won by J. Stuart Blackton in his *Baby Reliance III.*

Thorobred
See Red Wing Motor Co.

Thousand Island Boat & Engine Co.
Morristown, NY
Trade name: Lloyd

By 1908, the company was offering Lloyd two-cycle motors in ratings of 1½ to 20 hp.

Thrall Motor Company
Detroit, MI
Trade name: Thrall Refined

The Thrall was one of many small, three-port, two-cycle engines intended for small boats including canoes, launches, dinghies, and dories. It was a broad market, and in 1903 Thrall — then located at 9-31 21st St. — claimed it had sold 5,000 motors. If so, it is quite possible that a portion of this production was sold to other companies who re-badged the engines. The Brooks Boat Mfg. Co. of Saginaw was probably one of these resellers, and Brooks is also believed to have purchased DuBrie engines. While Thrall usually pictured in its ads a 3-hp, 60-lb. two-cycle model with a distinctive copper water jacket, it claimed to offer single- and multicylinder motors of up to 12 hp. By 1908 the company was still picturing the 3-hp model, but the name was now Thrall-Fishback. Whatever the ownership, Thrall was a small operation, employing only nine people in 1909, which tends to make one question its previous claims of high production volume or else that volume had steeply declined. In 1910 the Thrall received some detail changes. A 2-cylinder model at the Boston Show showed a timer position that had been raised (it was driven by enclosed bevel gears) and a new Thrall carburetor. "No valves, springs, triggers, cams, gaskets nor packing requiring attention," said the company. The address was then back to 21st St., and the ads once again read Thrall Motor Company. By 1911 or 1912 the company had become part of a group presided over by the Detroit Motor Car Supply Co., itself part of the Detroit Engine Works conglomerate. The Thrall did not appear in either 1911 or 1912 on the key lists of marine engine makers.

Thrall-Fishback Motor Co.
See Thrall Motor Co.

Tige Motor Company
Port Angeles, WA
Trade name: Silent Tige

Toledo Motor Boat & Power Co.
Toledo, OH
Trade name: Cranston

Toppan Boat and Manufacturing Company
Boston, MA
Trade name: Toppan Reversible

Arthur W. Toppan got his start in the boat industry in a small shop in Charlestown, Massachusetts. There, sometime between 1898 and 1900, Toppan built Swampscott-type dories intended primarily for use by local fishermen. Toppan soon developed a standardized production method to increase efficiency, and he began tinkering with the idea of converting his dories from oar and sail to power. According to Toppan himself, it took several years to get the idea into a commercially acceptable form. When that happened, Toppan moved from Charlestown to a larger plant on the Mystic River in Medford. "Here," said Toppan, "we built many lap-strake 26′ and 30′ fishing dories. They were arranged with a small cabin aft enclosing the motor and two berths. This was the start of the small motor boat for fishing." Later, Toppan introduced a carvel-planked "dory launch" for recreational customers.

Toppan ads claimed the boats were "the BEST ROUGH WATER BOAT in the World." These boats were available complete with engines that carried a Toppan nameplate but, in fact, were supplied to Toppan by Termaat & Monahan of Oshkosh, Wisconsin. After T&M's demise, when Toppan displayed two boats at the 1917 Power Boat Show in New York, each was powered by a Palmer Q1. Although Toppan's plant was located in Medford, the business office was in Boston. In the fall of 1916, Toppan moved from its original showroom on Haverhill Street to larger premises in the Trade Building at 87 Washington Street near Boston's famous Haymarket Square. There, in addition to rowboats and the familiar sailing dories, the latest Kermath and Universal engines (John Termaat's and Louis Monahan's newest engine endeavor) were on display. By the late '20s, Toppan was offering standardized runabouts (the Sea Dog), a 30′ cruiser (the Sea Crest), and a 32½′ Express Cruiser (the Champion).

Toquet Launch and Motor Company
Saugatuck, CT (Westport)
Trade name: "The Model" (later Toquet)

Entrepreneur and inventor Benjamin Louis Toquet gave his name to this company, which was organized on Feb. 2, 1901. It succeeded the Toquet Launch Company and was begun with $15,000 in capital, a generous enough amount for such an enterprise at that time. It is likely that company president George McKibbin, who gave his address as 29 Broadway, New York, was the principal investor. Toquet was officially listed as secretary. Three Toquet engines were offered including a single-cylinder 2-hp version (500 rpm), a 4½-hp twin (500 rpm), and an 8-hp (400 rpm) twin. According to Toquet, vibration was objectionable in a single-cylinder engine of more than 2 hp.

The Toquet was a sophisticated engine for its time. "The vaporizer and valves [check valve and needle valve] are all in one piece which is bolted to, and forms part of the motor, as does also the hot-air drum." The company brochure claimed the engine's "speed lever" effectively controlled rpm and that the engine could "be run at slow speed all day if desired." The make-and-break igniters were the company's own and relied on only a single spring. A battery and "sparking generator" (magneto) was the ignition equipment. In an era when cylinder head gaskets (packing) were widely recognized as potential trouble spots, the Toquet cylinder heads were "fitted on by a special process which makes an air tight joint without packing." Details were not provided. However, the company catalog of 1902 included several testimonials from satisfied owners.

Benjamin Toquet appears to have been the designer of the Toquet "Model" engine although no patent or other documentation for this conjecture exists. The Model engine was built in South Norwalk, presumably by the Specialty Manufacturing Co., which would later play a further role in Toquet's efforts. The engines were transported to Saugatuck for installation in the boats. As for the "launch" part of the company, a letter written by President McKibbin to Toquet in April 1901 suggests that the company planned to offer B. B. Crowninshield–designed launches and advertise Crowninshield's well-known name. "I think we would build as many boats of Crowninshield's design at a fancy price, as we would stock boats at a low price." It is not known if this plan was carried through. However, boats of 18′, 20′, and 24′ were available in 1901. They ranged in price from $285 to $785 at a time when the average American family had a yearly income of $750.

In 1902 the company hired a mechanical engineering graduate from Cornell named Chester B. DuBois. While DuBois's immediate assignment is not known, it is clear that the company was looking beyond marine engines. In January 1904 the company name was officially changed to the Toquet Motor Company. Soon, ads listed a New York address as the place to write for a brochure about the engines. Meanwhile, the Toquet Motor Car & Construction Company was organized in 1905. It built a number of high-grade automobiles with a T-head engine notable for its detachable heads and copper head gaskets. Was this the engine for which in March 1906 Toquet was informed he'd been allowed a patent? The details remain unknown. Only a limited number of Toquet cars were sold.

As for the marine engine, the Stanley Co. was publicly announcing that it produced the Toquet in 1908 at a plant in Lawrence, Massachusetts. Sometime in 1909 or 1910, Benjamin

1902 Toquet

1911 Kennebec, sometimes advertised as the "Torrey fishing engine"

1913 Kennebec

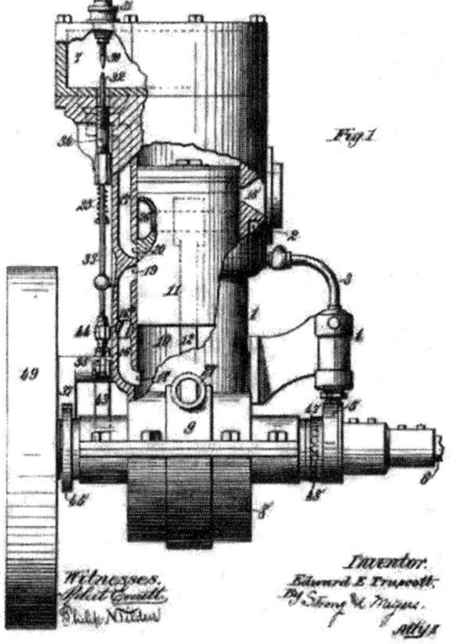

E. E. Truscott's patent drawing for his 1898 two-cycle engine. Mixture compressed in the base entered the combustion chamber through a valve forced open against spring pressure. The opening in the center of the crankcase is the check-valve-controlled air inlet port. Fuel vapor entered on the opposite side.

The 1906 Truscott heavy-duty model was an impressive four-cycle engine with a durable, one-piece camshaft, and combined intake/exhaust manifold.

1898 Truscott with removable shroud and reversing propeller

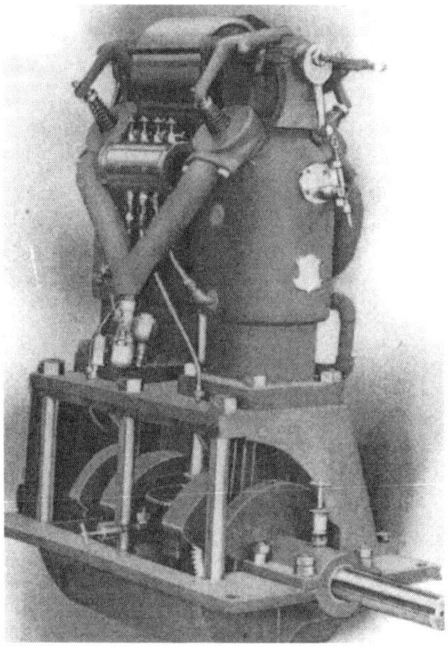

This cutaway shows Truscott's fascinating overhead cam 1906 engine.

1905 Tuttle port side

1911 Tuttle

Toquet moved on to apparently greener pastures with the production of an after-market carburetor for the Model T. This was produced under the moniker Toquet Carburetor Co. in Westport, Connecticut. Another business entity associated with Toquet was the Toquet Manufacturing Company, which may have been concurrent with the Launch and Motor Company and produced wire connectors and tools. The line of Toquet marine engines — no longer the original Model version — was taken over by 1910 by the Specialty Manufacturing Co. As for Benjamin Toquet, he received a patent for an engine timer (distributor) suitable to a 4-cylinder engine in 1921.

See also Specialty Manufacturing Co. *and* Stanley Co.

Torrens & Company
New York, NY
Trade name: N/A
Torrens offered a kerosene engine as of 1906. It was available in ratings of 1½ to 24 hp.

Torrey Roller Bushing Works
Bath, ME
Trade name: Kennebec
A Palmer engine is said to have been the model that Frank Lord used when asked to design a two-cycle engine for the Torrey Roller Bushing Works. The Works was begun in 1869 in a shop beside the Kennebec River. A brief note in *Power Boating* in 1911 stated that the engines had been made "for quite a few years" but was not specific as to what year production began. It does seem certain that the primary market was initially Maine fishermen; some ads referred to the engine as "The Torrey fishing engine." These "fishermen" engines had bore/stroke dimensions of 6¼″ × 7″ and were available as an 8-hp single or 16-hp double. They weighed about 70 lbs. per hp, a good indication of the emphasis on commercial-application durability rather than recreational use performance.

By 1911 Kennebecs were said to have been sold outside New England. The motors advertised in 1912 were available in 1-cylinder models of 2, 3, and 5 hp; doubles of 4, 6, and 10 hp; and triples of 6, 10, and 15 hp. These headless, two-port engines were equipped with make-and-break igniters of Torrey's own manufacture. Production ceased in 1917 when the forge that supplied crankshafts became too busy with war-related contracts to continue doing so. A surviving Kennebec at the Maine Maritime Museum, which appears to be original, is painted a dark red.

Traverse City Iron Works
Traverse City, MI
Trade name: Eclipse

Trebert, H.L.F. Engine Works
Rochester, NY
Trade name: Trebert Reliance
An engineering genius born in Pittsford, New York, Henry Trebert won a New York–to-Buffalo car race in 1901 in a vehicle of his own design. Before establishing the Engine Works, Trebert had been involved with a partner in an earlier effort building Brownell-Trebert engines, which were popular with racing-oriented boaters. Trebert's business activities went through enough evolutions to be confusing. In February 1906, for example, the company was advertised as Trebert Auto & Marine Motor Co., doubtless the hope being to appeal to both markets. But the emphasis was clearly on performance. "Our 34 h. p. motor has a record of winning eight cups in a boat of 34 ft. × 4 ft. 6 in. beam," said an ad. Pictured was a man, perhaps Henry himself, back straight, holding the engine in his hands. The Engine Works was in business from 1910 to 1916, and offered four-cycle engines with exceptional horsepower/weight ratios: the 40-hp, 4-cylinder weighed 760 lbs. These production models were pressure lubricated and offered in 1- to 6-cylinder models (cylinders cast in pairs) of 10 to 90 hp.

In 1912 three racing engines using separately cast cylinders were offered. These included 4-cylinder engines producing 40 hp and 90 hp, and the 4⅞″ × 5″ Trebert V-8. It developed 80 hp at a weight of just 540 lbs. and could achieve 3,000 rpm without load. A single piston-type valve performed both intake and exhaust functions. If all that wasn't enough, the Trebert V-8 was very compact, just 27″ wide and 27″ long. This Trebert was one of those astonishing devices that appeared rarely but periodically in the period before World War I, engines that suggested a far distant future but were too far ahead of their time to achieve meaningful sales for their creators.

Trego Motors Co.
New Haven, CT
Trade name: Trego

Treiber Engine Co.
Camden, NJ
Trade name: Treiber
O. D. Treiber had worked at Cooper-Bessemer until, at some point in the mid-'20s, he began developing a lightweight diesel for marine and electric generation applications. There was a V-type engine as well as inline models. In 1930, at the National Motor Boat Show in New York, Treiber displayed a trio of engines including a 100- and 150-hp six and a 300-hp V-12. Of course, solid injection was used.

Trenton Engine Co.
Trenton, NJ
Trade name: Reeves-Graef
In 1910 Trenton introduced a heavy-duty (300- to 375-rpm) engine designed by Ernest Graef. It was intended for cruisers and commercial boats. This was a sturdy-looking, fully enclosed, T-head engine with individual cylinders. Two-, 3-, and 4-cylinder models were offered, all with removable cylinder heads.

Trenton Malleable Iron Co.
See Rice, John V. Jr. & Co.

Tribune Engineering Co.
Oswego, NY
Trade name: Tribune

Trojan Launch & Motor Works
Green Island/Troy, NY
Trade name: Trojan
In 1901 the Trojan Launch Company was listed at the same address in Green Island (across the Hudson River from Troy) as the Grant-Ferris Company. It last appeared in 1904 under the category for Boat Builders. Almost certainly, Trojan was an effort by Grant-Ferris, similar to that of many other engine builders, to build and market its own line of boats powered by its own engines. It is possible that the boat builder continued for a period as the Trojan Launch & Motor Works after Grant-Ferris closed.
See also Grant-Ferris

Trout, H. G. Co.
Buffalo, NY
Trade name: Trout
The Trouts were two-cycle, heavy-duty oil engines with airless injection and pump scavenging. The smallest model in 1923 was an 8½″ × 12″ that produced 75 hp.

Trump Bros. Machine Co.
Wilmington, DE
Trade name: Pilot
One of at least two marine engines named Pilot, this one was a two-cycle motor with a removable cylinder head and jump-spark ignition. According to the company brochure circa 1913, the motor was available in a single-cylinder of 3 and 6 hp and a double of 6 and 12 hp. The original color appears to have been a medium red. The Pilot was just one Trump product — the catalog noted they built "high grade machinery for a variety of purposes" — but is a perfect example of yet another quirky, intriguing marine motor of great individuality.

Truscott Boat Manufacturing Co.
St. Joseph, MI
Trade name: Truscott
Founded in 1876 in Grand Rapids by English immigrant boat builder Thomas Truscott, the Truscott Boat Manufacturing Co. moved to St. Joseph in 1892. There, on the banks of the St. Joseph River, Truscott and his sons built, in addition to their boats, a pioneering line of two- and four-cycle marine engines, which were fascinating in conception and execution. Castings were produced in Truscott's own foundry.

Edward Truscott received a patent for a two-cycle "explosive engine" in 1899. The patent's chief claim involved control of the engine afforded by a check valve vertically mounted at the top of the cylinder head in its own "compartment." The valve controlled admission of vapor to the combustion chamber. It was opened by the pressure of air compressed in the base. A removable head was fitted, and a make-and-break igniter, operated by an eccentric on the crankshaft, was positioned in the area above the valve. This first Truscott had a feature not seen before or seldom, if ever, since on a marine engine — a removable polished aluminum jacket. The shroud gave this late 19th-century engine a decidedly 21st-century streamlined appearance. The shroud was intended to contain any grease or oil and "make it impossible to catch clothing," something that a thoughtful boat builder might well have recognized would be a selling advantage. As early as the Nov. 1898 issue of *The Rudder*, Truscott was advertising its boat and engine with the slogan "No Fire, Smoke or Heat."

Truscott took advantage of Buffalo's great Pan-American Exhibition in the spring of 1901 to show off its products, bringing several boats including an 18′ launch powered by a 2½-hp motor. The company's 18′ launch of that period was built with white oak frames, keel, stem, and deadwood and planked carvel-fashion with cypress. The hull was painted white, the bottom green. The sheer strake and cockpit coaming were varnished.

Edmund W. Roberts, the prolific engine designer, writer, and then editor of *The Gas Engine* magazine, paid a visit and wrote a detailed account of the Truscott's operation. Contemporary reports by an expert like Roberts are as rare as they are informative, and this one provides fascinating, contemporary information about an early gasoline engine and launch.

First of all, Roberts noted, "The vaporizer is in the bow at the tank, where a pure gasoline vapor is formed by the heat of the jacket water passing about it." (This suggests some sort of plumbing from the engine to the vicinity of the tank.) The vapor flowed through a hose or pipe back to the engine. There, said Roberts, "The mixture is formed directly in the crankcase, vapor entering one side of the crankcase through a swing check and air through a check valve on the opposite side of the crankcase. The charge passes from the crankcase to the top of the cylinder and enters the combustion space past a poppet valve." Throttling action on this 500-rpm engine was controlled by adjusting the lift of the aforementioned poppet valve. Roberts took care to note the Truscott's horizontally split base, and that the igniter and water pump were both operated by a single eccentric. Electrical current was furnished by a hydroxide battery.

Such a vaporizer system had easily identifiable demerits, and the Truscotts moved quickly to find something better. In fact, when the Wheeler-Schebler carburetor was introduced in 1901, it was Truscott who became the company's very first customer! The "second generation" Truscott two-cycle engine differed in its cylinder shape, and the igniter was redesigned as well. Current research suggests that ignition was make-and-break ("Hammer-break" was Truscott's term) through 1904, but the 1905 catalog listed 1-, 2-, and 3-cylinder, two-cycle engines with either jump-spark or make-and-break ignition.

As interesting and puzzling to today's collectors (because of nuances of details) as the two-cycle Truscotts are, the four-cycle engines are equally fascinating. The company was advertising engines of up to 40 hp in 1902, and the more powerful engines were four-cycle models. The first versions had an automatic overhead inlet valve

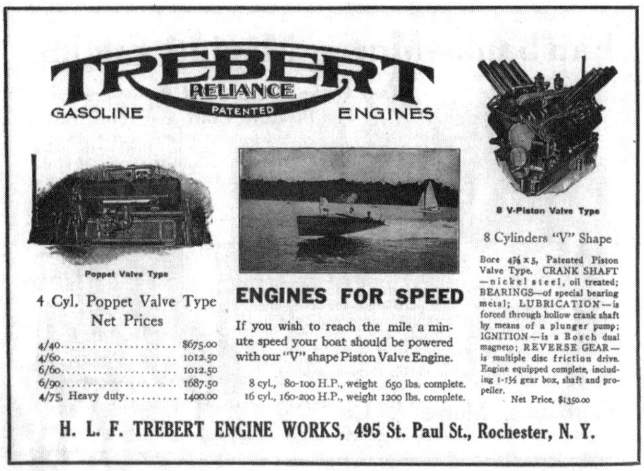

1913

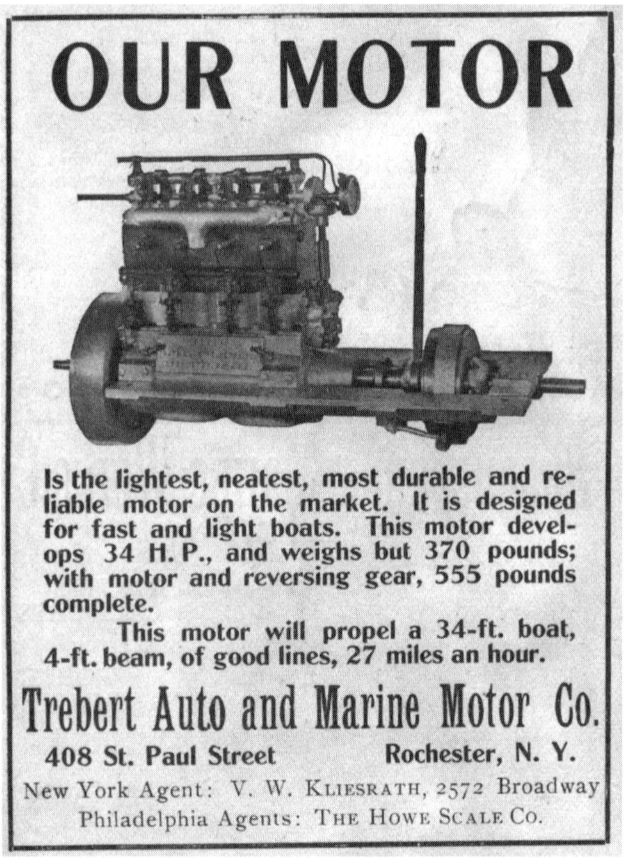

1905

1905

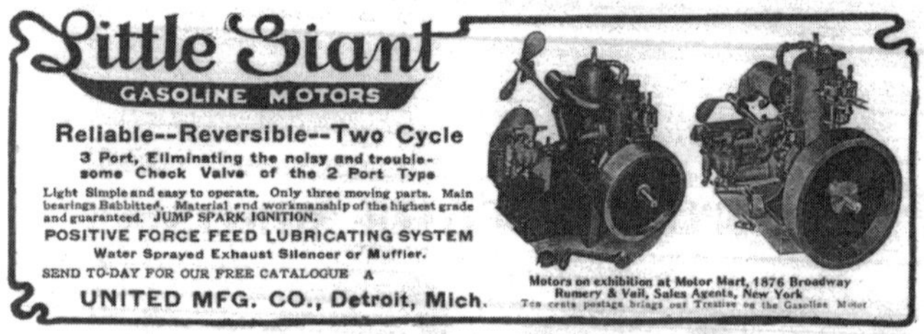

1908

This 1906 Trebert ad emphasized the engine's light weight

and side-mounted exhaust valve, both valves mounted together in a casting on the engine's starboard side. In 1906 a "heavy-duty" L-head Truscott line was unveiled. The intake/exhaust manifold was a single casting to warm the inflowing mixture, and ignition was make-and-break. Four- and 6-cylinder versions were offered. The engine's camshaft was a one-piece forging, an advanced technique in an era when many camshafts had their lobes cast separately, slid on, and mechanically attached by a "pin." By 1906 Truscott was said to be the largest employer in St. Joseph, with a staff of 700.

The two-cycle Truscotts shown at the 1910 Boston Boat Show were painted a dark color. However, their solid flywheels appeared to be painted white in the center with the rim either polished or painted another, lighter hue. Possibly, the engines were a navy blue or black. A photograph of the Truscott display at Boston shows what appears to be a V-type engine with overhead camshafts. This novel engine unfortunately, went unremarked in press coverage. At the New York Boat Show, however, inline Truscott production motors with an overhead camshaft were positioned around a 35′9″ flush-deck cruiser on display. The camshaft was operated by a vertical shaft that was driven by bevel gears in the center of the crankshaft. This was a novel idea, certainly, but one wonders about the ultimate wisdom of having rather large, spinning gears inside a crankcase. These overhead cam engines gave Truscott two model lines, the previously introduced heavy-duty motors and the "high speed" (750-rpm) line. In December 1910 Truscott published a photo of a 6-cylinder version of the overhead cam engine, now with a reversing wheel.

Two examples of Truscott-powered boats appeared for sale in *The Rudder* in 1910. They were a 1906 40′ × 10′ Toppan cabin launch with a 2-cylinder, four-cycle 15- to 20-hp Truscott and a 28′ × 6′ cabin launch, perhaps of Truscott's own build, with a 9-hp Truscott single.

The good times experienced in 1906 were destined not to last, and Truscott's decline and attempts to reverse its fortunes left a confusing legacy. In fact, even as the company was submitting pictures of its overhead cam, direct-reversing engine to the press in the fall of 1910, it was, together with others, seeking salvation under a new business model. This was the National Boat and Engine Company, announced that November.

See also National Boat and Engine Company *and* Truscott-Pierce Engine Co.

Truscott-Pierce Engine Co. (later the Truscott Boat Company)

St. Joseph, MI

Trade name: Truscott (two-cycle), La Salle (four-cycle)

At some point during or immediately after the decline of the Racine-Truscott-Shell Lake Boat Company, John M. Truscott and his brother Edward joined with remnants of the Pierce Engine Company to form the Truscott-Pierce Engine Company. Edward was president and John secretary. This probably happened in 1914 because Truscott was not listed among Racine-Truscott-Shell Lake managers in 1915. A surviving catalog refers to the company's product as the Pierce Marine Engine, a two-cycle machine. In May 1915 a 4-cylinder, 3¾″ × 4½,″ 15-hp at 900 rpm four-cycle "Truscott" was included in a list of motors suitable for Restricted Class Racing. It is not clear whether any were built — the engine was known officially as the La Salle — or sold. The Truscott-Pierce business was reported by *Michigan Investor* on April 3, 1915, to be in receivership. However, in Nov. 1915 *Power Boating* included the company in its list of marine engine makers. The published lineup included 10 two-cycle models — five singles and five doubles — ranging from 2 to 20 hp. Ignition was Bosch jump-spark. The carburetor was a Schebler. That December, the magazine reported there were "Hopes For Another Industry."

Current research suggests, inconclusively, that even as Truscott-Pierce was in difficulty, it was evolving into a different company. In May 1914 the Truscott Boat Company was said by *Michigan Investor* to be prospering. That said, the Truscott-Pierce entity was still listed in 1921, with Edward and John still listed as president and secretary. However, by 1930 the Truscott Boat Company was located in four buildings on Wayne Street. The company struggled through the Depression, was sold in 1940, built boats for the Government during World War II, and declared bankruptcy in 1948.

See also Pierce Engine Company *and* Truscott Boat Manufacturing Co.

Tuttle Motor Co.

Canastota, NY

Trade name: Tuttle

By circa 1902 or 1903, Tuttle had begun building what would become quite an extensive line of two-port, two-cycle engines. Most models were equipped with a carburetor of Tuttle's own design. By 1908, when the public became fascinated by engines of more than one cylinder, Tuttle promptly found a way to multiply his little 2-hp one-lunger and showed a 6-hp triple at that year's New York Boat Show. By 1911 the line had matured, with ratings of from 2 to 40 hp. The T series, available in 1, 2, or 3 cylinders, offered the largest displacement (6″ × 5½″). The 2- and 4-cylinder X models (4½″ × 4½″) produced 20 hp and 40 hp, respectively, and were the only models equipped with a Stewart carburetor. Engines of 14 hp or less were equipped with make-and-break ignition, the others with jump-spark. "All of the interior and exterior parts show a completeness of detail design which has never before been incorporated in any one motor," said company ads. Tuttle was also known for high-quality boats and usually took one or more, such as a 25′ × 7′ cabin cruiser or 22½′ open launch to boat shows.

Twice-Two Cycle Engine Co., The
Chicago, IL
Trade name: The Twice-Two
"The simplicity of the two-cycle with *more* than the efficiency of the four-cycle," said the company in 1908.

Tygard Engine
Plainfield, NJ
Trade name: N/A

U. S. Navy
Norfolk Navy Yard
Trade name: Norfolk et al.
Motivated by the prodding of forward-thinking officers, the Navy began replacing steam engines in its small craft with gasoline engines from a number of manufacturers including Standard, Sterling, Clifton, Craig, and Buffalo beginning around 1905. Its first submarines were powered by Otto 4-cylinder engines. The Navy also either contracted for exclusive engine designs to be built at its own facilities or, more likely, acquired rights to manufacture at the Norfolk Navy Yard. A few surviving photographs reveal a number of the engines once produced there. The earliest of these seem to have been a two-cycle, 2-cylinder labeled the GG and a 4-cylinder version, the KKW. A 20-hp, 4-cylinder Buffalo was built under license as the Norfolk KK. Later, the Navy adapted the Model N Van Blerck and, as an ambitious but not entirely successful "dieselization" program was begun about 1934, the Buda diesel. It was adapted by the Navy with American-designed Ex-Cell-O fuel injection rather than Bosch.

Ultra Six Marine Engine Co.
Chicago, IL
Trade name: Ultra Six

Union Gas Engine Company
San Francisco (later, after 1906 earthquake, Oakland, CA)
Trade name: Union
This company was the ultimate winner in the patent disputes between the Regan Vapor Engine Company and the Pacific Gas Engine Company. Incorporated on May 16, 1892, Union would become one of the country's longest-lived and most important engine builders. It began business as "Manufacturers of Regan Vapor Engines and Launches." An early brochure claimed that 500 of those engines were built in 1891 but doesn't specify under whose auspices. Because of Union's close connection to Regan, the company's claims about early sales figures and inventions should prompt caution for a researcher. Union claimed to have built the first marine gasoline engine — or at least the first one with electric ignition — in 1884 and to have patented a make-and-break igniter in June 1885. But these are references to Daniel Regan's work which, after the 1894 catalog, was seldom if ever explicitly referred to again.

A 25-hp, V-type engine was pictured in 1893. In 1901 Union claimed there were more than 3,000 of its engines in use. By the early 1900s, Union was publishing a thick brochure of which its marine engines — single-, double-, triple, 4-, and 6-cylinder models of up to 350 hp — were just a small part of the lineup. Stationary engines, water-pumping outfits, and hoisting engines were among the products shown. Ignition remained make-and-break. The connecting rods had shims to permit easy adjustment for wear. Crankshafts were forged from a solid billet before being turned and polished to a beautiful finish. The vaporizer had a petcock that permitted any water to be drained off, and could be disassembled, if necessary, without tools.

Union completed a 43′, 120,000 lb., 600-hp engine in 1915. By that time oil engines had been added to the line, and a diesel with Union's own patent injection system was placed on the market in 1923. By the 1930s Union 6- and 8-cylinder diesels were available in ratings of from 150 to 2,000 hp. Several of the eights were supercharged.

On its brochures Union always printed "1885" as its start date. Of course, that date referred to Daniel Regan's engines and electric ignition system. The company granted a license to its patents to the Pennsylvania Iron Works, makers of Globe engines in Philadelphia. Unions were painted a glossy black with gold trim.

Union Iron Works
Oshkosh, WI
Trade name: N/A

Unit Motor Co.
Kansas City, MO
Trade name: Unit Motor
The company was listed in 1922 as offering a line of two-cycle engines that ranged from 10 to 150 hp in models of 1 to 12 cylinders. Bosch ignition and Stromberg carburetors were specified.

United Engineering Works
San Francisco, CA
Trade name: United
In business well before the mid-teens, United made heavy-duty, hot-bulb engines with individual cylinders mounted separately on a base plate.

United Manufacturing Company
Detroit, MI
Trade name: Little Giant

United States Motors Corporation
Oshkosh, WI
Trade name: Doman, Bull Dog, Falcon
This company, an outgrowth of the H. C. Doman Company, was formed in September 1927. It had a very professional management staff including president E. H. Huesener, an electrical and mechanical engineer who had worked at Nash. His colleagues were equally qualified, and the whole staff was bolstered when engine pioneer J. D. Termaat emerged from retirement to join the company in January 1929. Its products included marine engines, stationary engines, and electric generating plants.

The Doman Heavy Duty marine series included the Model B 1-cylinder of 5 to 7 hp; the HM-2 2-cylinder, 12 to 15 hp; the HM-4 4-cylinder, 25 to 40 hp; and the TM-4 4-cylinder, 50 to 60

1929 United States Motors Corp. "Falcon"

Probably built sometime between 1914 and 1920, this 5-hp Union was rated at 400 rpm. It had an unusual F-head configuration. It is owned by the Columbia River Maritime Museum. (Stan Grayson photo)

1918 Van Blerck Model M, last of the classic T-head Van Blercks

1915 Universal Machine Co. Toledo one-lunger

1916 Van Blerck "Twin Six" overhead cam V-12 that developed 300-500 hp

Among Union's early marine engines was this fascinating 35-hp V-twin of 1895.

1914 Van Blerck B6

1949 Universal Flexifour

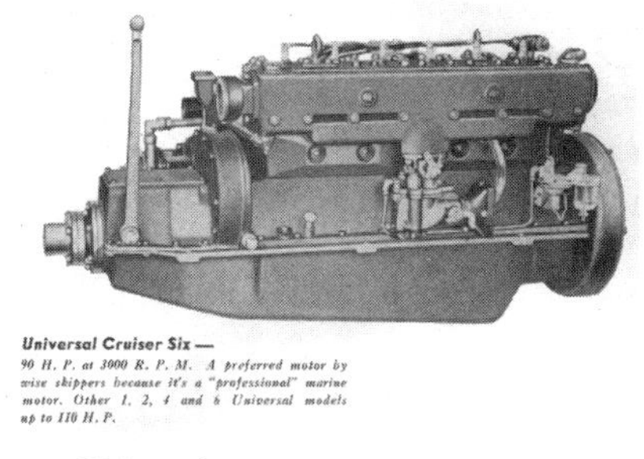

1949 Universal

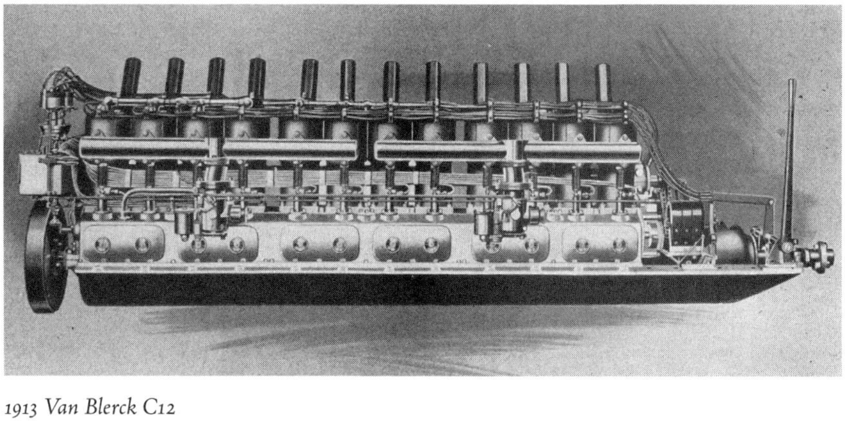

1913 Van Blerck C12

hp. The Medium Duty Bull Dog line included the U1K 1-cylinder 5 hp and the U2K 2-cylinder 10 hp. The Higher Speed line referred to the Falcon, available in 4-cylinder, 10 to 15 hp; a larger 40-hp version; and a 6-cylinder 50-hp version. Among the Falcon's key selling points were a counterbalanced crankshaft and full-pressure lubrication.

Universal Machine Company
Toledo, OH
Trade name: Toledo
Universal offered a 6-hp single and 12-hp double in 1912. These were two-cycle, 2-port motors with a 4½″ bore and 4½″ stroke. By 1915 the line had expanded to five models including a pair of singles (3 and 6 hp), two doubles (6 and 12 hp), and an 18-hp triple. These were jump-spark engines equipped with Red Head brand spark plugs and a Schebler carburetor or Detroit kerosene carburetor.

Universal Motor Co.
Oshkosh, WI
Trade name: Universal
In *MotorBoat* in April 1924, J. D. Termaat wrote: "... in 1913, Mr. Monahan and I resigned from active association in the Termaat & Monahan Company, the company, however, retaining this name, and we organized the Universal Motor Company. The automobile was setting new standards, and we knew that the marine engine had to meet these standards. The popular Universal, four-cylinder, four-cycle engine was the result of our efforts."

Termaat's brief account begs many questions. Chief among them is Termaat's failure to mention the apparent purchase by him and his partners of the Badger Motor Co., presumably because manager Lauson had developed a compact 4-cylinder engine. It is possible the engine was used in the little Woods Mobilette cyclecar of 1914. At any rate, in 1914 the Badger Mfg. Co. turns up in Oshkosh at 39 Ceape with Termaat listed as vice president. Although Louis Monahan's title is not given in the 1914 directory, by 1919 Monahan is listed as president and Termaat as vice president. Investor and inventor Emery H. Fahrney (who had by now taken to spelling his first name with a single "m") was then listed as treasurer.

Like Kermath, Universal became renowned as a company that had based its considerable success on a single model whose influence would be felt for well over half a century. This 4-cylinder, 6- to 10-hp, 800 to 2,500-rpm four-cycle engine was positioned to take advantage of a big slice of the market — that for 16′ to 24′ boats. What became known as the Flex-four or Flexifour (2¾″ × 4″) found a big market, indeed, and the engine was sold on a worldwide basis. Ads for the little Universal engine appeared in *MotorBoat* early in 1915 at about the same time as T&M ads for a similar-looking 7-hp engine named the "Little Special." Possibly, the "Little Special" was the one thing the partners regretted leaving behind when they quit T&M.

Universal was well established by 1925 when the Super-Four (3¼″ × 4½″) was introduced in models of from 15 to 50 hp and with direct drive or reduction gearing. The 80-hp VH 6-cylinder, and 110-hp 8-cylinder motors followed. A single-cylinder 8-hp WM was also added. In 1927 Termaat retired, and E. H. Fahrney became president. In 1928 a new 60,000-sq.-ft. factory was built just two blocks from Lake Winnebago. Served by both the Soo Line and Chicago & Northwestern, the factory was soon shipping its engines from this new plant, perhaps then the most modern in the industry. Castings were supplied by the Universal Foundry Co. established by Fahrney.

By the '30s Universal had matured. In 1932 the lineup included the following: for cruisers, the 99-cubic-inch, 25-hp AF (Flexifour "25"), the 149.3-cubic-inch, 35-hp GLS (Super-Four), the 148.3-cubic-inch, 40-hp AFS (Blue Jacket Six), the 260-cubic-inch, 60-hp CS (Six-260), the 347-cubic-inch, 80-hp CE (Eight 347); for runabouts, the 99-cubic-inch, 25-hp AF (Flexifour "25"), the 149.3-cubic-inch, 45-hp GLH (Super-Four), the 99-cubic-inch, 55-hp AR (Blue Jacket 125 Racing), the 99-cubic-inch, 47-hp AB (Blue Jacket Four), the 148.3-cubic-inch, 70-hp ADS (Blue Jacket Six), the 260-cubic-inch, 88-hp CS (Six-260), the 347-cubic-inch, 115-hp 8-115 (Eight 347); for utility, fishing, and auxiliaries, the 67½-cubic-inch, 8-hp WMG (Single), the 86-cubic-inch, 15-hp N-M (Flexifour "10-15"), the 99-cubic-inch, 20-hp A-F-M (Flexifour "25"), the 149.3-cubic-inch, 35-hp GLS-M (Super-Four), and the 148.3-cubic-inch, 40 hp, AFS (Blue Jacket Six). In addition, heavy-duty versions of the engines, from the single- to the 8-cylinder models, were available for heavy boats; they could be had with 2½:1 or 1¾:1 reduction gears.

After the war Universal acquired the Kermath 1- and 2-cylinder models built by Ballantine when that company stopped producing them, presumably in the early 1950s. In 1947 Universal introduced what would become famous as the Atomic Four. According to Universal specialist Robert Hess, the engine's nomenclature was as follows: UJ-5101 direct drive; UJR-5102, 2:1 reduction gear; UJVD- 5103, V-drive. Whatever the configuration, this L-head 64-cubic-inch, 30-hp engine became one of the most popular 4-cylinder gasoline marine engines of all for auxiliary power in sailboats. A Canadian version known as the Stevedore appeared in the mid-'70s. It is estimated that some 20,000 Atomic Fours were sold by the time production stopped in 1985. The familiar bronze-color paint of the Atomic Four is believed to have been consistent throughout its production life.

Although it may have ultimately proved the most popular of Universal's postwar engines, the Atomic Four was just one of many models. Universal acquired Nordberg's gasoline engines and offered a wide range of 6- and 8-cylinder models. The company was acquired by J. M. Nash in 1961, and when Nash became Medalist Industries,

1929 Van Blerck Jr.

1915 Watertown

1910 Weco Model F

1911 Vim

1910 Weco Model F

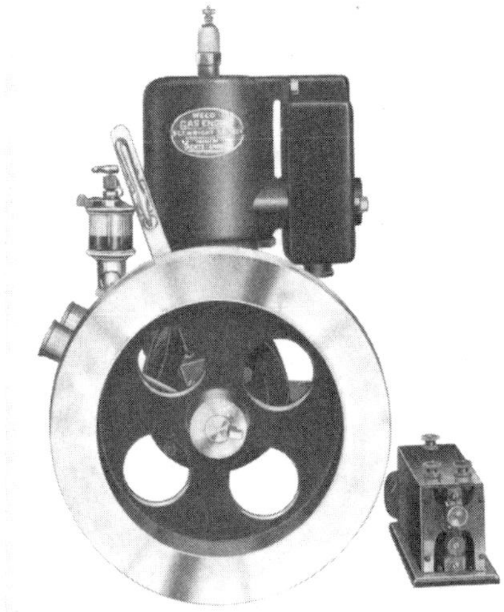

1910 Weco 3-hp Model A-1

1915 Vim

1913 Waterman

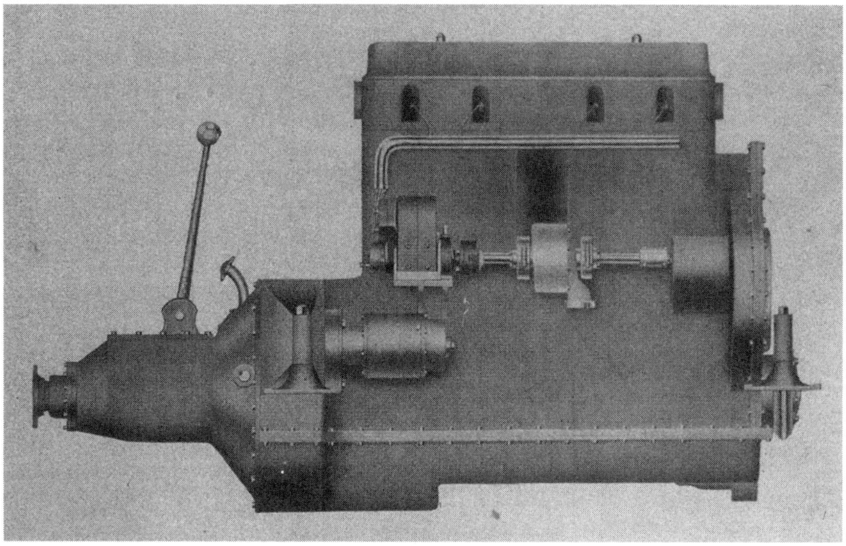

1920 J.V.B. (Joe Van Blerck): The engine was designed by Van Blerck and first built by Wellman-Seaver-Morgan. This modern overhead-valve engine featured a positive-acting sliding-gear transmission with a foot-controlled clutch. Beautifully clean-lined, the engine had full-pressure lubrication. A low-speed—600 to 900 rpm—model and a high-speed—1,000 to 1,450 rpm—type were offered.

Universal then became part of that company. A variety of automotive engines were then being converted for marine use. In the mid-'70s, with the Atomic Four being seriously challenged by emerging small diesels, Universal began marinizing Kubota diesels. In 1991, Universal was purchased by Westerbeke, Inc., of Avon, Massachusetts. Today, Universal is the only name that survives from the heyday of the American marine engine industry. Universal-branded diesels remain available, as do many parts for the Atomic Four.

Universal Products Co.
Oshkosh, WI
Trade name: Doman
See Doman, H. C.

V. J. Emery
See Emery, V. J.

Valentine, F. J.
Buffalo, NY
Trade name: Valentine
This short-lived company advertised an engine in 1908.

Valley Boat & Engine Co.
Saginaw, MI
Trade name: Faultless
In 1909, Valley became a successor to the American Gasoline Engine Co., which had offered both kits and completed engines. Valley offered engine castings and machined parts to be assembled by the "amateur mechanic" using an 8-inch lathe and a drill press capable of taking a 1-inch bit. The engine was, it is currently thought, a three-port, two-cycle with removable heads and jump-spark ignition designed by Emery P. Van Epps. Models up to 28 hp were said to be available. When the company moved to Michigan, it continued to offer its engine kits but also introduced a stylish tandem-seat cycle car named the Saginaw. It did no better than others of the genre, and production ceased in the same year it started, 1914. The company remained in business until 1919 when, according to the *Saginaw News*, "apparently something went wrong."

See also American Gasoline Engine Co.

Van Auken-Clevauc Co.
Yonkers-on-Hudson, NY
Trade name: Van Auken
In 1906 the company advertised a "new departure in gasoline engine construction." That departure was a horizontally opposed two-cycle engine in which two pistons worked in the same cylinder, their crowns forming the combustion chamber when both reached top dead center. Because each crankshaft was connected to an output shaft with a clutch, it was possible to drive and control the two shafts and their propellers independently, and the engine was advertised as the "twin-screw motor." The benefit, of course, was much-improved maneuverability. This engine, said to be (theoretically, one presumes) available in ratings of 8 to 300 hp, was the work of Clarence E. Van Auken. The Van Auken-Clevauc relationship apparently was a brief one, lasting little more than a year.

See also Van Auken Motor & Machine Works

Van Auken Motor & Machine Works
Bridgeport, CT
Trade name: Van Auken
In 1907 Clarence E. Van Auken established his own company at 205 Osborne St., listing the firm as "machinists & marine motor mfrs." The product remained the same ingenious engine Van Auken had hoped to build and market in his previous Van Auken-Clevauc effort. By 1909, Van Auken claimed to offer engines of 8 to 100 hp. The company naturally positioned itself as the one that offered the "advantages of twin-screw propulsion without the initial expense and subsequent extravagance of upkeep attendant upon the installation of two separate units of power." Van Auken's effort lasted through 1909; the company disappeared from the city directory in 1910, and Van Auken moved to New York. Still, the unusual engine lived on briefly.

See also Black Rock Machine Co.

Van Blerck Motor Co.
Detroit, MI, and Monroe, MI
Trade name: Van Blerck
Born in Holland in 1876, Joseph van Blerck emigrated to Detroit in 1902. There he pioneered engines known for their low weight per horsepower. Van Blerck quickly became an important figure in motorboat racing, working with Chris Smith and John Hacker, among others. His engines also were found aboard lavish, high-end motor yachts. In 1909, the Van Blerck Motor Company had its start in the area (at the foot of Hibbard) known generally as Motor Boat Lane. Van Blerck offered a line of 2-, 4-, and 6-cylinder T-head engines for both pleasure and racing use. Manganese bronze was used to produce a light, strong crankcase. The 650-rpm B models were intended for launches, cruisers, and commercial applications; the 1,000-rpm B-Specials and 1,200-rpm, lighter weight C-models for faster boats; and the 1,600-rpm C-Special models for racing. All had individual, cast-iron cylinders. On the higher-rpm C- and C-Special models, however, each cylinder's water jacket was cast with an open area that was then covered by a thin brass plate secured with numerous machine screws by a worker using a brace and screwdriver bit. These cutaway cylinders reduced overall weight by about 100 to 200 lbs. The cylinders were bolted to a robust aluminum crankcase (iron on the B) engineered for stiffness, a trait by then recognized as desirable for longevity.

Ever attentive to metallurgy, Van Blerck specified crankshafts of shock-resistant chrome nickel steel; connecting rods were drop-forged of "special" steel; pistons were turned of "best quality" iron ("semi-steel" on racing models); bearing shells were manganese bronze "lined with the best [they had a high percentage of nickel for wear resistance] Babbitt obtainable." The C- and C-Special models had force-feed lubrication, which Van Blerck realized early was a key to reliability for racing, and the system had a built-in strainer. This

1949 ad for Universal Atomic Four

1911

This 1915 ad heralded the start of the Universal Motor Co.

1913

feature was rare at the time on any but a few high-performance brands.

On the morning of June 7, 1912, the Van Blerck Motor Company was largely destroyed by fire. This potentially devastating blow was countered immediately by the finding of a new facility a mile or so inland on Leib St. With demand growing — 129 people were by then working at Van Blerck's — plans were now made to build a new facility 35 miles south of Detroit in Monroe. This factory, built of steel, concrete, and brick, was opened on Dec. 10, 1913. It contained five dynamometers — Van Blerck was perhaps the first manufacturer to certify horsepower — and was organized to permit efficient production of 1,000 motors a year.

An entirely new model line was introduced for 1915 consisting of the E (runabouts and express cruisers), E Special (higher compression and more advanced timing for competition), and EE (workboats and pleasure cruisers up to 65 feet). These engines had semi-steel cylinders cast in pairs (en bloc), permitting the previously external oil lines to be internal. An easily accessed strainer was built into the lubrication system to keep impurities from circulating through the engine. The crankcase on the EE was iron — on the E and E Special, manganese bronze for added strength. Only the oil pan was aluminum. By comparison to the previous models, the 1915 Van Blercks had larger crankshafts, increased from 1⅞″ to 2⅜″ in diameter to increase the safety factor of the main bearings. The camshafts were now innovative single-piece forgings that would "settle forever the trouble of loose and broken cams and defective timing." Like the engine, the reverse gear was pressure lubricated. It was a positive-acting, multidisc-type unit so the operator didn't have to worry about the boat creeping one way or another with the gear selector in neutral.

In 1916 Van Blerck unveiled the "Twin Six" (the same name used by Packard for its V-12). This was an engine of astonishing conception. It was just 8½′ long. Gone was T-head design, replaced by overhead valves driven by an overhead camshaft. Beautifully enameled, the engine looked particularly lean as the two spark plugs per cylinder were located inside the V. Twin high-tension magnetos were mounted at the aft end of the cylinders. Joe must have been working on this 2,545-cubic-inch V-12 as early as 1914 yet its overall design would remain contemporary through World War II. Tooling alone cost $12,500.

In January 1916 with orders flowing in at an increasing rate, the entire west side of the new factory was expanded and, for a time, both day and night shifts were necessary to meet demand. Joe Van Blerck continued to look ahead. In 1916 the company unveiled an overhead cam, V-12 aero engine that produced 185 hp and weighed only 580 lbs., a hp/weight ratio of 3.13:1. A nomenclature change occurred, probably for the 1917 model line, relating to the marine engines' bore/stroke dimensions. The 5½″ × 6″ "E" models were labeled the "I" and the 6″ × 6″, the "J". (The "J" was still on the U.S. Navy roster powering launches in the mid-'30s.) For 1918 the "M" models appeared. The most elegant of all Van Blercks, the "M" continued the T-head layout but added a counterbalanced crankshaft and an enclosed flywheel.

It was during this busy, productive period that the Van Blerck Motor Company took an exceedingly bad turn. On Oct. 1, 1916, vice president and general manager Charles Page quit and left the marine engine business. Van Blerck sales were then about $1 million. Now, things fell apart. The last year in which Joe Van Blerck's signature appeared on the company's annual report was 1917. Sometime in 1919 or 1920, Joe Van Blerck embarked on a two-decade odyssey as a famous but itinerant engine designer. He, his wife, and three children went first to Akron, Ohio. Joe carried with him plans for a thoroughly modern, 425-cubic-inch 4-cylinder marine engine, the J. V. B. This one was an overhead valve design. The $1,000 J.V.B. was built by the Wellman-Seaver-Morgan company and purchased by, among others, Elco for its 36-foot Cruisette.

Less than a dozen Detroit- and Monroe-built Van Blercks are known to exist. Any one of these motors would be the find of a lifetime for a collector, especially one of the 6- to 12-cylinder C-Special models. The same would be true for a Twin Six or an "M." However, not a single example of either motor is known to exist.

See also J.V.B. Engine Co., Wellman-Seaver-Morgan, *and* Elco

Van Blerck Motor Corporation and Van Blerck Motors

Plainfield, NJ, and Red Bank, NJ

Trade name: Van Blerck, Van Blerck Jr., Continental Van Blerck

In 1921 or 1922 Joe Van Blerck moved to New Jersey, where the Van Blerck Motor Corporation opened in Plainfield. Here Joe built a line of rugged L-head 2- to 12-cylinder medium-speed engines — the Model N. The N was adopted by the U.S. Navy, was used commercially, and also found its way into some lavish motor yachts of the Roaring Twenties. At some point, probably in 1928, Joe moved to Red Bank, New Jersey. There, Van Blerck Motors converted Continental engines for marine use and Joe built the impressively conceived, overhead cam, all aluminum, 33-hp, 165-lb. (with marine gear) Van Blerck Jr. The Junior existed in two versions. The first was 79.4 cubic inches and produced 25 hp. The later model displaced 90 cubic inches and developed 33 hp. These engines differed in detail. The 79.5-cubic-inch engine had an external oil pump and horizontal distributor. The later model had an internal oil pump and vertical distributor. The engine was able to be carried, thanks to a handle attached to the cylinder head. The Junior found its way into a variety of midget racing cars, and was even adapted to at least one mid-'20s French Ratier.

When Joe Van Blerck moved his family to Long Island, New York, he eventually settled down for good in Freeport where he owned a boatyard and built the building that housed Joseph Van Blerck & Son. The company produced a line of remarkably fabricated copper exhaust manifolds and also built the exhaust systems installed on the Packard V-12s that powered Elco PT boats. Joe Van Blerck died, at age 74, at his winter residence in Fort Lauderdale on September 5, 1949, leaving behind a fascinating legacy as one of the great pioneers of innovative, top-quality, performance marine engines.

See also Van Blerck Motor Co.

Van Epps Motors
Syracuse, NY
Trade name: Van Epps
In 1912 it is believed that inventor Van Epps became the sole proprietor of what had previously been known as Livingston & Van Epps. An ad for 2- and 4-cylinder Van Epps marine engines appeared in *Motor Boating* in Dec. 1912. It is likely that Van Epps Motors soon evolved into an auto repair facility.

Vanguard Engine Co.
Boston, MA
Trade name: Vanguard
In 1911 Vanguard produced a rather mystifying open-base, two-cycle engine. The bottom end was evocative of a steam engine while the cylinder heads were removable. Ignition was jump-spark.

See also George E. Belcher Machine Shop

Venn-Severin Machine Co.
Chicago, IL
Trade name: Venn-Severin

Verrier-Eddy Co.
New York, NY
Trade name: Verrier-Eddy Special
This company was the New York agent for Lathrop and others. At the New York Motor Boat Show in 1915 (the eleventh such show), Verrier-Eddy had on display, among other things, a two-cycle en bloc 3″ × 3″ motor with jump-spark ignition. Such a small motor would have been perfect for yacht tenders or little runabouts. The supplier of the motor was not mentioned by the company.

Viking
Seattle, WA
Trade name: Viking

Vim Motor Co.
Sandusky, OH
Trade name: Vim
The Vim was the work of George Holloway, who claimed to have 15 years of two-stroke development experience by 1912. Holloway applied his inventive nature to each component of the engine and, among other things, claimed to have perfected an anti-backfiring device as well as a "one adjustment automatic carburetor, the needle point which regulates the flow of gasoline being automatically controlled by the throttle so that the proper mixture is distributed into the cylinders at all speeds … The air is also automatically controlled by a number of floating balls. …" How well such language communicated with customers is unknown. The Vim's flywheel was corrugated for easy grasping and to eliminate the danger of a pull-out starter handle.

All these features paled by comparison to the Vim's port system, at least as it was described by its designer with regard to the high-speed models. Those Vims were said to have a "double system of separate and independent ports increasing the port area almost 100% over the regular type of 2-stroke engine." Peak power was said to be achieved at the very high speed of 1,200 to 1,300 rpm. Dual battery and magneto ignition was fitted.

Vim offered a full lineup of two-cycle engines in heavy-duty (450 rpm), medium-duty (700 to 800 rpm), and high-speed types (1,100 to 1,200 rpm). The latter were offered in 2-, 3-, and 4-cylinder versions fitted with dual carburetors. All Vims had individually cast cylinders and jump-spark ignition. Horsepower ranged from 3 to 55 in 1912. Mention of Vim ceased after 1919.

Vimalert Co., Ltd.
Jersey City, NJ
Trade name: Vimalert
Vimalert began marine conversions of the Liberty V-12 after World War I and continued to offer a series of marine-converted aero engines in the years that followed. In 1934 three models were offered. These included the 400-hp M-12 and a "Duplex Unit" consisting of two M-12s geared together. The P-2500 was a 700-hp 12-cylinder (6⅜″ × 6½″). By 1936, a 120-hp, 8-cylinder semi-diesel was listed.

Vulcan Engine Works
Philadelphia, PA
Trade name: Vulcan
A 1916 ad reveals that Vulcan focused on low-speed, 400- to 500-rpm engines, probably intended mainly for commercial applications. These were T-head models equipped with, as was usual, a single updraft carburetor. The smallest model was a 95-cubic-inch, 4-hp single with an 18-inch flywheel that weighed 300 lbs. The largest was a 2,526-cubic-inch, 75-hp 4-cylinder with a 26-inch flywheel that weighed 5,000 lbs.

W. C. Disbrow, Jr.
See Disbrow, W. C. Jr.

W. E. Schaefer
See Schaefer, W. E.

W. F. Safford Manuf. Co.
See Safford, W. F. Manuf. Co.

W. H. Mullins & Co.
See Mullins, W. H. & Co.

W. J. Young Machinery Co.
Lynn, MA
Trade name: Essex
See Young, W. J. Machinery Co.

W. M. Ryan, Co.
See Ryan, W. M. Co.

Walker, R. N.
Rome, NY
Trade name: Walker

This 1916 Winton L-head 6-cylinder gasoline engine was typical of the company's output. Beautifully finished and known for reliability, Wintons dominated the market for big motor-yacht power and were available in 80-, 125-, and 200-hp models. By the mid-'20s, engines like this were, however, being challenged by Winton's pioneering diesels.

1913 Wisconsin Machinery & Mfg. Co.

1913 Wonder

1913 Wonder twin

Circa 1932 Winslow Monovalve 2-cylinder diesel built by the American Diesel Engine Co.

This exotic racing engine was built in 1916 by the Wisconsin Motor Mfg. Co. It featured four valves per cylinder operated by chain-driven, overhead camshafts. With an aluminum crankcase, the engine weighed just 600 lbs. but produced 130 hp.

The 6-cylinder Walker engine had rotary valves. An ad appeared in *Motor Boating* in Sept. 1915.

Wambold Mfg. Co.
Milwaukee, WI
Trade name: N/A

Warren Mfg. Co.
Warren, MA
Trade name: Merry
In 1906, Warren listed four-cycle engines in 1-, 2-, and 4-cylinder models of 3 to 35 hp.

Washington Iron Works
Seattle, WA
Trade name: Washington-Estep
Founded in 1881, the Washington Iron Works was a prosperous supplier of equipment to the lumber industry when owners Frank and John Frink hired onetime Atlas-Imperial agent A. C. Estep to adapt his own diesel engine design for production at the Works. In the fall of 1922, the first Washington-Estep, a 3-cylinder, was installed in the American Tugboat Company's *Elmore.* The Washington-Estep was a four-cycle diesel. From the start the company emphasized the advantages of common rail injection versus high pressure air: "Simplicity and reliability of operation. Fewer delicate parts with a reduction in maintenance cost. Greater ease in obtaining suitable operators. More reliable ignition when starting or at extreme low speeds due to elimination of the refrigerating effect of the expanding injection air..."

The metallurgy and manufacturing techniques used at the Iron Works suggest an engine of very high quality. The Works brewed up its own unique iron alloy for the base, which was then rough-machined, heat-treated, annealed, and seasoned to eliminate any warping subsequent to final machining. The cylinders were fitted with heat-treated liners and had big water jacket spaces. Valves were built of heat-treated iron heads and nickel-steel stems. Twin valve springs were fitted to ensure long-lasting, smooth operation. The valves were mounted in cages, the best practice in those times when a valve might need to be removed and replaced with some regularity. It could be done on the Washington-Estep without draining cooling water or removing the cylinder head. As delivered, each engine came with at least one spare intake and exhaust valve with cage. These impressive engines were built to order, each one requiring 60 to 75 days to manufacture. Estep said that customers usually paid cash when their orders were placed.

The Iron Works offered a wide range of engine ratings. By 1928 33 different engines were cataloged. They ranged from a 44-hp, 2-cylinder to a 900-hp, 8-cylinder. The horsepower ratings, like those of most engines of that era, were conservative (a pair of 270-hp engines destined for two tugboats was tested; each proved to develop almost 400 horsepower). The smaller models were equipped with the company's own reversing gear while larger models were direct-reversing, the big camshaft shifted in position by the use of 150 to 250 lbs. of air pressure.

The most famous surviving Washington engine replaced a steam engine in the 120′ tugboat *Arthur Foss.* When launched in 1889 for the Portland Railway and Navigation Company, the *Foss* was equipped with a 122-hp, double-cylinder, compound steam engine built in San Francisco by the Union Iron Works. In 1934 this engine was removed, and a direct-reversing, 700-horsepower Washington-Estep was installed. It has a 2-foot stroke and the volume of its 6 cylinders totals 36,000 cubic inches.

Waterman Marine Motor Co.
Detroit, MI
Trade name: Waterman
In 1906 a Yale-educated attorney, Cameron B. Waterman, began production of the 2-hp Waterman Porto outboard he had invented. The shop was located at 1506-12 Fort St. Later, inboard models were offered. These included the K1 and K2 "canoe motors." The K-models all had 2¾″ × 3″ bore/stroke dimensions and included a 4-cylinder that produced 8 to 10 hp but weighed just 108 lbs. To expand the line, in 1908 Waterman began producing its Model X, an improved version of the Kowalsky that retained the headless cylinder design, had a rotary timer, and was available in 6- and 10-hp models. At the 1910 Boston Show, one of these fours was on exhibit in a Waterman "speed canoe." New for 1910 were the B models, with 4″ × 4″ dimensions and copper water jackets. The 5-hp at 1,000 rpm model weighed only 48 lbs. Such a motor could drive a 17′ canoe at 12 mph. Waterman also produced a line of "medium-duty" motors, the A models. These had cast-iron cylinders and were available in 1-, 2-, and 4-cylinder versions of from 4 to 30 hp. Waterman sold out to the Pausin Engineering Co. and was disolved in March 1918.

Watertown Motor Co.
Canastota, NY
Trade name: Watertown
Located in Watertown, New York, in 1908, the company built engines with a manganese bronze connecting rod with white-brass removable bearings. By 1914, the company was taken over by Gillespie-Boynton.
See also Gillespie-Boynton

Watkins Motor Company
Cincinnati, OH
Trade name: Watkins
Frank M. Watkins began engine production in 1901 in a shop at the corner of 6th and Baymiller as the Frank M. Watkins Mfg. Co. The Watkins engines were "canoe motors" in 1-, 2, and 3-cylinder models. These were a three-port design and had an aluminum base. By 1904, however, the line has expanded and included 1½- to 25-hp two- and four-cycle models. In 1908 the company name was modified to the Watkins Motor Co. At the 1910 Boston Show the two-cycle, 3-hp Watkins, with its copper water jacket, was noted as the lightest engine on display, weighing just 30 lbs. The 1911 lineup included two-cycle models of 3 to 20 hp. The com-

1913

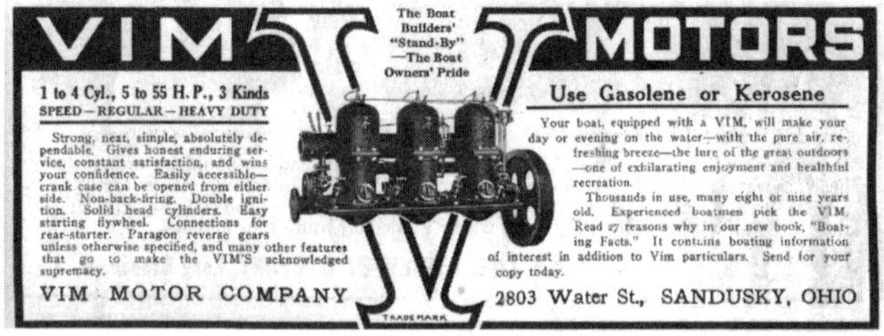

1913

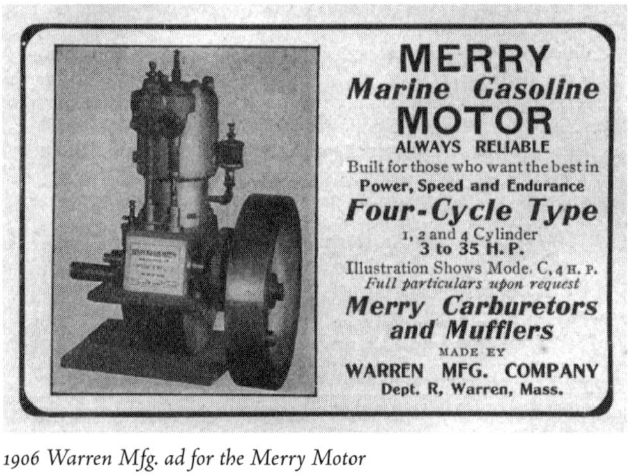

1906 Warren Mfg. ad for the Merry Motor

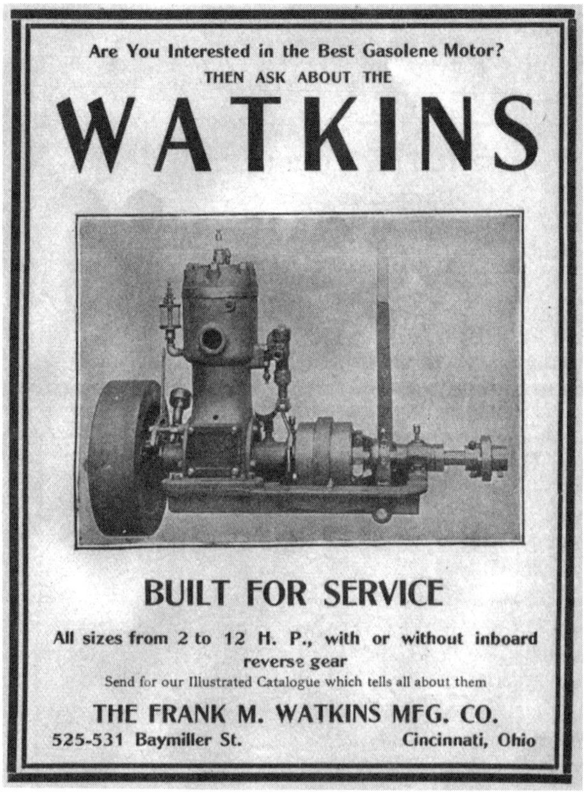

1905

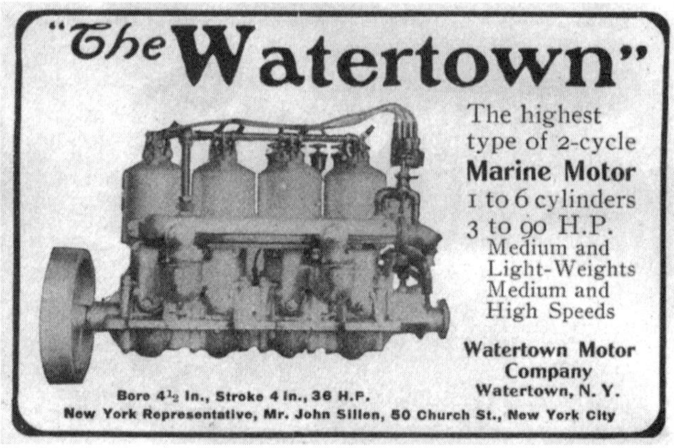

1911

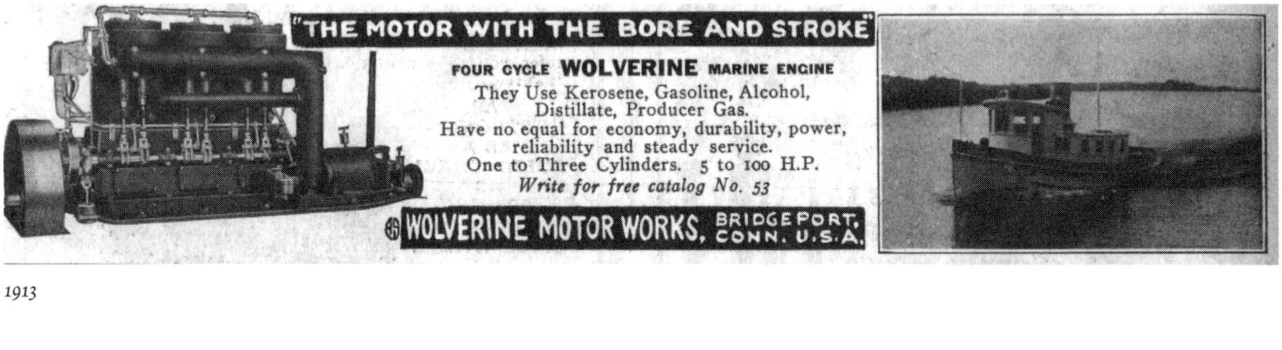

1913

pany continued to be listed in the City Directory until 1943 although engine production had ended many years earlier.

Waukesha Motor Co.
Waukesha, WI
Trade name: Waukesha
Waukesha, founded in 1906, built stationary and industrial engines. The latter were sometimes converted for marine use. Examples included the L-head Red Wing Arrowhead 50 gasoline engine of the 1940s and an Acadia version for fuel-oil operation.

Wayne Motor Co.
Detroit, MI
Trade name: Wayne
Wayne's two-cycle 4- and 8-hp motors were being advertised by 1911 with the claim that tests conducted by engineers at the University of Michigan proved horsepower was greater than claimed. This was probably true, but the same could be said for most marine engines of the period. The 4-hp one-lunger was said to drive an 18′ to 20′ boat at 6 to 10 mph. These were three-port engines, had jump-spark ignition and a Kingston or Standard carburetor.

Wellman-Seaver-Morgan Co.
New York, NY
Trade name: J.V.B. in 1920, later W-S-M
This firm produced the J.V.B., which had been designed by Joe Van Blerck after he left the company that bore his name in Monroe, Michigan, in 1918. The $1,000 J.V.B. was the first engine designed by Van Blerck for the broad market rather than the wealthiest boat owners. Among the engine's key design features were its valve-in-head layout (a dramatic change from the classic T-heads of most past Van Blercks), an en bloc cylinder casting with pressed-in liners, and a positive-acting sliding gear transmission. A high-speed (1000- to 1450-rpm) 45- to 60-hp model was offered and so was a medium-duty (600- to 900-rpm) 28- to 38-hp version. Among the first buyers was Elco, which used the J.V.B. in its 36-foot Cruisette. The W-S-M label was short-lived.
See also Sanderson-Cyclone Drill Co.

Wells & Adams
Bath, NY
Trade name: N/A

West Mystic Mfg. Co.
West Mystic, CT
Trade name: Universal
In business circa 1908, the company offered single- and double-cylinder, two-cycle engines with a removable head and jump-spark or make-and-break ignition. Lathrop would have been tough competition for this and other local hopefuls.

Westerfield Motor Co.
Anderson, IN
Trade name: Westerfield
"Engines for boats, Geared Pumps, Carburetors and other accessories," said this company's ad circa 1905. The engine was a 2-cylinder, horizontally opposed design available in 6-, 10-, 15-, and 20-hp ratings.

Western Gas Engine & Schaake Machine Works
Seattle, WA
Trade name: Western, Cowie
The 1910 Western was an open-base, splash-lubricated engine with overhead valves operated by a camshaft located high on the starboard side of the cylinders. Ignition was jump-spark. Two-, 3-, and 4-cylinder versions were offered in ratings of 25 to 75 hp.

The 1910 Cowie was available in models of 1 to 4 cylinders and ratings of 4 to 55 hp. Traditional valves were replaced by a "piston type [valve] operating in a cylinder and alternately covering and uncovering the inlet and exhaust ports."

Western Launch & Engine Works
Michigan City, IN
Trade name: Western
In business as of 1909, the company advertised "reliable high-grade Western Launches and Engines, built for use."

Western Machinery Co.
Los Angeles, CA
Trade name: Western

Western Motor Company
Logansport, IN
Trade name: Rutenber
Patented in 1905 by Edwin Rutenber, these four-cycle 25- to 40-hp, 1,200-rpm engines were primarily intended for automotive use, but they were advertised in *Motor Boating* in Dec. 1911.

White & Middleton Gas Engine Co.
Baltimore, MD
Trade name: White & Middleton
This company was justly famous for its horizontal-cylinder stationary engines whose patented governor made them exceptionally steady running and well-adapted for use as generating plants. They became very popular for electric lighting purposes. Occasionally, however, the engines were used in marine applications, and, as a matter of fact, they happened to garner quite a bit of publicity. In 1894, a Maryland farmer named Wickert equipped a launch with a White & Middleton and used the boat to carry produce down the Patapsco River to Chesapeake Bay. In 1897 inventor Simon Lake chose a White & Middleton to power his 36′9″ submarine, *Argonaut*. In 1901 Lake used a pair of the engines in a larger boat. That said, the company was a force in the stationary, rather than marine, engine market.

White, E. M. & Co.
Old Town, ME
Trade name: White

White Manufacturing Co.
See Goshen

Whitehall Motor & Boat Co.
Whitehall, NY
Trade name: Whitehall
The company was in business by 1911 building two-cycle engines.

Wilcox and Carlson Co.
Detroit, MI
Trade name: Cox
In 1910 this company's specialty was said to be a 2¼″ × 2½″ en bloc twin that weighed 64 lbs. and was intended for canoes and small skiffs. It was a jump-spark motor and was said to be the smallest of a line of twin-cylinder Cox models. So-called "canoe motors" were typically installed at the boat's aft end while the batteries and fuel tank were located in the bow to obtain an even balance.

Willamette Gas Engine
Portland, OR
Trade name: Webber
These heavy-duty gasoline engines were designed for fishing and work boats.

Willard, Charles P. & Co.
Chicago, IL
Trade name: Simplex
Willard was a steam engine and boiler company when, sometime in the late 1890s, it introduced a marine engine to be fueled by naphtha. Willard was one of several companies that decided Simplex would be a good name for an engine. The Simplex was a conventional two-cycle motor available in a 300-lb., 2-hp model and a 500-lb., 4-hp model. Both were rated at 300 rpm. It is likely the engines could also run on gasoline. The company did not become a factor in the marine engine industry, and operations ceased around 1900.

Willet Engine & Carburetor Co.
Buffalo, NY
Trade name: Willet
In 1908 Willet announced that its rotary valve, jump-spark engine would be available in 1-, 2-, 3-, 4-, and 6-cylinder models of 8 to 70 hp for marine or automotive use. A line of heavy-duty models was announced in 1910. These engines had iron cylinders cast en bloc in pairs and an aluminum base. A rotary valve replaced the more typical check valve. Two- and 4-cylinder models had a mechanical oiler. The company made its own carburetor.

Williamson Motor Co.
Philadelphia, PA
Trade name: N/A

Wm. L. Mantha Co.
See Mantha, Wm. L. Co.

Wilpen Company
Detroit, MI
Trade names: Brushmarine, Oakland
The 1-cylinder 3- to 5-hp Brush had been designed by Alanson P. Brush of Cadillac fame and was said to have as "little vibration as the ordinary two-cylinder engine." The Oakland was a 4-cylinder automotive engine but equipped with a marine gear. The company was in business by 1910 but may have ceased operation by 1912.

Wing, L.J. & Co.
New York, NY
Trade name: Wing
The company was founded circa 1896 and offered four-cycle engines. These included the usual vertical-cylinder configurations but also an inverted model. Production, whatever level it achieved, probably ended shortly after the 1900 models were produced.

Winkley, E. E.
Lynn, MA
Trade name: Winkley
Located on Union St., Winkley was in business from 1900 to 1912, advertising "high-grade gasoline engines for launches, tenders, dories and fishing boats."

Winona Machine & Boat Works
Winona, WI
Trade name: Winona

Winton Engine Works
Cleveland, OH
Trade name: Winton
Born in Scotland in 1860, Alexander Winton lived the American dream. After arriving in New York in 1879, he first applied his innate mechanical genius in a steam-engine shop and later became an engineer on ocean-going vessels. With his seagoing days behind him, Winton moved inland to Cleveland in the late 1880s and started the Winton Bicycle Company in 1891. The cycle and the timing were both excellent. Besides establishing a solid base of sales in the United States, Winton exported his bicycles, too, and he made some serious money. This he used, beginning in 1893, to fund experiments with a horseless carriage. Thus was planted the seed for what became, in 1897, the Winton Motor Carriage Company.

In April 1898 Winton sold his first car, a $1,000, single-cylinder machine. The sale of 21 more vehicles to wealthy customers made Winton the first man to commercially market an automobile in any volume. With the financial decline and market changes that followed World War I, Winton liquidated his car company in 1924 rather than introduce less expensive models.

Winton's success gave him the freedom to indulge his lifelong interest in the sea as a yachtsman. He spent some of his money on a steam yacht, the *La Belle*. His problem with the *La Belle* was that he had to let the professional captain know well ahead of time just when to have steam up. In 1911 Winton decided to have a big, gasoline-powered motor yacht constructed. Cox & Stevens designed a new, 133′ *La Belle*, and Winton began visiting marine engine companies to examine their products and make a selection. None were satisfactory, either because of noise or vibration. Instead, Winton designed a 6-cylinder engine with a 9″ bore and 12¾″ stroke that was rated at 150 hp.

Within four months, the patterns for the engine had been made, castings had been poured and machined, and three engines were built and installed in the *La Belle*. They proved to operate so quietly, so dependably, and so cleanly that Cox & Stevens, among others, suggested that Winton should manufacture them. The Winton Gas Engine and Manufacturing Company was organized in Cleveland in February 1912, and a new 100′ by 300′ factory was built on Cleveland's West 106th Street near the Winton automobile plant.

Winton's engines followed the general pattern of the *La Belle*'s. These were heavy-duty, 6-cylinder models that included the 80-hp W6, the 125-hp W5, and the Model 11. The latter's bore and stroke were increased in dimension over the *La Belle*'s engine and measured 9½″ by 14″ to provide 200 hp at up to 475 rpm. These Winton gasoline engines were mighty L-head machines that ranged in weight from the 5,400-lb. W6 to the 19,000-lb. Model 11. They remained in production with no major design change until the early 1930s. During World War I, the 8-cylinder W29 was introduced, a 300-hp model that was sold in some quantity to the U.S. Navy for use in minelayers and to the Russian Navy.

Exactly when Winton became fascinated by the possibility of diesel engines has gone unrecorded, but it must have preceded or else coincided with the building of the *La Belle*'s gasoline engines. "From that time until the day he died," according to Winton executive George Codrington, "Alexander Winton's life was devoted to the development of Rudolf Diesel's brainchild." The first Winton diesel, an engine that Winton and his colleagues believed was the "first American-designed internal combustion engine employing the Diesel principle of ignition and compression," appeared in the fall of 1913. This 6-cylinder engine was installed in the Winton automobile factory to generate electricity. A production version, the 175-hp W-16 was then introduced. Like other diesels of its era, it was a cast-iron machine with individual cylinders bolted to an iron upper crankcase. A three-stage air compressor was driven off the front of the crankshaft.

The six was followed by a 300-horsepower V-12 in 1915, a pair of which promptly took the place of the gasoline engines in the *La Belle*. These engines, which Winton designated as the W30, must almost certainly have been the first V-12 diesels ever made. In 1916 the W24 engine was unveiled, a 225-hp, 6-cylinder marine model with a 1215/16″ stroke and 18″ bore. It was followed by an 8-cylinder, 500-hp version, the W40.

Thus was launched a line of engines and a company that was destined to play a central role in American diesel-engine development. Winton diesels found homes in both big motor yachts and in a variety of big, auxiliary freight schooners ranging from 230′ to 255′ in length. The company was a successful innovator of diesel-electric drive, beginning with the installation in the 162′ Herreshoff-designed schooner yacht *Elfay* in 1919. This yacht, owned by R. A. Alger of Detroit, was going to be used for a world cruise. The vessel was taken to the George Lawley yard at Neponset (Boston) to have her steam engine removed and a diesel installed. It was a Model 59, 6-cylinder, 125-hp Winton driving a 75-kw generator that operated a 90-hp electric motor connected to the boat's propeller shaft. *Elfay* was said to have been the first diesel-electric yacht, and Winton soon found many more such applications.

An airless-injection system was introduced in 1928. It used a 3-cylinder, plunger-type pump to pump fuel into the manifold (common rail). From there, the fuel was injected into the cylinders by a camshaft-operated injector valve. A pair of these 700-rpm engines was promptly ordered for installation in a 94′ Mathis cruiser. They were handsome machines, fully enclosed, with a one-piece block cast of chrome-nickel iron that carried inserted cylinder liners.

By the mid-'20s, Winton was recognized as a premier supplier of diesel engines for yachts of 120′ to some 190′, as well as workboats. The company became a regular advertiser in boating publications, often showing a big photograph of the latest vessel to be equipped with a diesel electric or direct drive installation. Of the vessels classified as "yachts" by Lloyd's during the '20s, it was estimated that Winton powered some 57 percent of those equipped with diesels, and 61 percent had Winton gasoline engines. Approximately 500 of the latter were built before operating costs moved the market to diesels.

In May 1930 Winton was acquired by General Motors, which was especially interested in the expertise Winton had developed in production and fuel injection. The company was then renamed the Winton Engine Corporation. The engines continued to be known as Wintons. In 1934 the company became the Winton Division of General Motors and, in January 1938, the Cleveland Diesel Engine Division of General Motors. The name did not survive once GM began production of its own diesels during World War II.

Wisconsin Machinery & Mfg. Co.
Milwaukee, WI
Trade name: Wisconsin

In 1908, this company unveiled a twist on the three-port, two-cycle engine. The motor had "mechanical valves and in addition to the main exhaust valve a vacuum chamber is provided to suck out the tail-end of the exhaust and this thoroughly scavenges the cylinder." The complex explanation required to describe how this motor worked, coupled with the likelihood that it didn't work too well, led to a strategic change of plan. Wisconsin was advertising conventional two-cycle, two-port, dome-head marine engines in 1912. The engines had jump-spark ignition and a water-cooled muffler. Like a number of other firms, this one hoped to find buyers in the automotive market as well as the marine. It is doubtful that Wisconsin found much success in the former endeavor, but the company survived World War I to offer marine engines of 2 to 35 hp.

Wisconsin Motor Mfg. Co.
Milwaukee, WI
Trade name: Wisconsin Consistent, White Cap

Founded in 1911, this firm found success for a time by focusing on quality rather than quantity. Within just a few years, Wisconsin was offering attractive 4-, 6-, and 8-cylinder engines with en bloc cylinders and notably clean external design. "Decidedly handsome

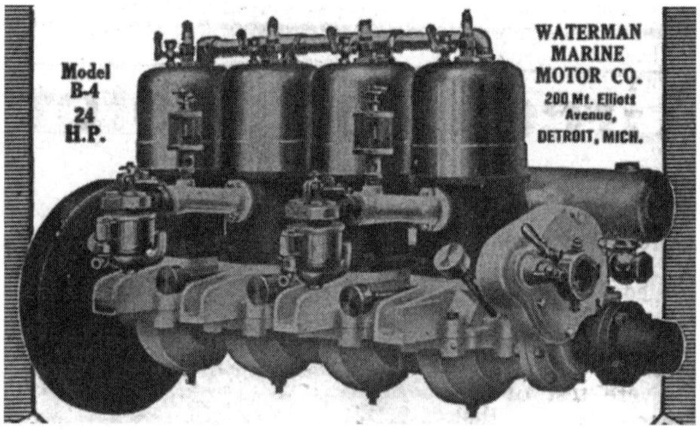

1915

1915

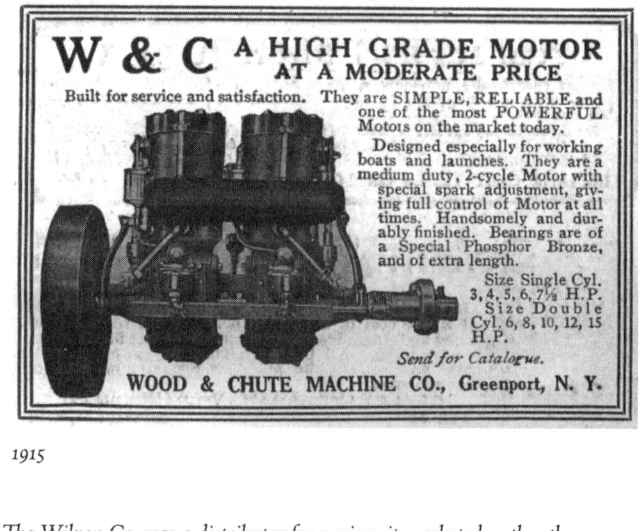

1915

The Wilpen Co. was a distributor for engines it marketed, rather than a manufacturer. The Brushmarine model, at least, was fitted with crankcase cover plates that carried the Wilpen name. This ad is from 1911.

in appearance," is how George Crouch described the motors in his report on the 11th New York Motorboat Show in the Feb. 1915 issue of *MotorBoat*. The 4-cylinder 22- and 25-hp models were L-heads; the others were T-heads with cylinders cast in pairs. Bosch dual ignition was available on standard and racing versions. An exceptionally advanced double-overhead cam racing engine of 1916 produced 130 hp at a weight of only 600 lbs. Racing models had chrome vanadium crankshafts, rods, and nuts and bolts. Bases were of cast aluminum.

Wisconsin touted its engines as "champions of world on road and speedway." In 1917, the company took the novel step of introducing profit-sharing for employees. When a Columbia-trained architect named N. W. Schenetzky took over the firm's management in 1922, he instilled a businesslike attitude that made Wisconsin one of the industry's stable producers for both marine and automotive applications. By the time of Schenetzky's arrival, the Wisconsin Consistent had reached a very high state of development. The upper and lower crankcase were both aluminum. The chrome nickel steel crankshaft had a tested strength of 140,000 lbs. psi. Cylinders and pistons were gray iron. After testing for three hours, each engine was taken down and the bores inspected and bearings adjusted as necessary. It then was finished in "battleship gray" enamel and was ready for shipping. Such an approach to motor manufacture worked until the late '20s, by which time the less expensive White Cap "4" and "6" models had been added to the line with the slogan "How Do You Prefer to Save?" Wisconsin disappeared from the city directory in 1928.

Wittmaack Machine Co.
Detroit, MI
Trade name: Premier

"The main feature is a mechanical disc valve on the crankshaft inside the base which opens and closes the intake port," claimed this company in a 1908 write-up. A removable copper water jacket helped ease maintenance. Although the company claimed to offer motors of 2 to 60 hp, nothing has emerged to indicate any level of production.

Woerner, Wm. J.
Rochester, NY
Trade name: The Midget

Wolverine Motor Works
Grand Rapids, MI
Trade name: Wolverine

After he sold his remaining interest in the Sintz Gas Engine Co. in December 1893, Clark Sintz promptly engaged in a new venture in 1894. Recognizing that he could not form another company that bore his own name without running afoul of the Sintz Gas Engine Co., Sintz decided to call the new firm Wolverine. In a brief and unsatisfactory article in *Power Boating* in January 1914, Sintz incorrectly attributed the name to "the state of Michigan" rather than the state's official animal. "I did not use my own name as it might cause mixing of the mails of the Sintz Gas Engine Co. and myself," he recalled in a rather offhanded manner. Sintz was assisted by sons Claude and Guy. The original patent, filed Dec. 14, 1894, was in Claude's name. The shop was located on Huron Street.

The Wolverine was a clever two-cycle engine touted as being exceptionally easy to start. While mixture was drawn into the base through a check valve and passed through a transfer port to the cylinder, a mechanically operated valve in a chamber on the side of the cylinder just below the removable head sealed the cylinder. Ignition was make-and-break. The Sintzes seem to have gotten this engine exactly right, and the Wolverine impressed an early and influential customer named Charles R. Snyder. A pioneer in the banana export industry in Panama, Snyder bought two Wolverines for installation in boats used to link his plantations. This effort was so successful that he ordered six more engines and visited Grand Rapids in 1895. Then he had Clark and Claude Sintz design a gasoline-powered, narrow-gauge locomotive for the banana plantations.

When Snyder and his brothers sold their business to the United Fruit Co., he used a portion of his new-found cash to buy Wolverine and became its president. Wolverine was incorporated on January 21, 1899, the shareholders having raised $20,000. (According to Clark Sintz, the incorporation took place in 1900 and he then left the company. However, a document on file at the Henry Ford Museum & Greenfield Village puts the year of Sintz's leaving at 1902.) Like so many early engine companies, Wolverine had a boat shop. In January 1901 Wolverine moved the boat-building part of its operation to Holland, Michigan, a first step toward phasing out this side of the business. Sometime in 1906 or early 1907, Charles Snyder decided to move Wolverine to Bridgeport, Connecticut, for easier access to export markets. Claude, who was the company secretary, was coaxed by Snyder to move east but he refused. "I can't do it," Claude Sintz remembered telling Snyder. "I like Michigan."

Snyder's strategic decision paid off, and the Wolverine eventually became known as "The World-Wide Motor." The company introduced a line of four-cycle, L-head engines and by 1911 offered 5- to 100-hp models to run on kerosene, gasoline, and, for a time, producer gas. Except for a 5-hp single and 7-hp double, all these Wolverines were 3-cylinder engines rated at between 280 rpm (the 12¼" × 14" 100-hp model) and 400 rpm. Ignition was make-and-break, using either Edison batteries or a Remy magneto. By 1918 a 6-cylinder (11" × 15"), 200-hp Wolverine was available. It weighed 16,630 lbs.

By 1922 the model line remained familiar, but Bosch ignition components were then used and Kingston carburetors were standard although a Schebler was optional on several of the smaller engines. A "heavy oil" engine had also been introduced although it was several years before the company began listing it as a "diesel." The engine was available in the standard color of "Egyptian gray" but could be custom-painted as desired. This fine American builder of heavy-duty marine engines remained in business until 1955.

See also Claude Sintz

Wonder Manufacturing Co.
Syracuse, NY
Trade name: Wonder
Another of the many upstate New York engine builders, Wonder was listed in Syracuse City Directories from 1907–1918, but the earliest Wonder, available in both stationary and marine versions was advertised in 1905. The earliest company ad dates to 1906 and lists the Wonder as being built in Syracuse by the R. M. Cornwell Company at 408 S. Salina St. A year later, the name had changed to the Wonder Manufacturing Co., and the shop had moved to Tallman St. Ray M. Cornwell remained the guiding force behind the enterprise. The Wonder was a two-cycle motor available by 1911 in single- and double-cylinder models of 2 to 40 hp. In addition, a 60- and even a 70-hp model were said to be available. The Wonder was a jump-spark engine with a separate cylinder head; some versions were equipped with one-piece intake/exhaust manifolds while others had separate manifolds. Either version would now make a quite desirable addition to any collection. The company had a New York City sales office and displayed at boat shows as distant as Chicago in 1910. In 1916 Ray Cornwell became the Syracuse agent for the Crow-Elkhart automobile. At some point thereafter, no later than 1919, the Wonder tooling was acquired by Henry F. and William P. King, successors to Barber Brothers.

Wonder Pump and Engine Co.
Vancouver, BC
Trade name: Wonder

Wood & Chute Machine Co.
Greenport, NY
Trade name: W & C
These sturdy two-cycle engines were intended primarily for the local workboat market. In 1915, the line included singes of from 3 to 7½ hp and doubles of from 6 to 15 hp.

Woodhouse Gasoline Engine Co.
Seattle, WA
Trade name: Olympic, Seattle Spirit
In 1920, the company offered a 4-hp single intended for small trollers, gillnet boats, skiffs, and tenders up to 22′. An 8-hp twin-cylinder version was intended for boats up to 30′ in length. Both engines could run on gasoline, distillate, or kerosene, thanks to, according to the company, the "unique manner" in which fuel was introduced through the cylinder head.

Worthington Pump & Machinery
New York, NY
Trade name: Worthington Diesel

Wright
Tacoma, WA
Trade name: N/A

Wright Aeronautical Corp.
Paterson, NJ
Trade name: Wright Typhoon
The Typhoon was an aircraft engine that was marinized on occasion. The Auto Engine Works' Capitol was the best-known conversion.

See also Auto Engine Works

Wright, C. T. Engine Co.
Greenville, MI
Trade name: Weco
The wealthy owner of the Greenville Implement Company with "vast business interests," Cass T. Wright claimed to have made four-cycle engines as early as 1896 before his conversion to two-cycle. Wright was not specific about his efforts to build four-cycle engines, and, given the nature of the C. T. Wright Engine Co., it seems likely that any four-cycle machine would have been the work of a now unknown inventor. As for Wright's two-cycle conversion, it was spawned by his exposure to the engines of William Gile, whose company Wright purchased in 1907. Wright's interest in the engine business seems to have been driven by the market opportunities he saw rather than any personal engineering or inventing accomplishments.

Based on the illustrations in Weco's catalog, it looks as if the Gile models with removable heads continued in production at Weco with no changes other than the nameplate, which now read Weco instead of Gile. Presumably the nomenclature changed as did color(s). In Weco nomenclature, models E (1 cylinder, 5 hp), F (2 cylinder, 10 hp), and G (2 cylinder, 15 hp) indicated the Gile designs with removable heads. The catalog claimed that the "gaskets of the Weco engine are positively guaranteed against leaking or blowing out." Regarding colors, the catalog noted, "Engines finished with eight coats of paint and varnish." Colors were brown, dark blue, gray, and olive green with polished brass. Other colors were available by order.

As for the dome-head Weco engines, these were not the same models that Gile had offered. These Wecos were the model A (1 cylinder, 3½ hp), B (2 cylinders, 7 hp), C (3 cylinders, 10½ hp), and D (4 cylinders, 14 hp). All were three-port, jump-spark engines. "Time and cuss words have driven the make and break system out of business," read the brochure. The origin of these Weco models is presently uncertain. The A-1, with its condenser exhaust manifold and 4-hole flywheel, looks much like the comparable Termaat & Monahan. However, the bore/stroke dimensions of the Weco and T&M are slightly different. It remains possible that the Weco A-1 was based on T&M components, and that the solid-flywheel Weco dome-head models were re-badged versions of another brand.

Like many other companies, C. T. Wright offered the Weco as a complete outfit including a New York coil, dry cell batteries, knife ignition switch, wiring and spark plug cables, shaft coupling and stuffing box. Buyers could select a steel or bronze shaft, the Weco weedless propeller, muffler, and a galvanized iron 6- or 12-gallon fuel tank. What's more, a complete boat could

This 1928 Wolverine diesel produces 100 hp @ 350 rpm. It is 11 feet long and 5 feet tall. This engine was restored by Larry Mahan and installed in his schooner Larinda. *(Ed Malitsky photo)*

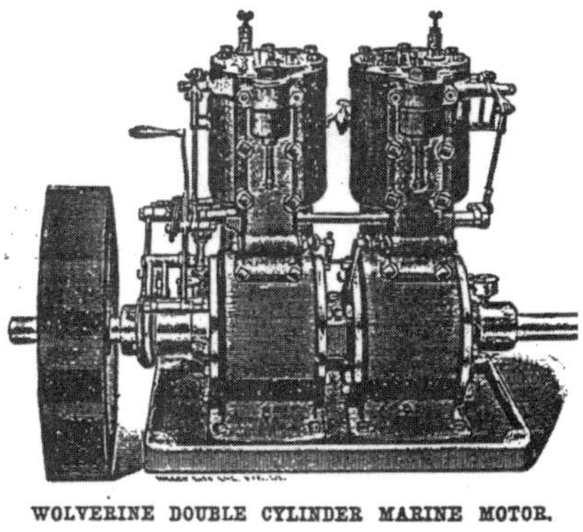

The first Sintz Wolverine, a two-cycle engine

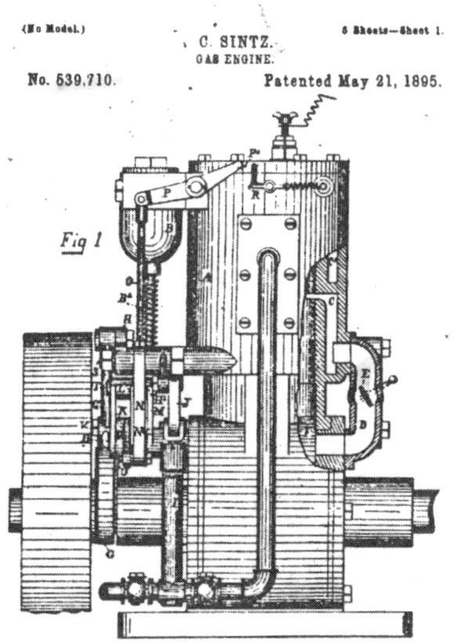

1895 Wolverine patent drawing

1923 Wright "Reliable"

be ordered. There were 16′, 18′, and 20′ launches. Production continued through 1917.

Wright Machine Co. (later Gunther-Wright)
Owensboro, KY
Trade name: Wright Reliable
Founded in 1880, Wright was building an impressive line of overhead valve kerosene engines with low-tension, make-and-break ignition by the early teens. These hefty 450- to 500-rpm engines were available in 3-, 4-, 6-, and 8-cylinder models. Wright claimed to have found a method for thoroughly "gratifying" kerosene, thus eliminating the problem of carbon buildup in the valves and lubrication system. Dealing with carbon buildup was a common problem of gasoline engines for decades, and owners of kerosene-fueled engines certainly faced the same challenge when it came to decarbonizing valves, seats, and cylinders. "The kerosene equipment we use on Wright Heavy Duty Motors gives just as much power as gasoline, and uses the same number of gallons of fuel," claimed company ads. The advantage was that kerosene cost half or less than half the price of gasoline. By 1922 the company had sales offices in New York, Seattle, and New Orleans.

Wright Motor Works
Buffalo, NY
Trade name: Wright
At a shop at 48 S. Division Street, Wright was building two-cycle motors as of 1905. The company introduced an automobile engine, presumably a four-cycle, in May of 1906.

Xargil Manufacturing Co.
Utica, NY
Trade name: Utica

Yacht Gas Engine & Launch Co.
Philadelphia, PA
Trade name: Baby Crown, Crown
Like many boatbuilding companies in the first decade of the 20th century, this firm saw a market for its own boats when powered by its own engines. In this case, the engine was quite unusual. It was an open-base, four-cycle design with water jackets made of aluminum alloy. Jump-spark ignition was used. Believed to have been introduced in 1905 or 1906, the Baby Crown is thought to have been a single-cylinder motor while the Crown was a double. The 2½-hp Baby Crown yacht tender motor had an aluminum alloy base and lignum vitae bearings. Neither the company nor its motors seem to have made much of an impact. Not a single boat from Yacht Gas Engine & Launch Co. was included in write-ups in either *The Rudder* or *Motor Boating*.

Young, W. J. Machinery Co.
Lynn, MA
Trade name: Essex
When the Essex Engine Co. stopped production, Young took over and built the engines in a shop on Union Street. Models ranged from 3 to 20 hp.
See also Essex Engine Co.

CANADA

A. D. Fisher Co.
See Fisher, A. D. Co.

Acadia Gas Engines Ltd.
Bridgewater, NS
Trade name: Acadia
The first Acadia catalog of 1907–1908 noted, "The designs of our engines are original and designed by our superintendent who for many years was connected with one of the first gas engine manufacturers in the United States and who for the past four years superintended construction and building of a well-known engine manufacturer in Maine. . . ." Who this man was or where he had previously worked is not mentioned. Might the Maine company have been Rice Brothers or Camden Anchor-Rockland Machine?

The Acadia was a classic two-port engine that was fitted with an "auxiliary third port air inlet" mounted in the transfer passage, "which greatly increases the power and speed." Both make-and-break and jump-spark models were offered. There was a light-duty line and one intended for heavy-duty commercial applications. Ratings ranged from 3 to 20 hp. Once the engines had proved themselves, no changes were made, and it was said that parts for a 1915 engine were the same as those being used when production finally ceased in the 1960s. The finishing process consisted of coating the cured casting with heat-resistant varnish, and applying a filler coat that was sanded smooth before the enamel finish coat was applied. Surviving engines built toward the end of Acadia production were painted a medium blue.

Before engines were shipped, those with make-and-break igniters were carefully adjusted so that when the flywheel hit 9 o'clock, the tripper would begin to make. At about 11 o'clock, it would break and the spark would be produced. It took about 2½ to 2¾ hours for a skilled engine fitter to complete a one-lunger and prepare it for operation in the test house. There, each engine was run to be sure it developed its full horsepower.

In 1919, president W. T. Ritcey hired Arthur E. Colchester to develop Acadia's first four-cycle engine. Colchester had been born in Windsor, Nova Scotia, in 1874 and developed an early interest in steam engines. He moved to Massachusetts, where he gained practical experience in gasoline engines in New Bedford and later opened a machine shop in Marblehead before returning to Nova Scotia. The four-cycle Acadia was a 40- to 60-hp (depending on rpm) 4-cylinder, overhead-valve engine with an aluminum upper crankcase and the lower portion of bronze. Atwater-Kent battery ignition and a Bosch magneto were the ignition components. The engine soon proved to be durable, but individual-cylinder engines were old technology by 1919. What's more, with Acadia grinding its own camshafts, cutting its own timing gears, and then assembling the engine by hand, the new Acadia was comparatively expensive, too expensive to become a real factor in the marine engine market.

As the years wore on, Acadia introduced a fuel-oil engine based on a Waukesha conversion, stationary engines, and such equipment as scallop and lobster pot hoists, and winches. Besides the facility in Bridgewater, a branch was established at St. John's, long the center of commerce in Newfoundland, where the two-cycle one-lunger remained in use for years after fishermen elsewhere had moved on to more modern engines.

As for Acadia, the big red factory in Bridgewater was closed, but equipment still remained in place in the late 1970s and thereafter. The possibility of selling one-lungers, or patterns to build the engines, to a third-world country didn't materialize, but the Acadia was built longer than any other example of the classic, two-cycle marine engine except for one. That was the Atlantic built in Lunenburg.

Adams Engine Co.
Penetanguishene, ON; Vancouver, BC
Trade name: Adams

After Peter Payette acquired Adams in 1914, the company name changed to the Adams Engine Company. By 1924 J. T. Payette was listed as the proprietor. The brochure included a full-size illustration of the make-and-break igniter, noting its easy adjustability and similarity (details weren't provided) to the igniters used on several pioneer brands of American-built engines including Globe, Standard, Craig, and Hercules. Adams made a special pitch to commercial operators based on its optional ($15) "Adams Universal Kerosene Carburetor." The advantage here, of course, was reduced fuel cost. Adams claimed expenses could be cut in half by equipping the engine with one of these devices. "No hot air drums, pipes or unsightly contrivances ... we have so perfected this Universal Carburetor that our 'SPECIALS' are started directly off kerosene or crude oil with one priming of gasoline. One quart of gasoline will start these engines for months." The Adams was available in 1- and 2-cylinder models, either gasoline or kerosene, in ratings of 2½ to 16 hp.

See also Payette, Peter

Atchison-Bower
Shelburne, NS
Trade name: Etherington

Barker McLean Motors Ltd.
Amherst, NS
Trade name: N/A

Barnes, E. F.
St. John's, NF
Trade name: Barnes

Eban Frederick Barnes was born in Boat Harbour, Newfoundland, in 1894. He became an apprentice mechanic in Nova Scotia and then a master mechanic in mainland Canada, after which he worked for a time in the U. S. He returned to Newfoundland in 1932 to open Barnes Machine Works in St. John's. Soon his company was making everything from elevators to ventilating fans, tools, and tanks. In 1945, with backing from the Newfoundland Industrial Development Board, Barnes introduced a 5½- and 7½-hp marine engine. Castings were supplied by United Nail & Foundry. Perhaps the most unusual thing about the Barnes was that it was a classic, two-cycle one-lunger introduced at a time when such engines were recognized as long out of date everywhere but in Newfoundland. In Newfoundland, the engine was celebrated as unique, a marine engine that was actually built in the province and used by Newfoundland fisherman who had for decades been an important market for other engine companies. By the spring of 1947, 350 engines had been sold, 650 by 1958. Finally, even the Barnes succumbed to more modern engines although parts were reportedly still available as of 1979. Barnes also built small steel fishing boats. He died in 1972. Today, the company is known as D. F. Barnes Ltd.

Barrie Gas Engine Co. (Canada Producer Gas Engine Co.)
Barrie, ON
Trade name: Barrie

Barrie offered gas and producer gas stationary models in 1910. Later, marine versions were introduced.

Bruce Stewart & Co., Limited
See Stewart, Bruce & Co.

Burrill-Johnson Foundry & Machine
Yarmouth, NS
Trade name: Burrill-Johnson

Byres Co.
Toronto, ON
Trade name: Honest Injun

Canadian Beaver Co.
Toronto, ON
Trade name: Canadian Beaver

The Canadian Beaver was a two-cycle engine available in three singles and three doubles as of 1915. The singles were rated at 3½, 6, and 8 hp. The doubles were rated at 9, 12, and 16 hp. The singles had make-and-break ignition, and the twins, jump-spark. Sherman-Cooper Co. was an important sales outlet in Toronto.

Canadian Gas Power & Launches, Ltd.
Toronto, ON
Trade Name: [badge listed company name]
This company's two-cycle engines were supplied by the Defiance Iron Works of Chatham, ON. The company ceased operations as of 1913.

Caron Engine Co.
Montreal, QB
Trade name: Caron
Alfonce Louis Caron and his brother manufactured this engine in a Bleury St. factory in downtown Montreal. The engine had been designed and patented by inventor Joseph Oscar Lavallee, who, like a number of others, sought to develop an effective two-cycle engine that didn't rely on the base for compressing the mixture. Lavallee's engine, for which he filed a patent in April 1920, was exceedingly clever. The piston descended onto a "plug" at the bottom of the cylinder. This plug forced the mixture into the transfer passage. "No Base Compression" was the ad slogan. Otherwise, the Caron was a three-port engine, the iron cylinder enclosed in a cast-aluminum water jacket. The crank case was also aluminum. The Model A Caron was a 3½″ × 2⅞″ engine that produced 4 hp and weighed just 90 lbs. Some of these engines were used by Dispro during the 1920s. A stationary version was also offered.

Champion Machine and Motor Works, Ltd.
St. John's, NF
Trade name: Champion
William H. and Cecil Trask, the sons of foundry owner L. M. Trask, began an engine-building company circa 1918. Their factory was equipped, in part, with some of the machinery previously used to manufacture artillery shells for the war effort. The prototype engine was a two-cycle, single-cylinder, 8-hp make-and-break. When cylinder castings from a local foundry proved too porous, the company contracted for cylinder castings from a paper mill in Grand Falls. The cylinders were then pressure-tested after machining. An 8′ test tank was constructed. A brochure printed in 1918 or 1919 described a comprehensive range of two-cycle and four-cycle engines. However, those reporting to work one day in 1919 (according to best recollections) found the factory locked, and a shipment of 200 crankshafts went unclaimed.

Cummings, J.W. Mfg. Co.
New Glasgow, NS
Trade name: N/A

Dickson
Port Carling, ON
Trade name: Dickson

Disappearing Propeller Boat Co. (Dispro)
Port Carling, ON
Trade name: Dispro
Several engines were used in the charming "dippy" skiff with its haul-up propeller shaft, including the Waterman K1 2 hp, and a Canadian version, the Fisher. Other engines used included the St. Lawrence and Caron.

Easthope Bros.
Vancouver, BC
Trade name: East Hope
Only a very few marine engine companies stayed in business from the early production era starting around 1900 until post–World War II. Easthope, was among them. Initially, the company was known as Easthope and Son. The inventive father was Ernest Easthope who brought his family from England to British Columbia in 1889. According to family lore, Vincent Easthope, the oldest of the sons, saw an ad for a Roberts two-cycle marine engine "around 1900." This prompted Ernest to make a drawing of a natural-gas-fueled, two-cycle engine he had seen in England. The ultimate result was a home-built 1½-hp two-cycle engine that was installed in a Peterborough canoe. A 3-hp followed, equipped with a homemade timer and mixing valve. The spark plug was fired by 6-volt dry cell and buzz coil. Friends and local Fraser River fishermen saw and wanted a similar engine, and the company was born. In 1905 growth was spurred by the purchase of quantities of engines for use in the Columbia River gill-net fleet of slender, double-ended fishing boats. In *Raincoast Chronicles* No. 5, Percy "Peck" Easthope, Vincent's younger brother, recalled the boats were powered in batches of five boats at a time. The largest of the two-cycle engines was a 6″ × 6″ rated at 30 hp. Sometime in 1910 or 1911, partners brought into the business forced the Easthopes out and then failed.

Although Vincent Easthope suffered an untimely death at age 23, brothers George and Peck reentered the engine business as dealers for the Connecticut-built Frisbee. The brothers had success with the Frisbee and eventually decided to design and build their own engine using Ford Model T parts and a home-built crankshaft and cylinder casting. The result was a 4-hp engine with a Gies reverse gear and Bemis or Cuno timer depending on availability. The Easthope had a balanced flywheel for smoother running. A Wico magneto was adopted for ignition. Adding a second cylinder produced an 8- to 12-hp engine, and a third resulted in a 15- to 18-hp version. The multicylinder versions had dual battery/magneto ignition.

Those Model T-based engines set the style for future Easthopes, although the later engines had larger bore/stroke dimensions and the model line became comprehensive. It ultimately included ratings of 2 to 60 hp. The cast-iron base mounted a reverse gear of Easthope manufacture. The engines remained always the simple, slow-turning type well known to fishermen, and sales continued strong even after World War II. Vince Easthope remembered the period 1945–1950 as some of the com-

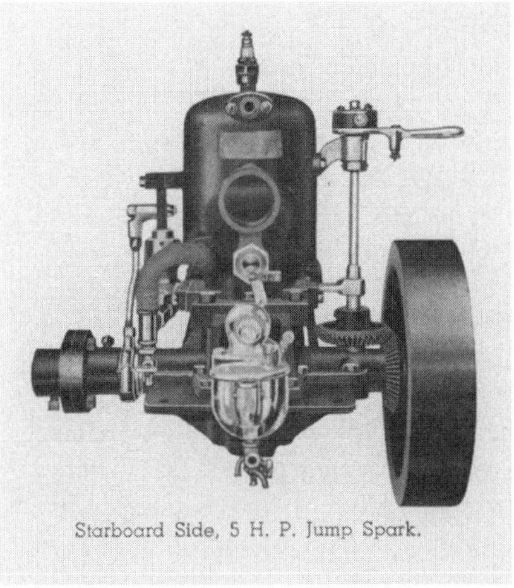

Acadia single, starboard side

Acadia single, port side

Acadia double, starboard side

Acadia 10-hp single

Acadia double, port side

Guarantee

1921 Bruce Stewart triple

1921 Bruce Stewart single

Circa 1923 Caron

pany's best years. Eventually, however, less expensive Chrysler Crown marine engines and converted Ford models with electric starting and higher operating speeds began cutting into Easthope sales. An effort to sell an Easthope, with enclosed valves and adapted to somewhat higher rpm, failed. Easthope Bros. Ltd. was sold in 1978 and became Easthope Sales & Service Ltd. New engines remained available until about 1983. Today, an Easthope is a fine example of this pioneering, family-run Canadian engine company whose product was known to generations of commercial fishermen and pleasure boaters.

Fisher, A. D. Co.
Toronto, ON
Trade name: Kingfisher
Fisher's 2-hp Model A-1, "Kingfisher," was used in Dispro boats.

Fleming Bros.
Halifax, NS
Trade name: N/A

Foreman Motor & Machine Co., Ltd.
Toronto, ON
Trade name: Foreman
Although Foreman's first models were two-cycle, the company transitioned to four-stroke, heavy-duty motors, probably during or after World War I. These were substantial engines with overhead valves, twin ignition, and a robust but unenclosed reverse gear. A 2-cylinder, 15- to 20-hp and 4-cylinder, 30- to 40-hp model were offered. Both were rated at 500 rpm.

Fraser Machine and Motor Co.
New Glasgow, NS
Trade name: Fraser

Gasoline Engine Co. Ltd.
Toronto Junction, ON
Trade name: Toronto Junction
The company was in business in 1901–1906 building a two-port, jump-spark engine. The engine was quite possibly a re-badged American brand.

Gilles
Carlton Place, ON
Trade name: Gilles

Guarantee Motor Co.
Hamilton, ON
Trade name: Guarantee
Founded in 1903, Guarantee was the brainchild of E. T. Krug and H. A. Crosby. Krug had learned the machinist's trade at Westinghouse in Pittsburgh and with the Rueger Gas Engine Co. in Buffalo. Crosby seems to have been involved primarily with the executive side of the business. Guarantee offered two- and four-cycle models by 1910, but five years later, only four-cycle engines were listed. These included 1-, 2-, and 4-cylinder side-valve models of from 4 to 20 hp. The single and double had individual cylinders, but the four was an impressive en bloc engine with a separate head and copper gasket. Jump-spark and make-and-break ignition were listed, but Guarantee advised "jump spark on account of freedom from mechanical parts, also more power from the same cylinder."

The company developed its own carburetor. A Guarantee brochure showed one of its two-cycle engine's crankshafts and that of a competitor. The Guarantee's crank was 2″ in diameter versus 1⁵⁄₁₆″ for the competitor. Guarantee achieved some measure of success for it was reported that the engines were shipped to British Columbia, Newfoundland, and Bermuda. A local history recorded that 200 Guarantee motors were working in boats in Hamilton Bay alone. When motor production ended, perhaps in the 1920s, the shop turned to automobile repair.

Hamilton Motor Co.
Hamilton, ON
Trade name: Hamilton
Hamilton offered 1-, 2-, and 3-cylinder, two-cycle engines from circa 1902–1910.

Hawboldt Gas Engines
Chester, NS
Trade name: Hawboldt
Born in 1876 in the tiny village of Marriotts Cove, Nova Scotia, Forman Hawboldt was a merchant's son who became a jeweler and then, in 1900, embarked on a career that would make him one of Canada's marine engine pioneers. Hawboldt designed a single-cylinder, two-cycle, jump-spark engine, built the needed patterns, and, with the help of a Scotsman named John Douglas, set up a foundry in his barn. The first engine emerged in 1906. The company was still going in 1932 when Hawboldt's son Frank joined the business after he graduated from Nova Scotia Tech in 1932. He recalled that there was little to improve on the classic Hawboldt although the igniter points and some other parts were changed to stainless steel from ordinary steel. Newfoundland fishermen were an important market for Hawboldt until the early '30s when fishing practices took the boats ever farther from shore. "They had to have more power," Frank Hawboldt recalled, "and the marine engines of that time would have cost too much. They started converting car engines then, and they did the trick just as well." Faced with the downturn in demand for their two-cycle marine engines, the Hawboldts began making hardware and other castings although the occasional engine was still built until 1954. Today, Hawboldt Industries manufactures a variety of marine equipment in its 3,700-sq.-ft. Chester facility.

J. W. Cummings Mfg. Co.
See Cummings, J. W. Mfg. Co.

Lloyd's Manuf. & Foundry Co., Ltd.
Kentville, NS
Trade name: Minas

Lunenburg Foundry Ltd. (now Lunenburg Industrial Foundry & Engineering Ltd.)
Lunenburg, NS
Trade name: Atlantic
The foundry was begun as the Lunenburg Iron Company in 1891 to make cast-iron stoves. In 1907, after the foundry burned and was rebuilt, it be-

came a family-owned business. Initially, Lunenburg mayor John James Kinley was president. He was succeeded by his personable son, M.I.T.-trained Jim, in 1965. Peter Kinley, Jim's son and a naval architect, followed and is today the president and CEO.

In 1908 the foundry built its first gasoline engine, which had been designed by Dan Young who had started work at the company as a pattern maker in 1899. Although Young's formal schooling had ended after 4th grade, he enrolled in an International Correspondence School Course and learned mathematics and mechanical drawing. Young may have had less intuitive knowledge of gasoline engine technology than many men who became involved in the business, but he had confidence in his ability to design an engine, make all the needed patterns, and oversee its construction. In 1907 he purchased and disassembled for study a 2-cylinder Palmer, primarily seeking guidance for port layout and compression ratio. The result was a two-port, single-cylinder 5-hp engine. It was purchased by a stove fitter at the foundry who used it for 30 years.

What followed was a beautifully executed series of drawings for seven engines ranging from 3 to 10 hp. It is thought that these were two-port engines. In 1910 a three-port was developed, and the design of the "Atlantic" engine thereafter became standardized. The carburetor was mounted under the exhaust manifold where its vapor would be heated. Crankcase compression was increased by balancers on the crankshaft that filled excessive volume in the base. An auxiliary air valve was mounted in the base.

Most two-cycle marine engines were overbuilt, and that was especially true of the Atlantic. Dan Young's son Charlie reported, "If a casting was supposed to be an inch and a half thick, my father would say, 'Well, better make it 1⅝" just to be sure.'" Like most other marine engines, the Atlantic was for years made of an alloy known as "gray iron." However, at some point, chromium nickel was added to the molten alloy to increase its hardness. The engines were also unusual in that, probably during the late 1930s, the pistons were made not of iron but of "saltwater" aluminum. The result was corrosion resistant while offering one-third less weight than iron, which translated into an rpm increase of from 500 or 600 rpm to 800 to 900 rpm.

Nearly every component of the engine was made within the foundry except for the Bemus timers on jump-spark models and the Schebler carburetor. Later, however, the Bemus was replaced by a plunger-type, crankshaft-driven timer made by the foundry. After the Schebler Model D was discontinued, the foundry made its own version. The parts for the make-and-break igniter were made in-house, of course, of mild steel hardened by immersion in potassium cyanide, the deadly fumes of which the workmen had to take care to avoid. The bought-out components included coils, water pump check valves, the cylinder petcock, and priming cup.

The largest two-cycle Atlantic was a 54-hp, 7" × 7½" 3-cylinder designed in 1928 and fed by a 2" Schebler carburetor. Five of these engines were built. Charlie Young remembered them well, saying they were "monstrous things and I was afraid of them." Young cited the biggest single change in the Atlantic engine as occurring in the late 1930s when the engine's circular base, so typical of the first generation of marine engines, was changed to the more familiar base split longitudinally. Throughout the '20s and into the 1930s, Atlantic engine sales were brisk. At one time, Newfoundland alone accounted for 400 engines a month.

After World War II, demand for the two-cycle engine dwindled, and the foundry introduced marinized versions of Chrysler and Buda 6-cylinder engines and a variety of V-8s. An English Ford diesel was also converted for marine applications. Efforts were made to continue production of the Atlantic in developing countries. In 2008 president and CEO Peter Kinley recalled, "We discussed our engine technology with representatives of the National Research Council in Ottawa. It was determined that there were great advantages to using the two-cycle engine in developing countries. With the help of the Canadian International Development Agency we set up a joint venture in Colombia, South America, where they were to be produced under license. We duplicated a set of patterns for the 5-hp model and shipped them to our joint venture." This project, however, soon proved unfeasible. So did the idea of manufacturing in China.

Although the Atlantic was still being produced into the early 1990s, Peter Kinley reported that "the demographic of our customers was changing. Instead of communities of fishermen using them to feed their families we saw more and more collectors and engine buffs who wanted them for posterity rather than for their use. I met a gentleman from San Francisco who was taking delivery of one of our engines and asked him what type boat the engine was going to be installed in. It was a bit of a revelation that he admitted that it was not going to go in a boat but was instead going to be put in a glass case in his living room."

At about that time, it was becoming harder to get forged cranks and coils and other parts at a reasonable price. What's more, reported Peter Kinley, "The old guys that built the Atlantic engines did so without much concern for writing down their procedures and methods. Some of the young diesel mechanics would lend their hands at building engines, but it took time to relearn the old ways and it was clear that from an economic basis our per-unit costs were going up."

All these factors and others combined to force the closing of the company's iron-casting operation. The last Atlantic Engine produced was for the Engineering School at Memorial University in St. John's, Newfoundland, where it is used to teach students something about the technology that changed the coastline of Canada.

1914 Guarantee fitted with heater for intake mixture and both jump-spark and make-and-break ignition

1914 Guarantee 4-cylinder four-cycle, available in 20-25– and 30-35–hp models with magneto, coil, or coil and-tension magneto ignition.

1923 St. Lawrence

1914 Guarantee single-cylinder engine: "solid head cylinder with no gaskets to blow out," float-feed carburetor, water-cooled exhaust, jump-spark for maximum power, ribbed flywheel to eliminate starting knob or handle.

1921 Canadian-built Michigan Marine

1921 Semmelhaack "Maple Leaf"

This 1914 Guarantee was built by the Hamilton Motor Co. of Ontario, Canada. This two-cycle engine is equipped with a friction clutch, governor, and winch to haul lobster pots.

Hawboldt

It's estimated that more than 12,000 Atlantic engines have been built. The 4-hp Atlantic with make-and-break ignition was among the most popular in the latter years of production. In 1952 engineers at Canada's National Research Council subjected the 4-hp single and an 8-hp twin to scrupulous dynamometer tests. They found the 4-hp actually produced 5.8 hp at 650 rpm while the 8-hp model developed a maximum power of 17.2 hp at 950 rpm — 115 percent more than the manufacturer's rating.

Matheson & Co. Ltd.
New Glasgow, NS
Trade name: N/A

McKeough & Trotter Ltd.
Chatham, ON
Trade name: McKeough & Trotter
The company offered two- and four-cycle engines of 1½ to 16 hp in 1921.

Michigan Marine Motor Corp.
Ford City, ON
Trade name: Mich. Marine Motor
See Michigan Marine Motor Corp., Detroit

Midland Engine Works
Midland, ON
Trade name: Hanly Marine
This company was the outgrowth of an iron foundry begun in 1895 by J. Bruce Hanly. Around 1900 or 1901, Hanly's younger brother joined him, and they founded the engine works and produced two-cycle marine engines. It is possible that a stationary version was also offered. The first engines were headless, but later models had removable cylinder heads. The model line included a little 1½-hp canoe engine (3″ × 2½″), a 4″ × 4″ 3-hp single, a 5″ × 5″ 4-hp single, a 5¾″ × 5½″ 6-hp single, a 4″ × 4″ 6-hp double, a 5″ × 5″ 8-hp double, a 5¾″ × 5½″ 12-hp double, and a 5¾″ × 5½″ 18-hp triple. The power of a big one-lunger is evident in the company price list, which noted, "A 6-h.p. single cylinder motor requires a stronger clutch than a three-cylinder 12 h. p., as the blow from the explosion is much stronger in the 6 h. p. than in the 12 h. p." These were jump-spark engines and came equipped with six Columbia dry cell batteries (12 for a 2-cylinder), 4′ of secondary wire and 10′ of primary wire, a stuffing box, timer, Eclipse carburetor, and other equipment needed to complete the installation and run the engine. A wide variety of reverse gears was offered: Paragon, Gies, Baldridge, Carlyle Johnson, Bull Dog, and Ball.

After a fire in 1907, the company rebuilt, this time contracting for a cement building. During World War I, Midland supplied munitions. The extent of postwar engine production is unknown, but engines may well have been built until at least 1923. The company continued until 1937 when, after Bruce Hanly's death, it was closed down.

Moir, W&A
Halifax, NS
Trade name: N/A

N.S. Steel & Coal Ltd.
New Glasgow, NS
Trade name: N/A

Pacific Standard Motor Works
Vancouver, BC
Trade name: Pacific Standard

Payette, Peter & Co.
Penetanguishene, ON
Trade name: Payette
Peter Payette established his machine shop and foundry in 1880. The Payette engine was designed by Stanley R. Du Brie, possibly before 1900. It was among the earliest two-cycle marine engines built in Canada. In 1914 the company acquired the Adams Engine Co. and continued manufacture under that name.

See also Adams Engine Co.

Phoenix Foundry
St. Johns, NB
Trade name: Essex

Pictou Foundry & Machine Co.
Pictou, NS
Trade name: N/A

Reid Gasoline Engine Co.
Hamilton, ON
Trade name: Reid
The Reid engine was a 2-hp, four-cycle.

Roberts Motor Works
Lachine, Quebec
Trade name: Roberts
The company was in business as of 1911. It is plausible to suggest the engine was designed by Edmund Wilson Roberts.

Schaake Machine Works, Limited
New Westminster, BC
Trade name: Yale
Schaake built a heavy-duty, open-base four-cycle engine in sizes of 20 to 75 hp. This engine was quite possibly the same model that had previously been built in Seattle by Western Gas Engine & Schaake Machine Works.

Schofield-Holden
Toronto, ON
Trade name: Schofield-Holden

Semmelhaack-Dickson, Ltd.
Montreal, Quebec
Trade name: Maple Leaf
The Maple Leaf was touted as being "Canadian-Made" and was available in two models. These included a 45-lb., 2½- and a 72-lb., 5-hp twin. Both had copper water jackets and jump-spark ignition. The company was in business as of 1921.

St. Lawrence Engine Co.
Brockville, ON
Trade name: St. Lawrence
A bike shop owner named H. W. Going together with engineer C. T. Wilkinson founded St. Lawrence in 1905 on St. Andrew Street in Brockville. Local sources reported that the two-cycle St. Lawrence was based on the Rice engine, built in Bordentown, New Jersey, and the distinct, removable dome head of the Rice is clearly evident on the St. Lawrence. The St. Lawrence two-cycle

motor remained in production into the 1930s, but it was joined by converted Ford engines starting in the '20s. The Model T, Model A, and V-8 were all adapted by St. Lawrence. One local historian claimed the two-cycle St. Lawrence was still being built in 1948. "Every dollar invested in the Company is Canadian and every employee is a British subject." By the early 1970s, St. Lawrence had become a general engineering shop with a marine specialty.

Stewart, Bruce & Co. Ltd.
Charlottetown, PEI
Trade name: Imperial
The "Motor That Makes the Mark" was a headless two-stroke available in two model ranges. The Model 'A' was offered as a 5-hp single or 10-hp double intended for pleasure boats. The Model 'C' was a heavy-duty engine designed for commercial fishing and work boats and cruisers. Single- and double-cylinder models were available in 10-, 20-, and 30-hp ratings.

Swift Motor Car Co. of Canada
Chatham, ON
Trade name: Swift

Sydney Foundry & Machine Co.
New Glasgow, NS
Trade name: N/A

Sylvester Mfg. Co. Ltd.
Lindsay, ON
Trade name: Sylvester

Trask & Company
Halifax, NS
Trade name: Trask

Tudhope Foundry
Orillia, ON
Trade name: Tudhope

Vivian Gas Engine
Victoria, BC
Trade name: Vivian
Sometimes called the Henry Ford of the Canadian diesel industry, Will Vivian grew up in Vancouver, left school when he was just nine years old, and went to sea as a cabin boy before returning to work on fishing boats and, it is believed, at the Easthope works then in Coal Harbour. Vivian pursued further experience at Imperial in San Francisco. He built his first engine — a slow-turning, heavy-duty, 6-hp four-cycle — in Vancouver in 1909 at age 19. What did heavy-duty mean as it related to such an engine? According to David Conn in *Raincoast Chronicles* No. 9, a production 5-hp Vivian had a 150-lb. flywheel, and the engine accounted for at least another 150 lbs. Later, beginning around the late '20s or 1930, the focus was on heavy-duty diesels, the first Canadian diesel marine engines.

By 1938 the lineup ranged from 10 to 240 hp. Vivian prospered during World War II and Vivian Diesels & Munitions was incorporated in 1942. Vivian 240-hp diesels were installed in approximately 240 64′ harbor tugboats built in Ontario by the General Bridge Co. In 1945 Vivian licensed a 4-cylinder model to Australia's Commercial Steels & Forge in Sydney and showed a 10-cylinder diesel in 1949. Vivian sold the company that year, and the brand ceased to exist in the early '50s. Will Vivian died in 1965. Like the records of nearly all the pioneer diesel-engine companies, those pertaining to Vivian disappeared along with remaining parts and patterns. In this case, it is believed the materials were controlled by Canadian Car and Foundry, which divested itself of them in the mid-1970s.

Wonder Pump and Engine Co.
Vancouver, BC
Trade name: Wonder

Bibliography

Magazines and Papers

In addition to the *New York Times*, *Detroit News*, and *San Francisco Chronicle*, local newspapers were consulted whenever possible in the cities and towns in which specific marine engine companies were located. Other publications used include the following:

American Industries (various issues)

Canadian Motor Boat (various issues)

Claude Sintz Interview, 1952, Owen Bombard, Ford Archives

Diesel and Gas Turbine Progress, 1948

The Gas Engine (various issues, principally 1898–1915)

Gas Engine Magazine (various issues)

High Lights of the Life of Clark Sintz, Ford Archives

"Historical Development of Internal Combustion Engines on the West Coast," Aug. 17, 1960, SAE paper, Charles A. Winslow

"A History of Gray Marine Motor Co.," John W. Mulford, 1961

Horseless Age (various issues 1900–1915)

Michigan Marine Engine Manufacturer Notes, Scott M. Peters, March 2003

Michigan Tradesman (Dec. 12, 1894, June 8, 1904, Jan. 13, 1907, April 27, 1910, etc.)

Motor Boating (various issues 1911–1940)

MotorBoat (various issues 1912–1940)

Pacific Powerboat (various issues 1915–1930)

Power Boating (Jan. 1914, Nov. 1915, etc.)

The Rudder (various issues 1898–1955)

Scientific American (various issues 1900–1912)

The Society of Naval Architects and Marine Engineers (1905–1910)

"Tests of Atlantic 8 H. P. Two-Cycle Marine Gasoline Engine," National Research Council of Canada, 1952

Toquet Launch and Motor Co., papers, Westport, CT, Historical Society

Yachting (various issues 1932–1955)

Books/Booklets

City Directories were routinely consulted

Gas Engine Troubles and Installation, J. R. Rathbun, reprint edition by Ron Lachniet, July 1977

I Kept No Diary, F. R. Banks, Airlife Publishing Ltd., Shrewsbury, England

Marine Engines, Their Construction and Management, Carl H. Clark, D. Van Nostrand Co., 1914

Motor Boats, Construction and Operation, Thomas H. Russell, 1910, The Charles C. Thompson Co., Chicago

750 Manufacturers of Inboard Marine Engines, Max F. Homfeld, 1991

The Standard Catalog of American Cars, Beverly Rae Kimes

Brochures

Acadia Gas Engine Co., Adams Engine Co., American Engine Co., Atlas-Imperial Engine Co., Auto Engine Works, Bessemer Gas Engine Co., Bridgeport Motor Co., Buffalo Gasolene Motor Co., Camden Anchor, Rockland Machine, Campbell Motor Co., Caron Engine Co., Charles Brady King Engine Co., Clark Sintz (Springfield, Ohio), Cleveland Diesel Engine Division/GM, Cooper-Bessemer Engine Co., Corliss Engine Co., C. T. Wright Engine Co., Detroit Engine Works, Du Brie Motor Co., Eagle Co., Elbridge Marine Engines, Electro-Motive Division of General Motors, Erd Motors Corp., Fay & Bowen Engine Co., Ferro Machine & Foundry Co., General Machinery Co., Goshen Motor Works, Gray Motor Co., Guarantee Motor Co., Hall Gas Engine Co., Hallett Manufacturing Co., Loew Manufacturing Co., Lozier Motor Co., Niagara Motors Corp., Palmer Brothers Engine Co., Pausin Engineering Co., Peerless Marine Engine Co., Pierce-Budd Engine Co., Regan Vapor Engine Co., Roberts Motor Company, Sintz-Wallin, St. Joseph Engine Co., Standard Engine Works, Standard Gas Engine Co. (San Francisco), Stanley Co., Syracuse Gas Engine Works, Toquet Launch & Engine Co., Torrey Roller Bushing Works, Union Engine Co., Van Blerck Motor Co.

Index